201

Topics in Current Chemistry

Springer-Verlag Berlin Heidelberg GmbH

Carbon Rich Compounds II

Macrocyclic Oligoacetylenes and Other Linearly Conjugated Systems

Volume Editor: A. de Meijere

With contributions by
S. C. Brand, U. H. F. Bunz, F. Diederich, L. Gobbi,
M. M. Haley, S. I. Kozhuskov, A. de Meijere, J. J. Pak,
U. Scherf

 Springer

This series presents critical reviews of the present position and future trends in modern chemical research. It is addressed to all research and industrial chemists who wish to keep abreast of advances in the topics covered.

As a rule, contributions are specially commissioned. The editors and publishers will, however, always be pleased to receive suggestions and supplementary information. Papers are accepted for "Topics in Current Chemistry" in English.

In references Topics in Current Chemistry is abbreviated Top. Curr. Chem. and is cited as a journal.

Springer WWW home page: http://www.springer.de
Visit the TCC home page at http://www.springer.de/

ISSN 0340-1022

ISBN 978-3-662-15610-0   ISBN 978-3-540-49451-5 (eBook)
DOI 10.1007/978-3-540-49451-5

Library of Congress Catalog Card Number 74-644622

© Springer-Verlag Berlin Heidelberg 1999

Originally published by Springer-Verlag Berlin Heidelberg New York in 1999
Softcover reprint of the hardcover 1st edition 1999

Cover design: Friedhelm Steinen-Broo, Barcelona; MEDIO, Berlin
Typesetting: Fotosatz-Service Köhler GmbH, 97084 Würzburg

SPIN: 10649238   66/3020 – 5 4 3 2 1 0 – Printed on acid-free paper

Preface

Not too long ago, graphite and diamond were the only two known modifications of carbon. That changed dramatically with the discovery of C_{60} in 1985 and all the higher fullerenes soon thereafter. Nevertheless, this breakthrough did not stand alone in paving the way to the new era of chemical and physical research into carbon rich compounds that we are now enjoying.

The last 20 years have witnessed the development of a powerful repertoire of new carbon-carbon bond forming processes, especially metal-catalyzed and metal-mediated ones. These, together with other useful organic synthetic methodologies, including important protection and deprotection procedures, have made the synthesis of new targets possible that were inconceivable previously. The whole range of modern acetylene chemistry and numerous new materials simply would not exist without these new methodologies. Along with the fast growing set of tools goes an ever increasing number of publications concerning impressive arrays – cyclic, two-dimensional and even three-dimensional – of acetylenic and diacetylenic units linked by aliphatic, aromatic or organometallic connectors. Nowadays there seems to be no limits to achieving what the fantasy of the chemist can come up with.

It appeared timely to compile recent developments in modern oligoacetylene chemistry and other new materials – such as polyparavinylenephenylenes and the like – as these carbon rich compounds and their properties bear relevance to our modern understanding of the importance of organic chemistry as well as the emerging field of organic materials science. The term "carbon rich" is used here and in the preceding volume on this topic (TCC Vol. 196) to refer to everything that has a carbon to hydrogen ratio of $1:(<1)$.

Göttingen, Januar 1999   Armin de Meijere

Contents

Contents of Volume 196
Carbon Rich Compounds I

Volume Editor: A. de Meijere
ISBN 3-540-64110-6

Contents of Volume 197
Dendrimers

Volume Editor: F. Vögtle
ISBN 3-540-64412-1

Contents of Volume 198
Design of Organic Solids

Volume Editor: E. Weber
ISBN 3-540-64645-0

Macrocyclic Structurally Homoconjugated Oligo-acetylenes: Acetylene- and Diacetylene-Expanded Cycloalkanes and Rotanes

Armin de Meijere* · Sergei I. Kozhushkov

Institut für Organische Chemie der Georg-August-Universität Göttingen, Tammannstrasse 2, D-37077 Göttingen, Germany. * E-mail: ameijer1@uni-goettingen.de

There is ample evidence for the significance of modern acetylene chemistry and its role for the future development in several areas of organic chemistry. In this chapter, the successful syntheses and attempted approaches to macrocyclic structurally homoconjugated oligoacetylenes and oligodiacetylenes, as well as some of their physical properties, are discussed. Such compounds, which are formally derived from permethylated cycloalkanes by insertion of $-C\equiv C-$ or $-C\equiv C-C\equiv C-$ fragments between each pair of adjacent sp^3-hybridized carbon atoms, have been termed [n]pericyclines and expanded [n]pericyclines, respectively. These and analogous hydrocarbons formally derived from [n]rotanes (perspirocyclopropanated [n]pericyclines and "exploded" [n]rotanes) as well as heteroanalogues of [n]pericyclines are presented. In addition, the preparations and properties of macrocycles with mixed ethyne and butadiyne expanders as well as the attempted syntheses of perspirocyclopropanated "exploded" [n]rotanes are also covered. In view of the reported properties, the question of cyclic homoconjugation and homoaromaticity in these unconventional compounds is discussed. Finally, some chemical transformations of these macrocyclic oligoacetylenes and oligodiacetylenes, e.g. the conversion of diacetylene-expanded [n]rotanes into crowns of thiophenes, are presented.

Keywords: Alkynes, Coupling reactions, Macrocycles, Pericyclines, Expanded pericyclines, Cyclopropanes, Exploded [n]rotanes, Heteropericyclines, Homoconjugation, Homoaromaticity, Strain energy, Oligothiophenes.

Topics in Current Chemistry, Vol. 201
© Springer-Verlag Berlin Heidelberg 1999

1
Introduction

The syntheses of macrocyclic oligoacetylenes with a capacity for cyclic conjugation and the investigation of their properties have once again become an active
area of acetylene chemistry. After the rather quiet period following the classical
work of Sondheimer et al. [1], it has gained tremendous momentum in recent
years. Undoubtedly, the drastically advanced methodology of metal-catalyzed
and metal-mediated cross-coupling reactions [2] has contributed to the rapid
development in modern acetylene chemistry [3]. Macrocyclic, structurally
homoconjugated oligoacetylenes are of special interest to evaluate the potential
effects of homoconjugation and cyclic homoconjugation, particularly in neutral
molecules. This account reviews the synthetic efforts that have been made
towards the assembly of macrocyclic oligoacetylenes and oligodiacetylenes in
which each pair of ethyne and butadiyne moieties is separated by a quaternary
carbon atom, a CH_2 group or a heteroatom. Such macrocycles can also be viewed
as expanded carbo- or heterocycles, in which each single bond is replaced by an
ethyne or a butadiyne fragment. In addition to the synthesis of such compounds,
their relevant spectroscopic and structural properties will be discussed. The few
chemical transformations and insights into the chemical reactivity of these
macrocycles, especially the macrocyclic oligodiacetylenes, will also be presented.

2
[n]Pericyclines

The preparation and chemical transformations of [n]pericyclines have recently
been reviewed by Scott et al. [4]. This section is therefore to be considered as an
update of that excellent review. The term "[n]pericyclines" was advanced in the
pioneering work of Scott et al. [5, 6] to describe molecules containing n $-C \equiv C-$
units distributed symmetrically around the perimeter of a cycle with n vertices,
i.e. a cycloalkane with an ethyne moiety inserted into each single bond (Fig. 1).

The first members of this family – cyclonona-1,4,7-triyne and cyclododeca-
1,4,7,10-tetrayne 2 ($n=3, 4$) – as well as their substituted analogues remain
illusive to this day. In order to overcome preparative complications associated
with labile hydrogens on the doubly propargylic carbon atoms, the higher
[n]pericyclines 3–6 were prepared only as the fully or almost fully methylated
derivatives [6].

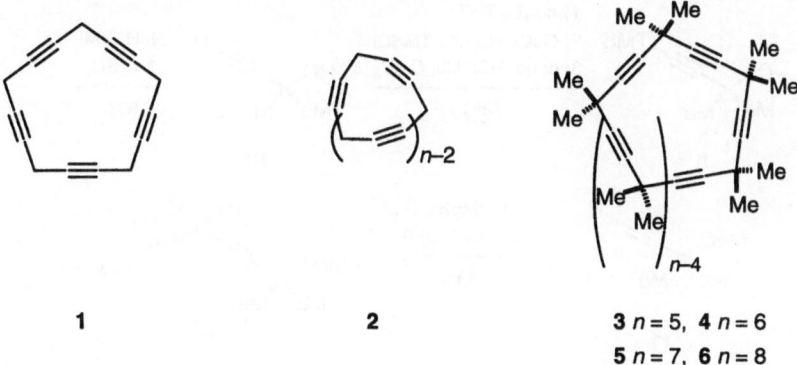

1　　　　　　**2**　　　　　　**3** $n = 5$, **4** $n = 6$
　　　　　　　　　　　　　　　　　　5 $n = 7$, **6** $n = 8$

Fig. 1. [5]Pericycline **1**, pericycline **2** of order n, and permethylated [n]pericyclines **3–6** ($n = 5–8$)

2.1
Synthetic Routes to [n]Pericyclines

[n]Pericycline ring systems **2** can be prepared by a stepwise assembly of an appropriate acyclic precursor **7** with the same number of carbon atoms and its subsequent cyclization (one-component approach, pathway A) or by twofold alkynylation of a difunctional C_1-building block with an acyclic homoconjugated [n]oligoyne **8** (two-component approach, pathway B in Scheme 1).

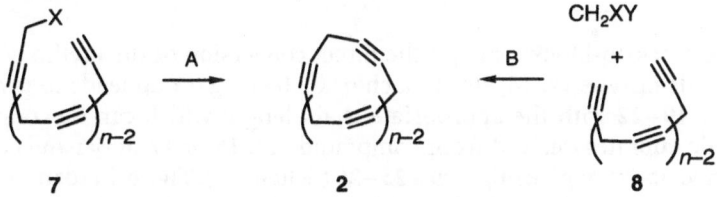

7　　　　　　　　**2**　　　　　　　　**8**

Scheme 1. Conceivable synthetic approaches to [n]pericyclines

Following the first of these two strategies, the whole series of permethylated [n]pericyclines **3–6** ($n = 5–8$) has been synthesized starting from a single building block – dimethylpropargyl alcohol **10** – which is commercially available and can easily be prepared from acetone and acetylene (Scheme 2) [4, 6].

The subsequent chain extension can be accomplished by the pedestrian step-by-step homologation sequence via the acyclic diyne **12** or by a more efficient block-to-block strategy. The step-by-step approach includes protiodesilylation of diyne **12** followed by coupling with the propargyl chloride **9** following the same protocol as for the preparation of **12** from **11** and subsequent repetitions of protiodesilylation and alkylation with chloride **9** to reach stages **16** and **18**, respectively (Scheme 3).

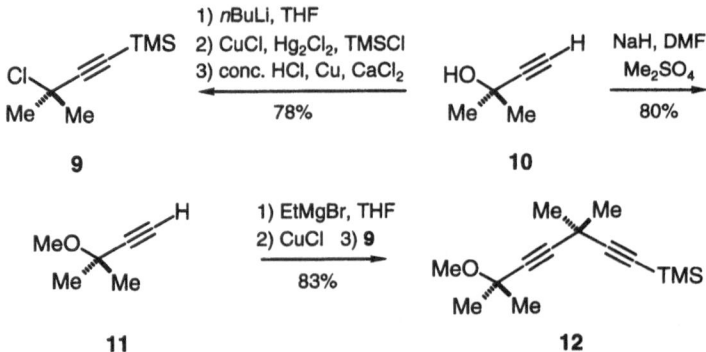

Scheme 2. The main building blocks for [n]pericyclines

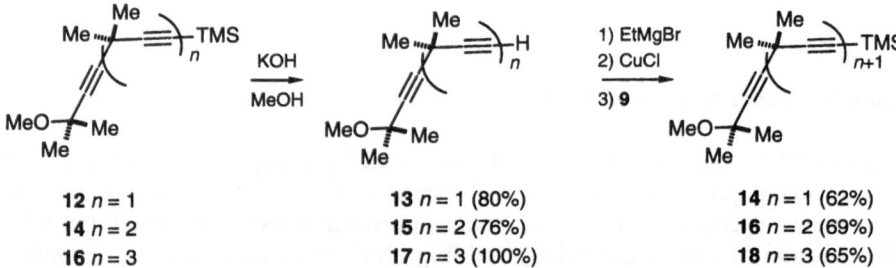

Scheme 3. Regioselective step-by-step chain lengthening of the precursor **12** to [n]pericyclines

In the block-to-block strategy, the direct conversion of the methoxy group in compounds of type **12**, **14**, etc., to a chloride leaving group leads to propargylic chlorides **19–22** with the appropriate chain length which can be coupled with acetylenic cuprates derived from compounds **13**, **15** or **17** under the conditions mentioned above to give oligoynes **23–26** (Scheme 4). These, in turn, are converted to the corresponding chlorides **27** in 82–93 % yields. Accordingly, a set of only two to three standard repetitive operations was used in both types of preparations of the acyclic oligoacetylenic compounds for eventual cyclization [4–6].

The acyclic oligoynes **23–26** can be cyclized under Friedel-Crafts conditions, i. e. by treatment with AlCl$_3$ in CS$_2$, which presumably proceeds via the intermediate tertiary propargylic **27** and β-silyl-substituted vinylic carbocations of type **28** (Scheme 5).

The yield of the macrocycle drops steadily as the ring size increases from 35 % for **3** ($n=5$) to 22 % for **4** ($n=6$), 6.2 % for **5** ($n=7$) and 1.5 % for **6** ($n=8$). By an analogous sequence, octamethyl[5]pericycline **30** has also been prepared, albeit in lower yield than the permethylated analogue **3** (Scheme 6) [6].

The two-component approach has been elaborated for the preparation of [n]pericycline derivatives with $n=4$ and $n=8$, and 10. These compounds are not easily or not at all accessible along the one-component route. In principle, the appropriate acyclic precursors can be prepared in the same manner as the ones for

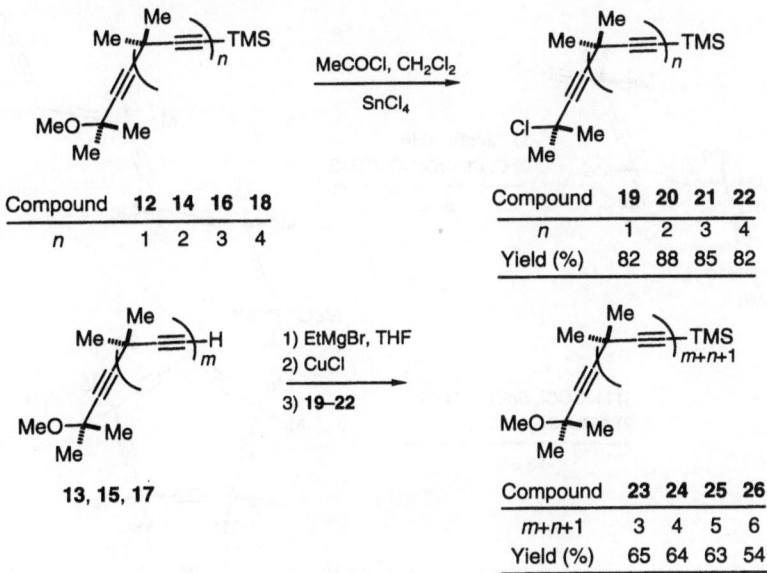

Compound	12	14	16	18
n	1	2	3	4

Compound	19	20	21	22
n	1	2	3	4
Yield (%)	82	88	85	82

13, 15, 17

Compound	23	24	25	26
m+n+1	3	4	5	6
Yield (%)	65	64	63	54

Scheme 4. Starting materials for [n]pericyclines

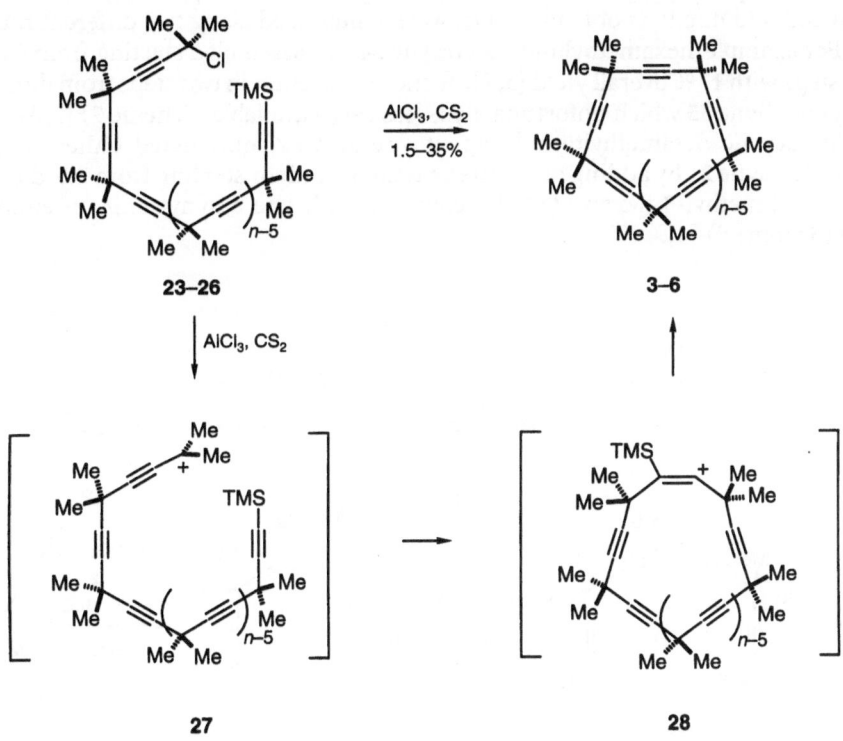

23–26

3–6

27

28

Scheme 5. Synthesis of [n]pericyclines and possible mechanistic rationalization

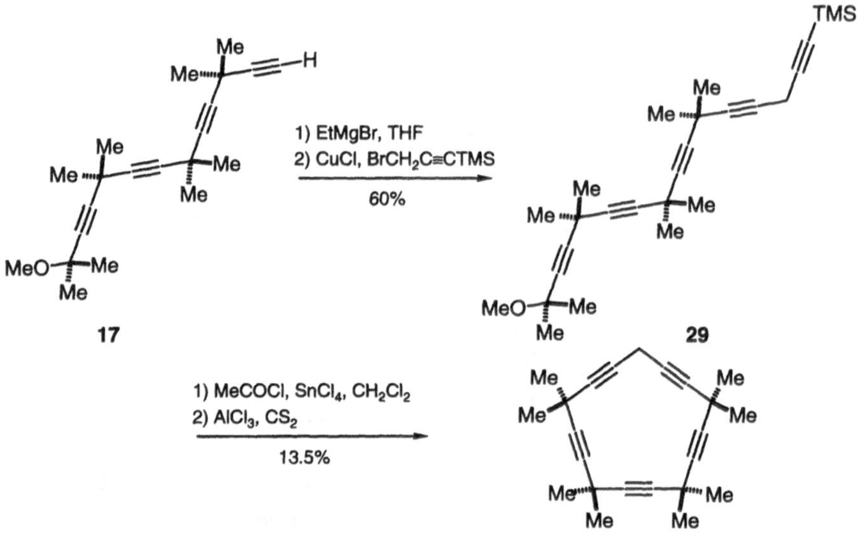

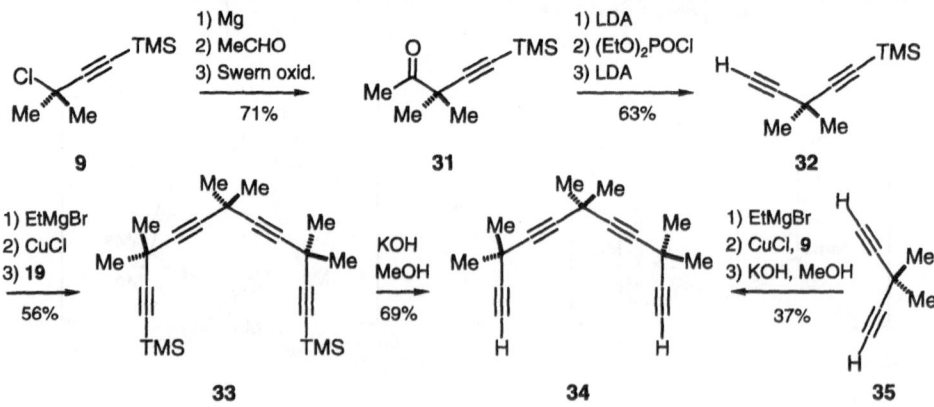

Scheme 6. Preparation of octamethyl[5]pericycline **30**

30

the one-component approach from the building block **9**, yet acyclic oligoynes with even and odd numbers of triple bonds were synthesized along two different routes. For example, hexamethylundecatetrayne **34** was assembled starting from **9** in five steps with 17% overall yield [3, 7], or, more efficiently, in two steps from dimethylpentadiyne **35** which unfortunately is less easily available (Scheme 7) [7, 8].

The acyclic octamethyltetradecapentayne **36** was constructed either from chlorotetrayne **21** by adding one dimethylethynyl unit or starting from 2,5-dichloro-2,5-dimethyl-3-hexyne (**37**) by coupling with one ethynyl unit on either side (Scheme 8) [4].

Scheme 7. Starting materials for [*n*]pericyclines via a two-component approach

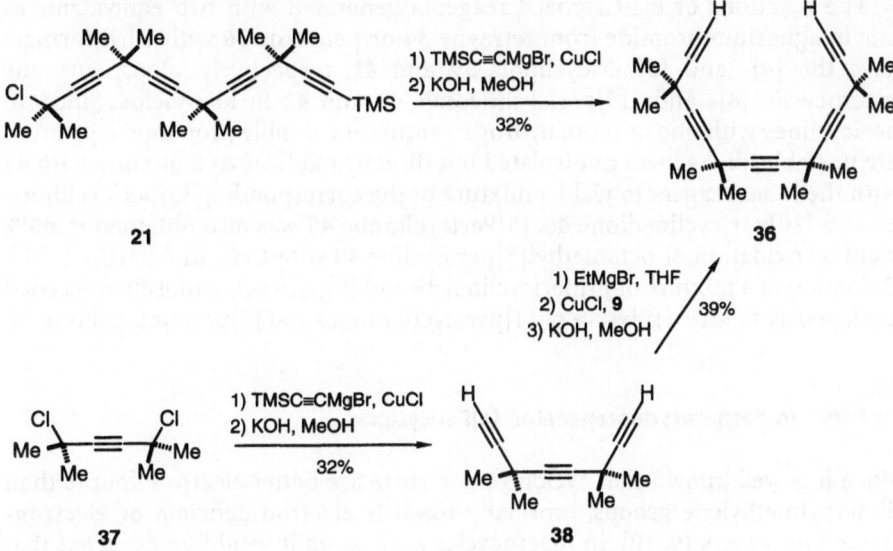

Scheme 8. Starting materials for [*n*]pericyclines via a two-component approach

39 *n* = 4 (10%)
41 *n* = 5 (27%)

40 *m* + *n* = 8 (6%)
42 *m* + *n* = 10 (not isolated)

Jones oxid.

43 *n* = 4 (54%)
45 *n* = 5 (18%)

44 *m* + *n* = 8 (32%)
46 *m* + *n* = 10 (2.2% from **36**)

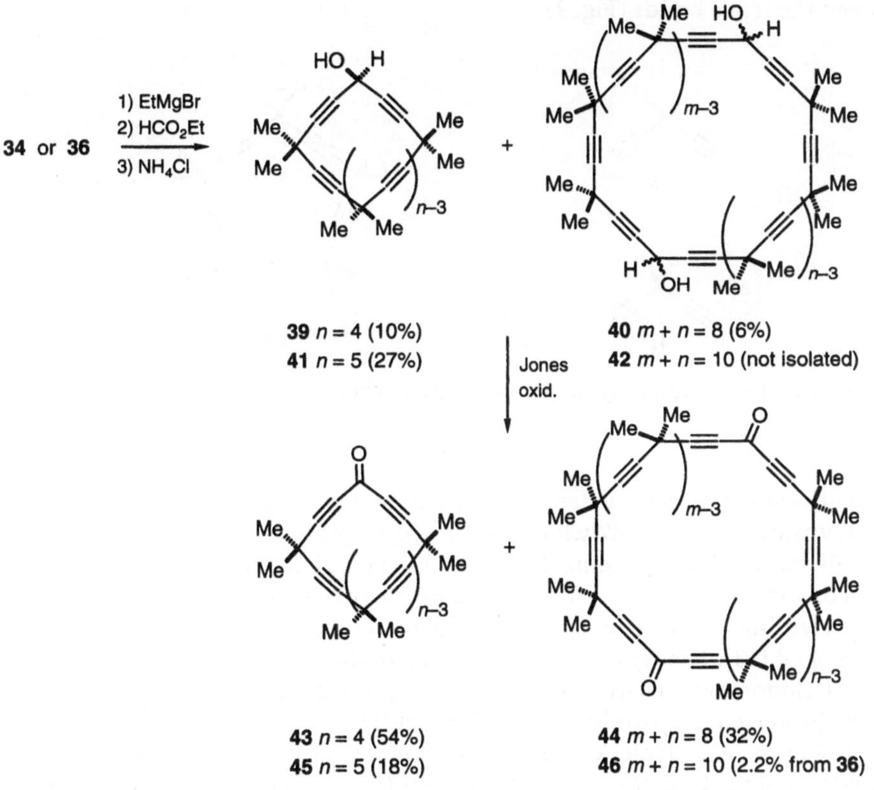

Scheme 9. Pericyclinols and pericyclinones via a two-component approach

The reactions of bis-Grignard reagents generated with two equivalents of ethylmagnesium bromide from tetrayne **34** or pentayne **36** with ethyl formate gave the [4]- and [5]pericyclinols **39** and **41**, respectively, along with the macrocyclic [8]- and [10]pericyclinediols **40** and **42** in low yields. Since all pericyclines with one or more hydrogen atoms at a doubly propargylic position are unstable, diol **42** was not isolated but directly oxidized as a mixture with **41** with the Jones reagent to yield a mixture of the corresponding [5]pericyclinone **45** and [10]pericyclinedione **46**. [5]Pericyclinone **45** was also obtained in 63% yield by oxidation of octamethyl[5]pericycline **30** with CrO_3 in AcOH/p-TsOH. Oxidation of a mixture of [4]pericyclinol **39** and [8]pericyclinediol **40** proceeded analogously to give a mixture of [4]pericyclinone **43** and [8]pericyclinedione **44**.

2.2
En Route to Perspirocyclopropanated [*n*]Pericyclines

Since it is well known that cyclopropane rings are better electron donors than dimethylmethylene groups, especially towards electron-deficient or electron-attracting centers [9, 10], in macrocycles such as **48** it would be expected that the spirocyclopropane rings rather than the dimethylmethylene groups in [*n*]pericyclines would more efficiently transmit the electronic interaction between the triple bonds (Fig. 2).

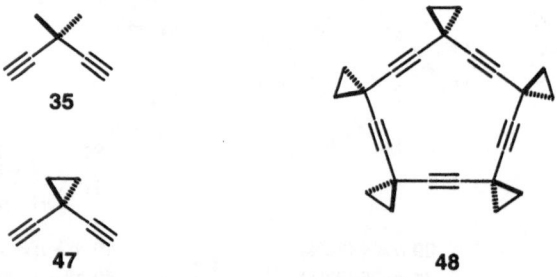

Fig. 2. Conceived perspirocyclopropanated [5]pericycline **48**

This anticipation is evidenced in the photoelectron spectrum (PES) of 1,1-diethynylcyclopropane **47** which shows a π,π-split of 1.4 eV compared to only 0.6 eV for 3,3-dimethylpenta-1,4-diyne **35** [11, 12]. Synthetic methodology for the construction of appropriate acyclic precursors for perspirocyclopropanated [5]pericycline **48** (Scheme 10) and analogous compounds has been successfully developed [13]. The regioselective coupling of the lithium phenylthiocuprate **50** with 1-(iodoethynyl)-1-(trimethylsilyl)cyclopropane (**51**) gave the bis(trimethylsilyl)-protected dehydrodimer **52** (49% yield) which was selectively protiodesilylated at the acetylenic terminus, deprotonated and iodinated with elemental iodide in almost quantitative overall yield (97%). The resulting iododiyne **53** was then treated with heterocuprate **50** to give the dehydrotrimer **54**. Two more

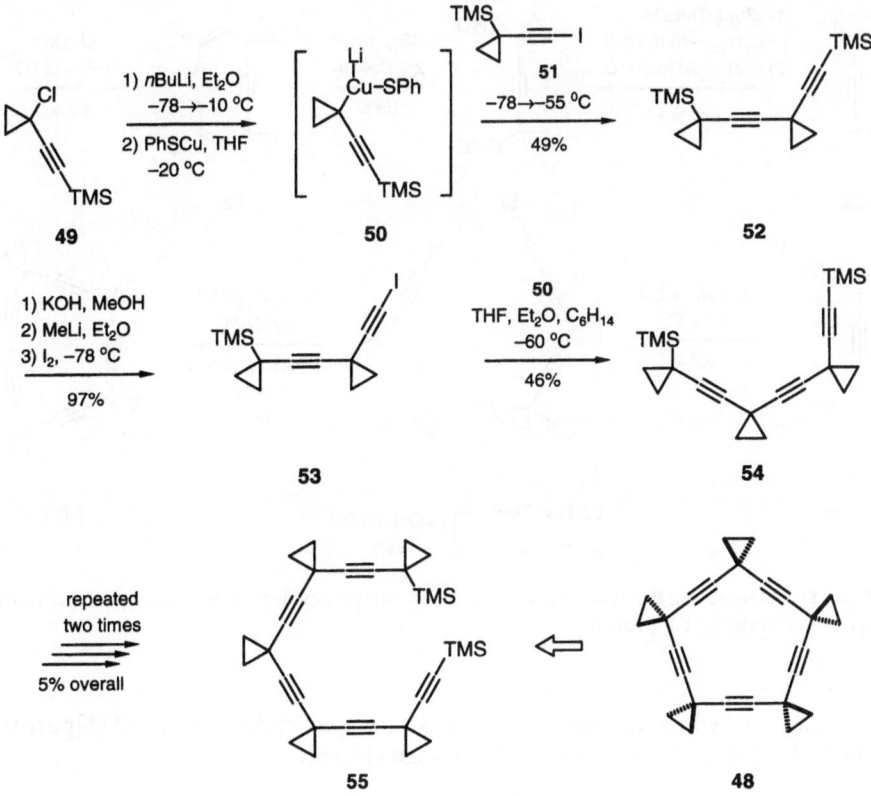

Scheme 10. Assembling the synthetic precursors to perspirocyclopropanated [5]pericycline 48 via a one-component approach

iterations of this same sequence led to the bis(trimethylsilyl)-protected dehydropentamer 55 in 5% overall yield. Although 55 could be protiodesilylated at the acetylenic terminus and at both ends, no appropriate cross-coupling method was available at the time, and a desperate attempt to achieve the coupling by transformation of the desilylated 55 into a derivative bearing an iodine atom on the cyclopropane ring followed by treatment with palladium acetate in DMF did not lead to a detectable amount of 48 [14].

A second, more convenient approach was tried starting with dicyclopropylacetylene 56 [15–17] which, after deprotonation with *tert*-butyllithium at both propargylic positions, yielded the bisaldehyde 57 upon treatment with dimethylformamide. The latter was converted to the bis(dibromoethenyl) derivative 58, and this in turn to the dibromotriyne 59 by standard methods (Scheme 11) [18]. The dibromide 59 was coupled with the ethynylcyclopropylheterocuprate 50 to give the terminally bis-protected pentayne 60.

Although an attempted twofold alkynylation of a difunctional C_1-building block with the deprotected pentayne 61 could not be achieved, intramolecular acetylene coupling under oxidative conditions was successful and gave the cyc-

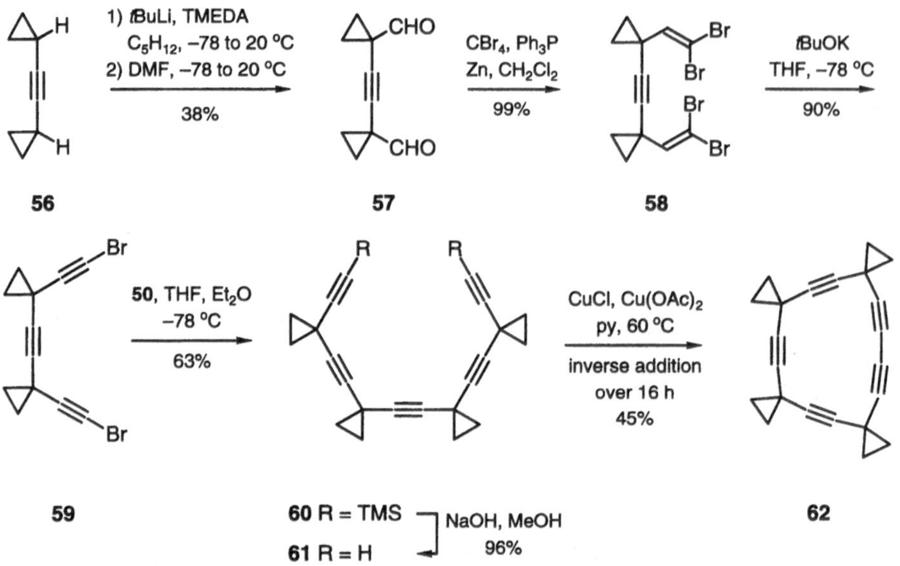

Scheme 11. Assembling the synthetic precursors to perspirocyclopropanated [5]pericycline 62 via a two-component approach

lic pentayne **62** which can be considered as a perspirocyclopropanated [5]pericy-cline lacking one spirocyclopropane linkage [15,18].

2.3
Heterocyclic Analogues of [*n*]Pericyclines

Heterocyclic analogues of [*n*]pericyclines of general formula **64** in which $X = S$, SiR_2, PR or any combinations with dimethylmethylene ($X = CMe_2$) linker units can conceptually be approached by applying three main synthetic strategies (Scheme 12): Pathway A is analogous to that developed for pericyclinones and pericyclinedinones (see above, Scheme 9); along pathway B, two or more frag-

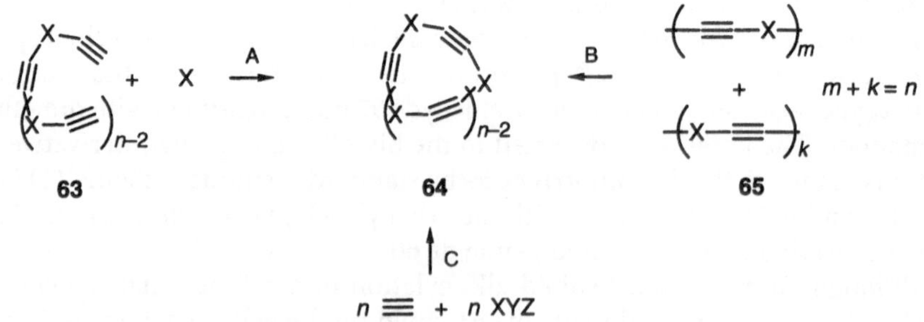

$$m + k = n$$

Scheme 12. Conceivable synthetic approaches to hetero[*n*]pericyclines **64**

ments, each containing heteroatoms and acetylene moieties, would be linked to each other – and, last but not least, a more random access from n acetylene and n heteroatomic fragments ("shotgun" approach, pathway C) can be applied.

In contrast to the series of hydrocarbons, heterocyclic analogues, even of cyclonona-1,4,7-triyne 2 ($n=3$), are known. This is because heteroatom linkers more easily adopt smaller bonding angles and thereby provide some relief of overall angle strain. Since heteroatoms have different electronic properties to carbon atoms, the perceived homoconjugative and homoaromatic effects might be expressed more pronouncedly in heterocyclic [n]pericyclines.

The first two principles were successfully applied by Scott et al. for the preparation of tri- and tetra-*tert*-butylphospha[3]- and -[4]pericyclines **67** and **68**, respectively (Scheme 13) [19]. While **67** was obtained as a single compound, the *trans*-diastereomer, the tetraphospha[4]pericycline **68** was obtained as a mixture of four diastereomers.

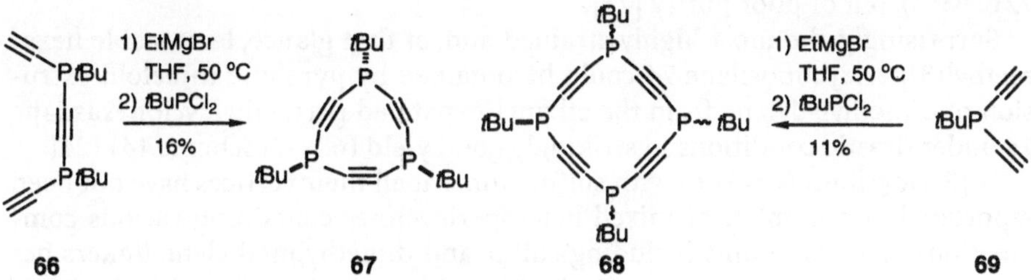

Scheme 13. Preparation of phospha[n]pericyclines

The feasibility of strategies B and C has been demonstrated particularly for the synthesis of the so-called pericyclinosilanes (previously also named cyclosilethynes [20]) **70–72** (Fig. 3). Historically, an approach of type C was the first one ever, in that [4]pericyclinosilane **70** was prepared by simple pyrolysis of a mixture of calcium carbide and dichlorodimethylsilane in a molten salt mixture of KCl/NaCl at 400 °C, albeit in low yield (actual yield not reported) [21]. A

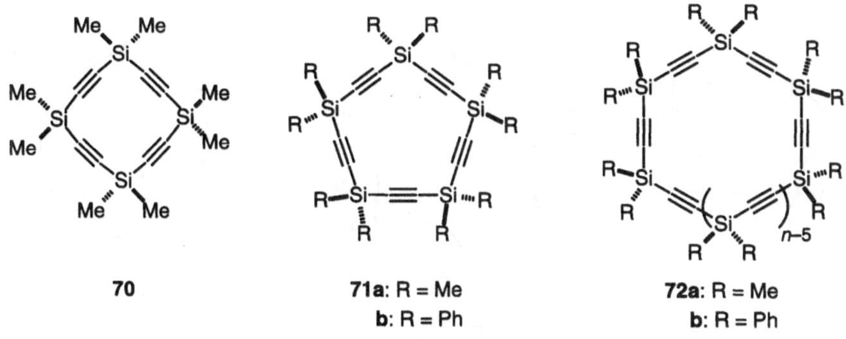

Fig. 3. The so-called pericyclinosilanes (sila[n]pericyclines) **70–72**

mixture of [n]pericyclinosilanes 72a (n=3–12) was obtained when dichlorodimethylsilane and acetylene were added simultaneously to a suspension of K/Na alloy in THF, and [4]- and [5]pericyclinosilanes 70 and 71a were isolated from this mixture in low yields (actual yields not reported) [22]. Under the same conditions, dichlorodiphenylsilane gave [5]pericyclinosilane 71b as the main product in 30% crude yield.

Following strategy B in Scheme 12, two variants have been successfully executed. In the first one, the reaction of dimethylbis(dimethylchloroethynyl)silane (ClMe$_2$SiC≡C)$_2$SiMe$_2$ with the oligo(dimethylsila)oligoynebismagnesium bromide BrMgC≡CSiMe$_2$(C≡CSiMe$_2$)$_l$C≡CMgBr (l=1,2) provided compounds 70 and 72a (n=6) in 2.4 and 5% yield, respectively. The corresponding [7]- (n=7) 72a (3.8%) and [8]pericyclinosilanes (n=8) 72a (5%) were also prepared along this route [20]. In the second variant, bislithiodiacetylides R$_2$Si(C≡CLi)$_2$ were treated with the corresponding dichlorodiorganylsilanes R$_2$SiCl$_2$ to give [6] pericyclinosilanes 72a (n=6) and 72b (n=6) in high crude yields (72a: 95%; 72b: 93%), but of poor purity [23].

Surprisingly, the most highly strained and, at first glance, least stable hexamethyl[3]pericyclinosilane 74 could be obtained by pyrolytic threefold extrusion of dimethylsilylene from the ethynyl-expanded permethylcyclohexasilane 73 under drastic conditions in strikingly good yield (68%) (Scheme 14) [24].

[n]Pericyclines (n=3–6) with sulfur atoms at all their vertices have not been reported, but a number of mixed heteropericyclines containing various combinations of heteroatoms including sulfur and dimethylmethylene linkers between the ethynyl units such as 75–77 (Fig. 4) have been prepared. This chem-

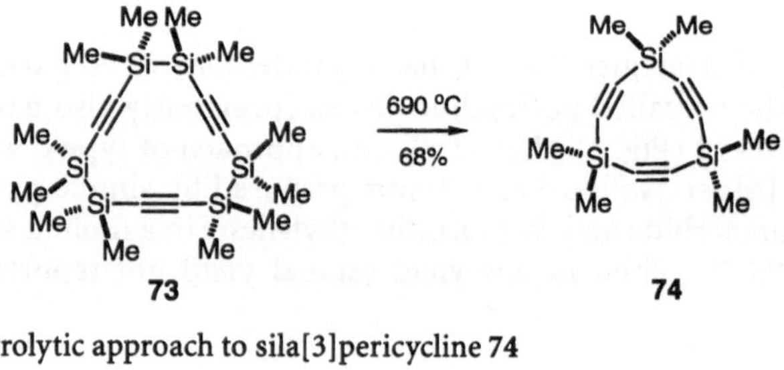

Scheme 14. Pyrolytic approach to sila[3]pericycline 74

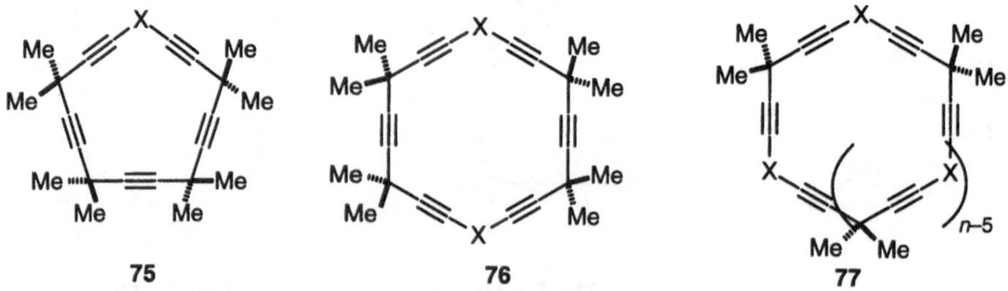

X = S, Me$_2$S, tBu$_3$P; n = 5, 6

Fig. 4. Some mixed heteropericyclines 75–77

istry has been excellently reviewed by Scott et al. [4] and, since the majority of these results were not published as original papers, there is no need to repeat this work here.

2.4
Structural Features of [n]Pericyclines and the Quest for Homoconjugation as well as Homoaromaticity

The X-ray crystallographic structures of [n]pericyclines 3 – 5 ($n = 5, 6, 7$) resemble those of the corresponding cycloalkanes with the same number of sp^3 carbon atoms: The acetylene-expanded permethylcyclopentane 3 adopts an envelope conformation, the homologue 4 exists in a chair conformation, and the expanded permethylcycloheptane 5 has a tub conformation [4,25]. Octamethyl[5] pericycline 30 adopts a non-planar conformation with C_{2h} symmetry [25], whereas the ring in hexamethyl[4]pericyclinone 43 is perfectly planar [4]. No significant deviations from the normal values for C–C and C≡C bond lengths were detected in 3 – 5, and in all cases the acetylene moieties are slightly bent outwards, e.g. the average C–C≡C angle for 3 was reported to be 177.3°. The [5]pericyclinosilane 71a also exists in an envelope conformation, and all the bond lengths also exhibit normal values, yet the acetylene moieties are bent inwards to the same extent (with an average Si–C≡C angle of 177.2°) as those in 3 are bent outwards, apparently because the ≡C–Si–C≡ angle prefers to be smaller then the ≡C–Si–C≡ angle in [5]pericycline 3 [25,26]. The 12-membered ring in [4]pericyclinosilane 70 is planar and the overall structure has D_{4h} symmetry [26] with the acetylene moieties being bent outwards with an average Si–C≡C angle of 173.1° and a ≡C–Si–C≡ angle of 103°. Even in 70 the bond lengths do not show any significant deviations from normal values. The X-ray crystallographic structural analyses of tri-*tert*-butylphospha[3]- 67 and tetra-*tert*-butyl-phospha[4]pericycline 68 confirm the significant relief of bond angle deformation at the acetylenic carbon atoms brought about by the heteroatoms: the endocyclic bond angles at the phosphorus atoms are 91 and 96°, and the P–C≡C angles are 163 and 174°, respectively [4].

    As aptly expressed by Scott et al. [4] who conceived the [n]pericyclines, the quest for cyclic homoconjugation and neutral homoaromaticity in these hydrocarbons has a similar connotation as that for these same phenomena in triquinacene: some experimental evidence appears to confirm the existence of such effects, but other experimental data and a number of computational results indicate that such effects are negligible, at least for these neutral species. Indeed, the main experimental result in the case of decamethyl[5]pericycline 3, as for triquinacene, interpreted in terms of a small, yet significant homoaromatic stabilization, was its heat of hydrogenation ($\Delta H^\circ_{hydr} = -340.7$ kcal/mol) [27]. Comparing the heats of hydrogenation in a series of analogous acyclic oligoynes of type $Me_3C(C≡C-CMe_2)_nMe$ ($n = 2 - 5$) allowed the derivation of an enthalpy increment value of –69.8 kcal/mol for each single ethynyldimethylmethylene moiety in an acyclic homoconjugated oligoyne assembly. Applying an additivity Scheme with this increment, a value of –349 kcal/mol was calculated for the heat of hydrogenation of 3, and the difference of about 6 kcal/mol in the experimental value was attributed to a homoaromatic stabilization of the cyclic system. A significant degree of electronic interaction between the acetylene moieties in 3 and 4 was also revealed by their photoelectron spectra [25]. Furthermore, electron transmission spectroscopy of 3 and 4 demonstrated a large splitting (ca. 1.6 eV) of the nonbonding orbitals of these compounds, and their LUMOs being stabilized by 0.4–0.7 eV relative to the LUMO of acetylene [4,25]. Additional evidence has also been extracted from UV spectra [4, 6, 25].

Criteria which do not support or even contradict any significant effect due to neutral homoaromaticity in such hydrocarbons are on the one hand the geometric data obtained for permethyl[5]pericycline 3 which reveal no shortening of C–C single bonds or lengthening of C≡C triple bonds as would be expected in the case of cyclic delocalization (the values 1.190(5) Å for C≡C and 1.480(7) Å for ≡C–C bonds are absolutely normal for such bonds) [25], and on the other the NMR spectra which show no evidence for ring currents resulting from cyclic homoconjugation [6].

Computational studies have been performed for [5]pericycline 3 at various levels of theory. In the early stages, MNDO calculations suggested that interactions between the triple bonds in such compounds would be hyperconjugative in the π-system and homoconjugative in the σ-system [28]; however, MM2 force-field and ab initio STO-3G calculations later indicated that no special stabilization would arise from such electronic interactions, but they would influence the spectroscopic properties of [n]pericyclines [25]. In line with these results an ab initio HF/3-21G study concluded that "the magnitude of aromatic stabilization, whether positive or negative, is probably not larger..." than 1 kcal/mol independent of the accuracy of the method used [29]. Eventually, the DFT calculations performed at a reasonably high level of theory by Schleyer, de Meijere et al. (Becke3LYP/D95(d)/Becke3LYP/6–31G*) demonstrated the total absence of any evidence for homoaromaticity in [5]pericycline 1, and decamethyl[5]pericycline 3 as far as geometric, energetic and magnetic criteria were concerned, and that they are not homoconjugated despite their intriguing structures (Fig. 5) [30].

Since the highest occupied molecular orbitals (HOMOs) of a cyclopropane ring [9, 10] are much closer in energy to the π-MOs of an acetylene unit than are the σ-MOs of a gem-dimethylmethylene group, perspirocyclopropanated [5]pericycline 48 ought to exhibit considerably stronger homoconjugative effects, and this expectation appeared to be supported by the experimental observation that the energy splitting between the π-MOs parallel to the plane of the ring in 1,1-diethynylcyclopropane 47 is much larger (1.4 eV) than in 3,3-dimethylpenta-1,4-diyne 35 (0.6 eV). [11, 12]. Nevertheless, even for perspirocyclopropanated [5]pericycline 48, Becke3LYP/6–31G* calculations indicated no exceptional ground state properties [30]. Thus, calculated geometrical parameters for 48 (1.211 Å for C≡C and 1.449 Å for ≡C–C$_{cycl}$ bonds, 114° for ≡C–C–C≡ angle) differ only slightly from those experimentally determined for 1,1-diethynylcyclopropane 47 (1.197, 1.442 Å and 115.1°, respectively) [31]. Based on the homodesmotic equation: $5 \times 48 + 5 \times HC≡CH = 5 \times 47$ (using experimentally determined values of $\Delta H_f^\circ(g) = 54.5$ kcal/mol for acetylene [32] and 128.7 kcal/mol for 1,1-diethynylcyclopropane 47, [33]), a $\Delta H_f^\circ(g)$ for 48 of 371.2 kcal/mol can be

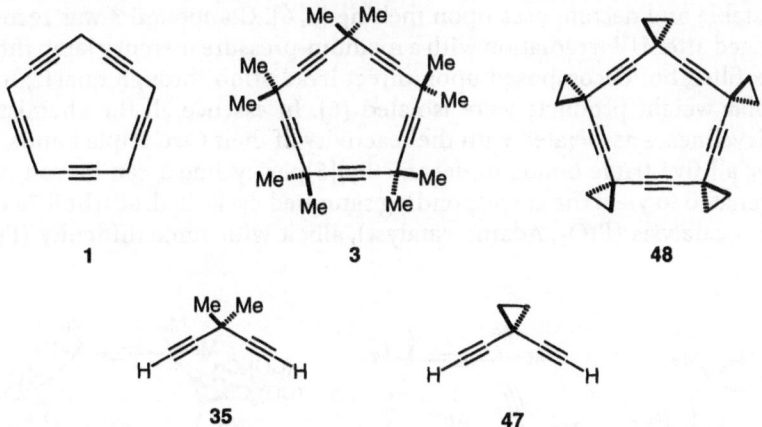

Fig. 5. [5]Pericyclines with conceived cyclic homoconjugation and neutral homoaromaticity

calculated. However, the calculated difference between the strain energy of pericycline **48** and the sum of strain energies from increments for the subunits such as **47** was only 0.6–0.9 kcal/mol (Becke3LYP/D95(d)/Becke3LYP/6–31G*) confirming that perspirocyclopropanated [5]pericycline **48** does not experience any significant homoaromatic ground state stabilization [30].

Practically the same conclusion has been reached for the [4]-, [5]- and [3]pericyclinosilanes **70, 71 a** and **74** [26,34]: AM1 calculations indicate a strong conjugative interaction between the C≡C bonds and the silicon π-MOs which leads to an overall splitting of the corresponding MOs of 1.73 eV; however, the total energy is not different from that of a non-conjugated model [26]. A strong conjugative interaction in [3]pericyclinosilane **74** was derived from the PE spectroscopic data [34]. The very high value of the silicon–carbon coupling constants in the NMR spectra, as well as UV spectroscopic data for **70, 71 a**, could not be interpreted beyond doubt as a result of cyclic homoconjugation, and structural experimental data do not confirm the presence of cyclic homoconjugation [26].

All this is in line with the most recent finding for triquinacene, for which the direct determination of its $\Delta H_f^\circ(g)$ from its experimentally measured heat of combustion finally corroborated the results of the most advanced computational studies that triquinacene is not homoaromatic [35]. Evidently, heat of combustion measurements should also be carried out for some representative [n]pericyclines to finally settle the quest for their neutral homoaromaticity.

2.5
Chemical Properties of [n]Pericyclines and Analogues

The [n]pericyclines **3–6** are all colorless, crystalline, light- and air-stable solids which do not exhibit any shock sensitivity. They exhibit sharp melting behavior without showing signs of decomposition. However, octamethyl[5]pericycline **30** is less stable and decomposes upon melting [4, 6]. Compound **3** was recovered unchanged after UV irradiation with a medium-pressure mercury lamp through a Pyrex filter, but decomposed upon direct irradiation through quartz; no low molecular weight products were isolated [6]. In essence all the chemistry of [*n*]pericyclines is associated with the reactivity of their C≡C triple bonds.

Thus, all five triple bonds in decamethyl[5]pericycline **3** can be completely hydrogenated to yield the corresponding saturated cyclic hydrocarbon **78** under platinum catalysis (PtO$_2$, Adam's catalyst), albeit with some difficulty (Fig. 6).

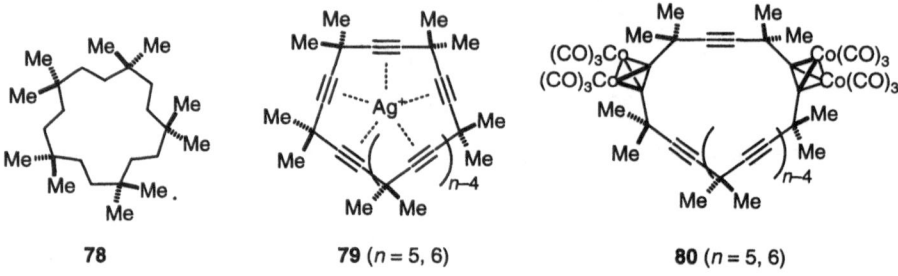

Fig. 6. Products of chemical transformations of pericyclines **3** and **4**

[5]- and [6]-Pericyclines **3** and **4** form 1:1 complexes with silver triflate in 74 and 75% yield, respectively. The NMR spectra of these complexes revealed their structures to be similar to **79** with the silver cation placed in the center of each ring. With Co$_2$(CO)$_8$ the penta-and hexaynes **3** and **4** form typical mono- and bis(hexacarbonyl)dicobalt complexes **80** (Fig. 6). The hydrocarbons **3** and **4** were regenerated quantitatively from the cobalt, as well as from the silver, complexes. [*n*]Pericyclines reacted rapidly with bromine to give a mixture of a large number of products [6].

Stability towards heating, irradiation and exposure to air was also reported for [*n*]pericyclinosilanes **70–72** [20, 21], and even the highly strained **74** [24]. But, in contrast to the hydrocarbons, the pericyclinosilanes did not react with bromine [21]. On the other hand, the presence of the silicon atoms brings about a specific reactivity of these macrocycles making them labile and capable of changes in ring size. Thus, when a mixture of compounds **70–72a** (*n*=6) was

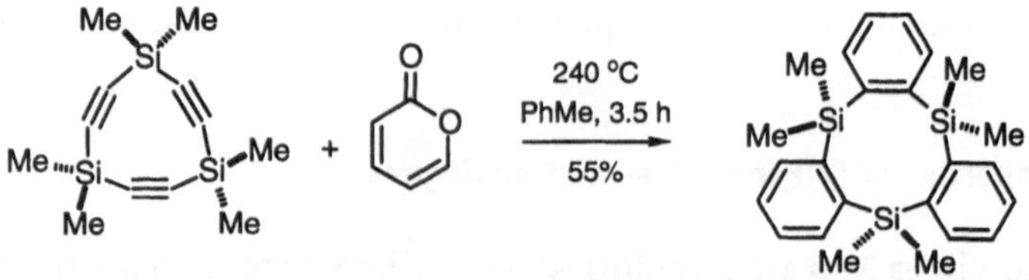

Scheme 15. Thermal cycloaddition of α-pyrone to sila[3]pericycline **74**

stirred in THF in the presence of a catalytic amount of BuLi at 20 °C over a period of 6 d, the [8]pericyclinosilane **72a** ($n = 8$) was isolated in 50% yield [26]. Moreover, a randomization of substituents was observed when a mixture of **72a** ($n = 6$) and **72b** ($n = 6$) was treated with alkyllithium reagents [23]. The highly strained [3]pericyclinosilane **74** was reported to react with three molecules of α-pyrone to produce 1,1,4,4,7,7-hexamethyltribenzo-1,4,7-trisilacyclonona-2,5,8-triene (**81**), presumably via Diels–Alder reaction of all three triple bonds followed by extrusion of carbon dioxide (Scheme 15) [24].

3
Acetylene-Expanded [*n*]Pericyclines and Butadiyne-Expanded [*n*]Rotanes

Although the [*n*]pericyclines do not exhibit the perceived homoaromatic stabilization and cyclic conjugation, they are non-the-less interesting and challenging compounds. Even more challenging are the expanded [*n*]pericyclines **84** which are formally derived from the [*n*]pericyclines **3–6** by insertion of an additional acetylene unit between each two dimethylmethylene groups or from permethylcycloalkanes by insertion of 1,3-butadiyne spacers between the vertices (Fig. 7). By analogy, butadiyne-expanded [*n*]rotanes (expanded perspirocyclopropanated [*n*]pericyclines) **85** can be conceived. Although the angle strain in such compounds would be distributed over a larger number of *sp*-hybridized carbon atoms, thus making even the smaller members of this family like **82** and **83** accessible, the heat of formation per carbon atom is more positive than that

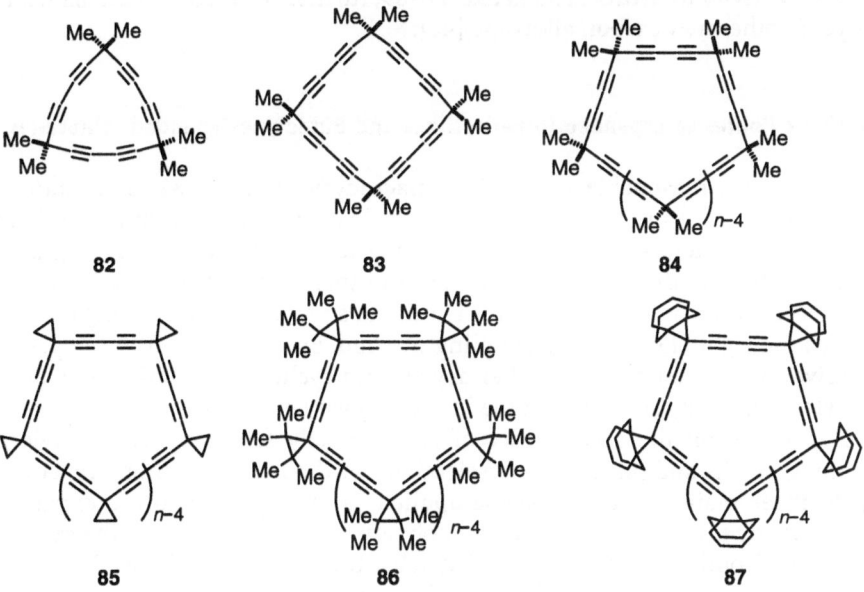

Fig. 7. Expanded [*n*]pericyclines and "exploded" [*n*]rotanes

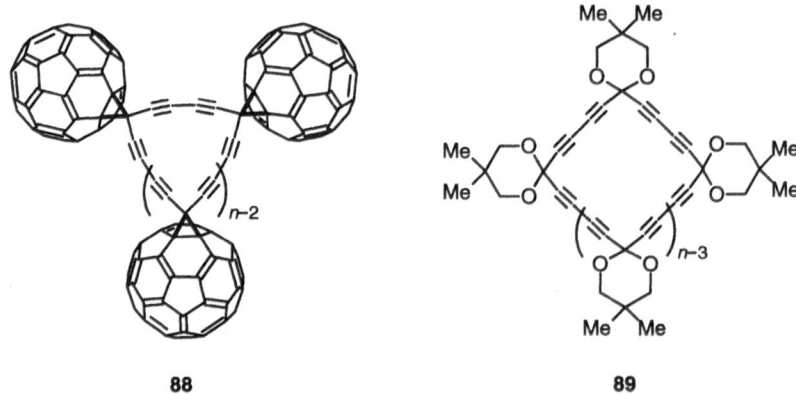

Fig. 7 (continued)

of the [n]pericyclines and thus the expanded systems of types **84** and **85** ought to be more labile than the corresponding parent compounds like **3–6** and **48**. Since, on the other hand, butadiyne units can conveniently be created by oxidative coupling of terminal acetylenes, a broader range of such macrocycles including types like **86** and **87** with fully substituted cyclopropane, even **88** with methano-C_{60}-fullerene or **89** with acetal moieties at all the vertices, can be approached. Interesting spectroscopic and structural features, as well as thermal and chemical properties, can be expected from such intriguing macrocycles [7, 36–39]. The strain energy of the cyclopropane ring brings an element of additional reactivity to hydrocarbons like **85–89**; furthermore, compound **88** would be yet another new carbon allotrope [40].

3.1
Synthetic Routes to Expanded [n]Pericyclines and Butadiyne-Expanded [n]Rotanes

Obviously, any reasonable approach to macrocycles of types **82–89** containing butadiyne moieties makes use of one or more of the well-known methods for acetylene–acetylene homo- or cross-coupling according to Glaser [41], Eglinton [42], Hay [43], Cadiot-Chodkiewicz [44] and their modern modified variants. The success then depends on the availability of an appropriate building block as well as the right choice of the coupling method. Even the solvent often plays a decisive role, as some of the higher acyclic and cyclic dehydrooligomers of diacetylenes are only sparsely soluble in the majority of solvents.

Depending on the accessibility of appropriately functionalized acetylenic building blocks and on the chemoselectivity of their cross-coupling reactions, up to three main strategies can be applied as an approach to the expanded [n]pericyclines (Scheme 16). The most obvious and simplest one is the random cyclodehydrooligomerization of 3,3-disubstituted-1,4-pentadiynes **90** ("shot-gun" approach, pathway A). The second strategy (pathway B) is more directed towards achievable ring sizes, as [n]macrocycles are prepared by oxidative

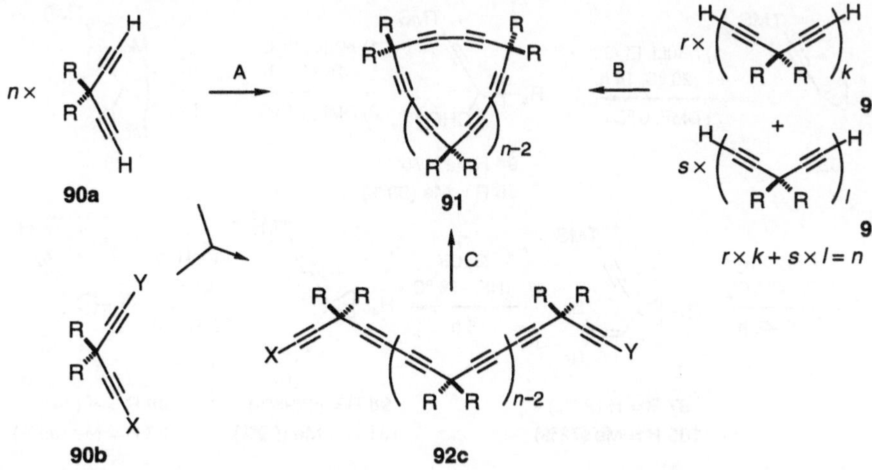

Scheme 16. Conceivable synthetic approaches to expanded [n]pericyclines

acetylene coupling of two or more acyclic dehydrooligomers **92** of the same or different chain lengths. These precursors, in turn, have to be synthesized by a stepwise directed assembly of appropriate monomers like **90a** and **90b**. The third and most target-oriented strategy (pathway C) calls for the ring closure of an acyclic dehydrooligomer **92c** of the corresponding chain length either by oxidative or some other cross-coupling method.

Although there is no general procedure for the preparation of the monomeric building blocks, the 3,3-disubstituted 1,4-pentadiyne derivatives of types **90a** and **90b**, a limited number of standard procedures have been used in most cases. The synthesis of 1-(trimethylsilyl)-3,3-dimethyl-1,4-pentadiyne **32** is summarized above (Scheme 7) (for the preparation of deprotected diyne **35** see also [45, 46]). The starting materials for the butadiyne-expanded [n]rotanes **85** and their permethylated analogues **86**, bromodiacetylenes **99** and **102**, were obtained in a similar manner starting from (trimethylsilylethynyl)cyclopropane **93** [17] and its permethylated 1-chloro derivative **95** [47], respectively. Deprotonation of **93** (or lithiation by halogen–metal exchange in the case of **95**) with n-butyllithium and subsequent treatment with dimethylformamide followed by Corey–Fuchs olefination and dehydrobromination with potassium $tert$-butoxide in THF gave 1-bromo-5-trimethylsilyldiynes **98** and **101** each in 66% overall yield. These diynes were cleanly protiodesilylated to give the 1-bromo-1,4-diynes **99** and **102** in 74 and 99% yield, respectively (Scheme 17) [36, 37, 48].

This approach provides a far greater versatility than one starting with the previously reported synthesis of the parent 1,1-diethynylcyclopropane **47** [49], and it was also applied for the preparation of 7,7-diethynyldispiro[2.0.2.1]heptane (dispirocyclopropanated diethynylcyclopropane) **108** [48] starting from the ester **103** [50] (Scheme 18).

The basic building block for the protected expanded [n]pericyclinones **89** [39] was obtained by simple acetalization of 1,5-bis(trimethylsilyl)penta-1,4-diyne-

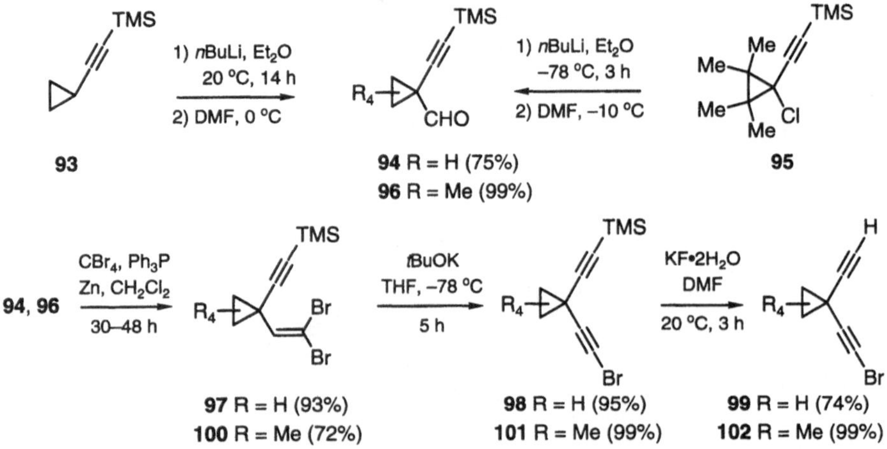

TMS

1) nBuLi, Et₂O
20 °C, 14 h
2) DMF, 0 °C

93

TMS

1) nBuLi, Et₂O
−78 °C, 3 h
2) DMF, −10 °C

R₄
CHO

94 R = H (75%)
96 R = Me (99%)

TMS

Me
Me
Me
Me
Cl

95

94, 96

CBr₄, Ph₃P
Zn, CH₂Cl₂
30–48 h

TMS

R₄
Br
Br

97 R = H (93%)
100 R = Me (72%)

tBuOK
THF, −78 °C
5 h

TMS

R₄
Br

98 R = H (95%)
101 R = Me (99%)

KF•2H₂O
DMF
20 °C, 3 h

H

R₄
Br

99 R = H (74%)
102 R = Me (99%)

Scheme 17. The main building blocks for butadiyne-expanded [*n*]rotanes and their permethylated analogues

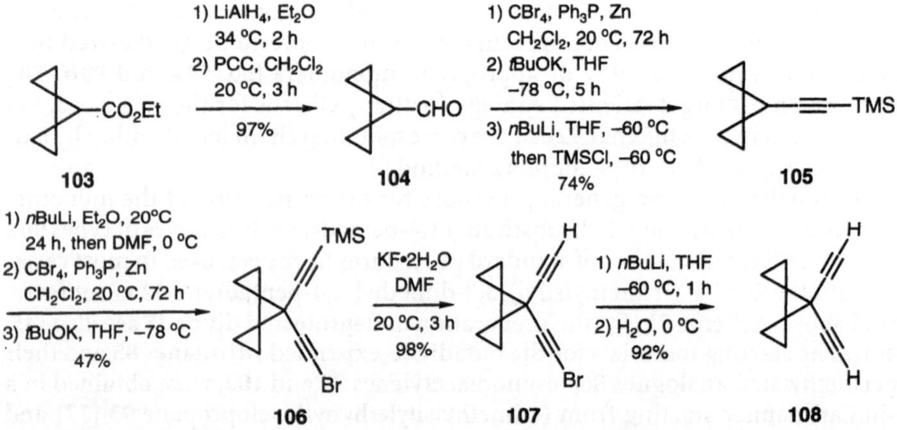

1) LiAlH₄, Et₂O
34 °C, 2 h
2) PCC, CH₂Cl₂
20 °C, 3 h
97%

—CO₂Et

103

—CHO

104

1) CBr₄, Ph₃P, Zn
CH₂Cl₂, 20 °C, 72 h
2) tBuOK, THF
−78 °C, 5 h
3) nBuLi, THF, −60 °C
then TMSCl, −60 °C
74%

≡—TMS

105

1) nBuLi, Et₂O, 20 °C
24 h, then DMF, 0 °C
2) CBr₄, Ph₃P, Zn
CH₂Cl₂, 20 °C, 72 h
3) tBuOK, THF, −78 °C
47%

TMS

Br

106

KF•2H₂O
DMF
20 °C, 3 h
98%

H

Br

107

1) nBuLi, THF
−60 °C, 1 h
2) H₂O, 0 °C
92%

H

H

108

Scheme 18. The main building blocks for perspirocyclopropanated expanded [*n*]rotanes

trione **109** (prepared by oxidation of 1,5-bis(trimethylsilyl)penta-1,4-diyne) [51]), followed by protiodesilylation (Scheme 19). Finally, even compounds as exotic as diethynylmethanofullerene **115** and substituted diethynylmethanofullerene **118** have been prepared via diethynylcarbene addition onto C₆₀-fullerene **112** and its derivative **116**, respectively, with subsequent protiodesilylation, and have been suggested as synthetic precursors for expanded [*n*]rotanes of type **88** with fused C₆₀ moieties at every three-membered ring (Scheme 19) [38, 52].

Despite its simplicity, the one-step "shotgun" synthesis of expanded [*n*] pericyclines has found only limited application because of its low selectivity and the difficulties associated with the separation of the product mixtures. For

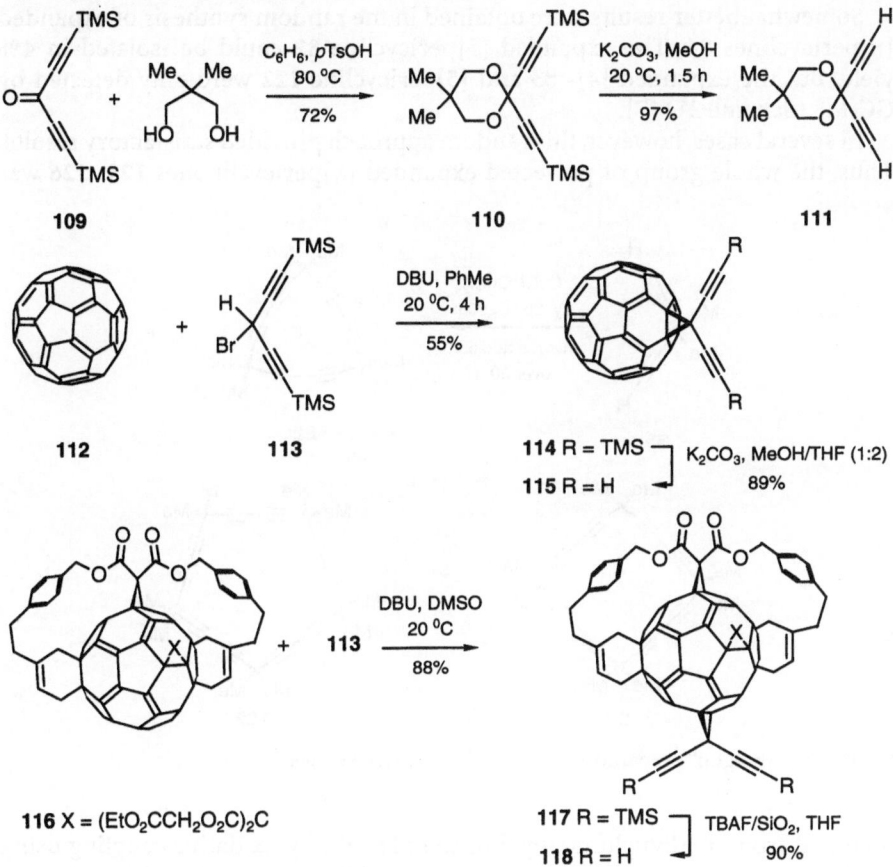

Scheme 19. The main building blocks for expanded [*n*]rotanes and [*n*]pericyclinones of types **88, 89**

example, the attempted oxidative coupling of 1,1-diethynylcyclopropane **47** led to a mixture of the open-chain dehydrooligomers **119 – 121,** consisting of two to four 1,1-diethynylcyclopropane units, in low yields (Scheme 20) (higher dehydrooligomers may also have been formed, but were not isolated and characterized) [37].

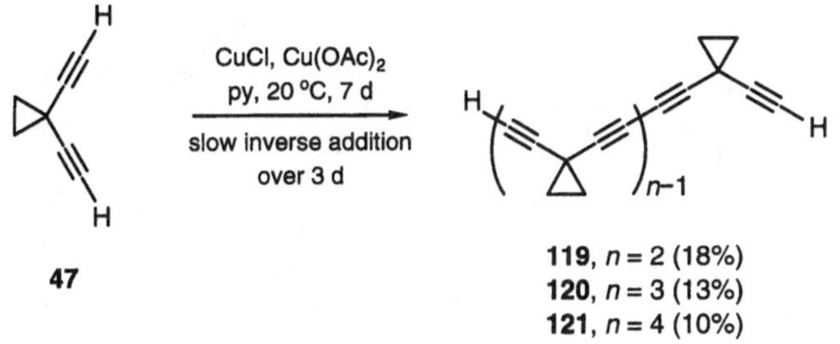

119, *n* = 2 (18%)
120, *n* = 3 (13%)
121, *n* = 4 (10%)

Scheme 20. Attempted "shotgun" synthesis of expanded [*n*]rotanes

Somewhat better results were obtained in the random synthesis of expanded [*n*]pericyclines 84: The expanded [3]pericycline 82 could be isolated in 4% yield, but the expanded [4]- 83 and [5]pericycline 122 were only detected by GC-MS (Scheme 21) [7].

In several cases, however, this random approach provided satisfactory results. Thus, the whole group of protected expanded [*n*]pericyclinones 123–126 was

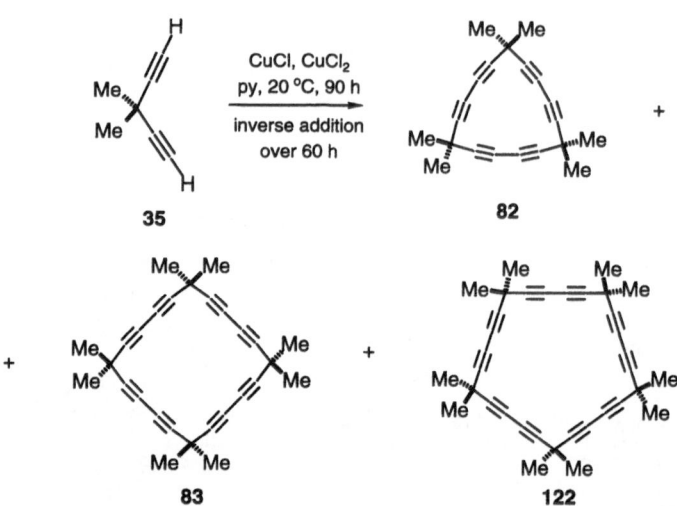

Scheme 21. "Shotgun" preparation of expanded [*n*]pericyclines

prepared from diethynyldimethyldioxane 111 [39] by oxidative coupling using modified Eglinton conditions [53] (Scheme 22). Macrocycles 123–126 were separated by preparative gel permeation chromatography (GPC). Under Hay conditions, the yield of compounds 123–126 was negligible [39].

In some rare cases, the "shotgun" method appears to be the only feasible one. A typical example is the cyclodehydrooligomerization of diethynylmethano-

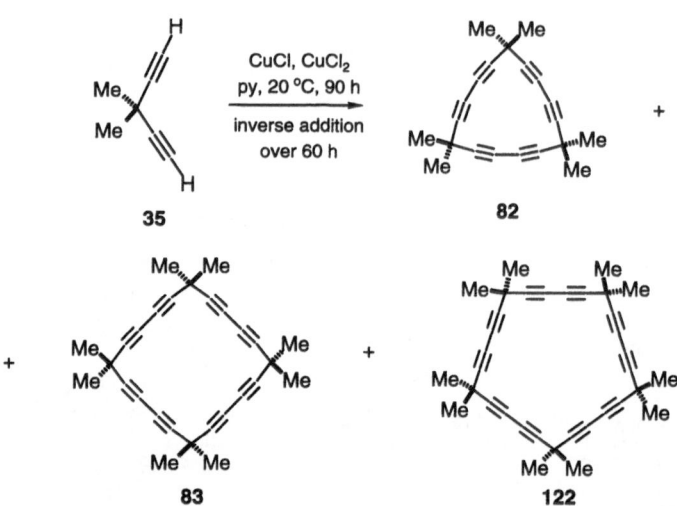

123 *n* = 4 (6%)
124 *n* = 5 (13%)
125 *n* = 6 (9%)
126 *n* = 7 (3%)

Scheme 22. Random preparation of protected expanded [*n*]pericyclinones

fullerenes **115** and **118** (Scheme 23) [38], for which the stepwise assembly seems to be less advantageous (although possible [52]) and the problem of solubility of the products, as well as of the starting materials, was of prime importance. The attempted oxidative coupling of compound **115** led only to a mixture of insoluble products [38]. Oxidative cyclooligomerization of the highly substituted diethynylmethanofullerene **118** under Hay conditions failed to produce isolable amounts of the cyclic dehydrooligomers, but the application of Eglington coupling conditions led to remarkable success: Compounds **127** and **128** were isolated in 53% combined yield [38]. Larger cyclic dehydrooligomers were not detected; their formation may have been suppressed due to the steric bulk of the substituted C_{60}-fullerene moieties.

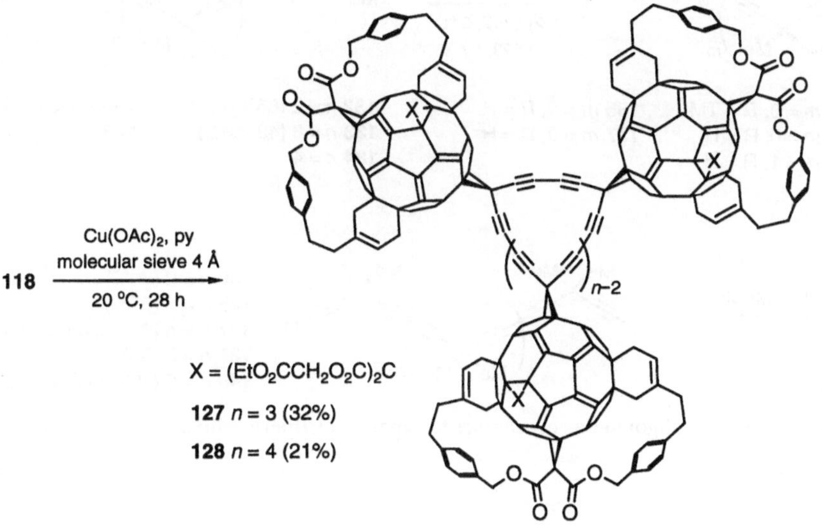

$$118 \xrightarrow[\text{20 °C, 28 h}]{\substack{\text{Cu(OAc)}_2,\text{ py} \\ \text{molecular sieve 4 Å}}}$$

$X = (EtO_2CCH_2O_2C)_2C$

127 $n = 3$ (32%)

128 $n = 4$ (21%)

Scheme 23. "Shotgun" synthesis of C_{60}-fullerene-annelated expanded [n]rotanes

The second and third strategies for the assembly of macrocyclic oligodiacetylenes, consisting of oxidative cyclizing of one, two or more open-chain dehydrooligomers, have been widely used for the preparation of acetylene-expanded [n]pericyclines and butadiyne-expanded [n]rotanes like **82–86**. The starting building blocks for these syntheses can easily be prepared from the corresponding monomers like **32, 35, 98, 99, 101, 102,** and **129**. Using only two standard operations in a repetitive way, namely the Cadiot–Chodkiewicz coupling via intermediate copper derivatives and protiodesilylation under various conditions, it was possible to construct – step by step, if necessary – any open-chain permethylated [4, 7] as well as cyclopropanated [36, 37] dehydrooligomer of any desired chain length (Schemes 24 and 25).

This reaction sequence was also found to be convenient and versatile for preparing the potential synthetic precursors to permethylated **86** and spirocyclopropanated exploded [n]rotanes (Scheme 26) [48].

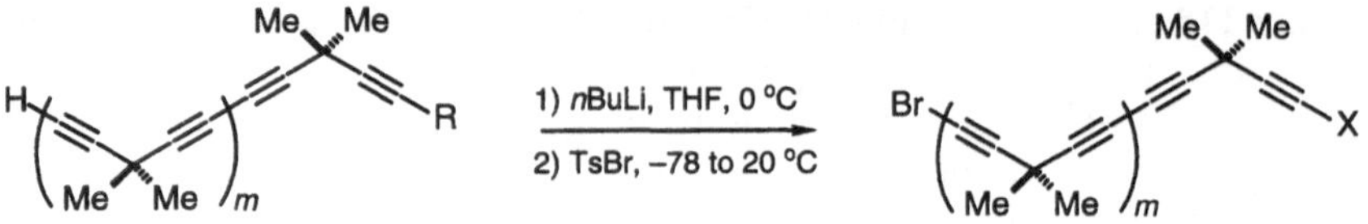

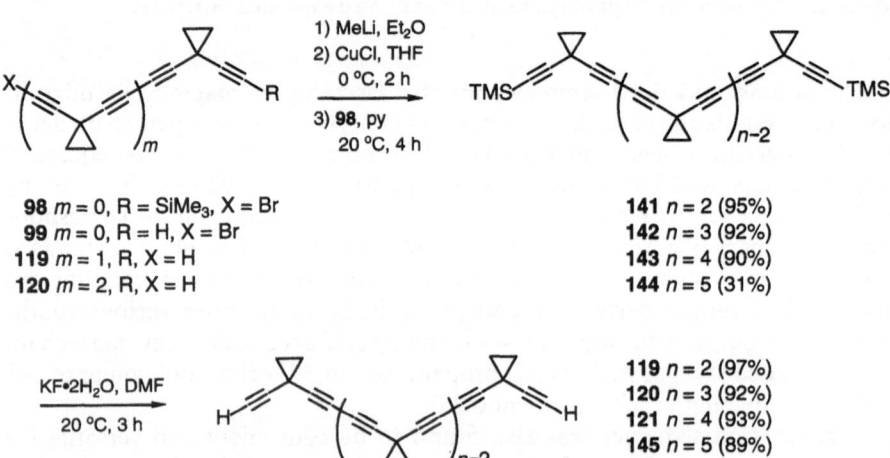

32 m = 0, R = TMS
130 m = 1, R = H

1) nBuLi, THF, 0 °C
2) TsBr, −78 to 20 °C

129 m = 0, X = TMS (76%)
131 m = 1, X = Br (69%)

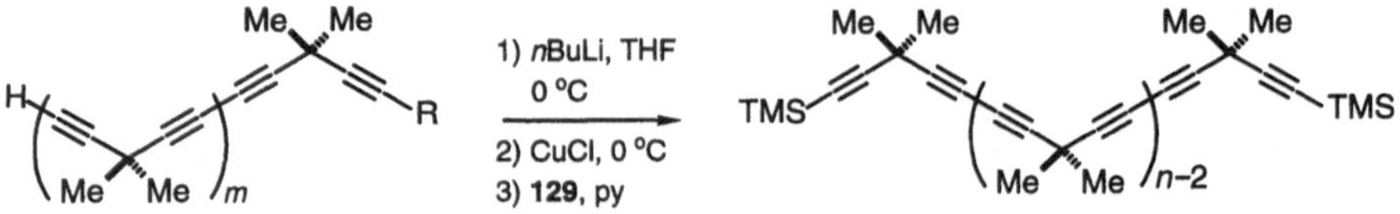

1) nBuLi, THF 0 °C
2) CuCl, 0 °C
3) **129**, py

32 m = 0, R = TMS
35 m = 0, R = H
130 m = 1, R = H

135 m = 2, R = H
137 m = 3, R = H

132 n = 2 (86%)
133 n = 3 (42–53%)
134 n = 4

136 n = 5 (45%)
138 n = 6

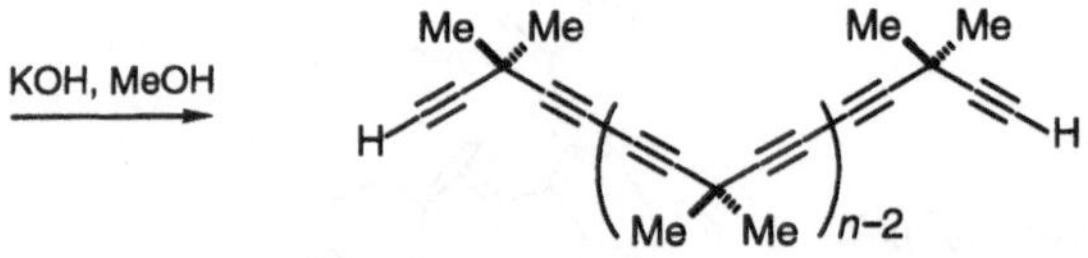

KOH, MeOH

130 n = 2 (88%)
135 n = 3 (87%)
137 n = 4 (73% over 2 steps)
139 n = 5 (63%)
140 n = 6 (46% over 2 steps)

Scheme 24. Acyclic oligoynes as precursors to expanded [n]pericyclines

1) MeLi, Et₂O
2) CuCl, THF 0 °C, 2 h
3) **98**, py 20 °C, 4 h

98 m = 0, R = SiMe₃, X = Br
99 m = 0, R = H, X = Br
119 m = 1, R, X = H
120 m = 2, R, X = H

141 n = 2 (95%)
142 n = 3 (92%)
143 n = 4 (90%)
144 n = 5 (31%)

KF•2H₂O, DMF
20 °C, 3 h

119 n = 2 (97%)
120 n = 3 (92%)
121 n = 4 (93%)
145 n = 5 (89%)

Scheme 25. Acyclic oligoynes for the preparation of expanded [n]rotanes

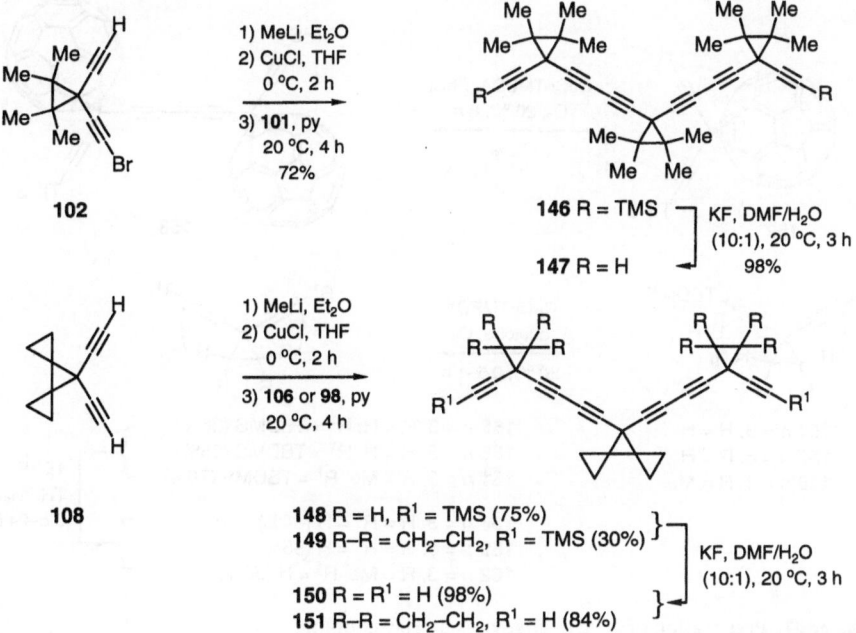

Scheme 26. Acyclic oligoynes for the preparation of permethylated and spirocyclopropanated expanded [*n*]rotanes

Potential starting materials for the syntheses of exploded [*n*]rotanes via approaches B and C containing an even number of cyclopropane units may also be prepared by applying the Hay coupling procedure (Scheme 27) [48, 52].

The conversion of the dehydrotrimer **135** into the corresponding bis-cuprate followed by coupling with dibromide **131** (Cadiot–Chodkiewicz conditions) gave the expanded [5]pericycline **122** in 53% isolated yield (Scheme 28) [4]. The more versatile approach by simple oxidative cyclooligomerization of dehydro-oligomers of type **135** under high dilution conditions as shown in Scheme 28 provided the acetylene-expanded [3]- **82**, [5]- **122** and [6]pericyclines **163** in reasonable to excellent yields [4,7].

Analogously, when solutions of the acyclic precursors **119–121**, **146**, **156** and **159** or mixtures of two of them in pyridine were added over a period of 3 d to a slurry of cuprous chloride and cupric acetate in pyridine [54], and the reaction mixtures were stirred at ambient temperature for an additional 4 d, the cyclic dehydrooligomers of 1,1-diethynylcyclopropane **47** could be isolated in relatively good yields by column chromatography and/or recrystallization (Scheme 29). A 1:1 mixture of the short-chain starting materials **119** and **120**, and the dehydrotrimer **120** alone, gave rise to the cyclic dehydropenta- **165**, dehydrohexa- **166**, dehydrohepta- **167**, dehydroocta- **168**, and dehydrononamer **169** which resulted from all the possible simple combinations of ($n \times$ [3] + $m \times$ [2]) couplings. Similar results were obtained with a mixture of **120** and **121**. A direct ring closure of the

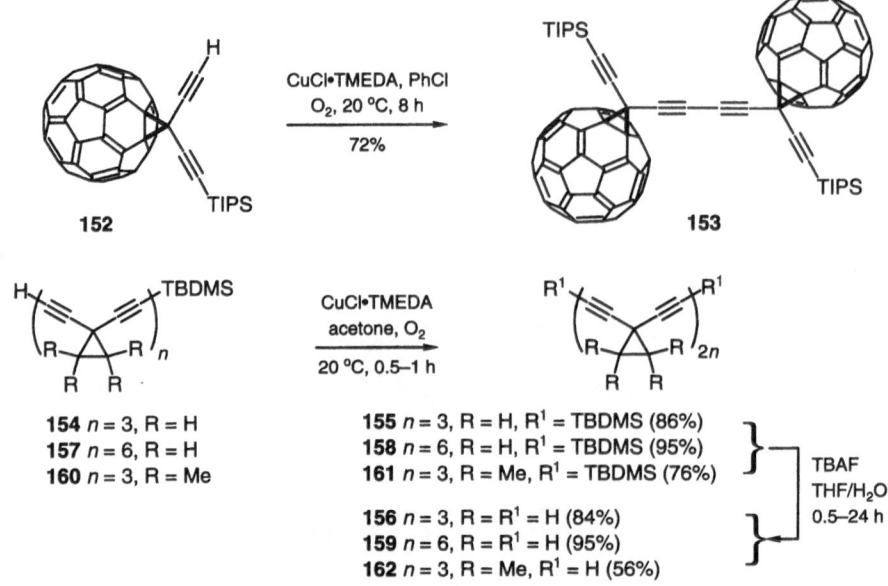

Scheme 27. Preparation of acyclic oligoynes by Hay coupling

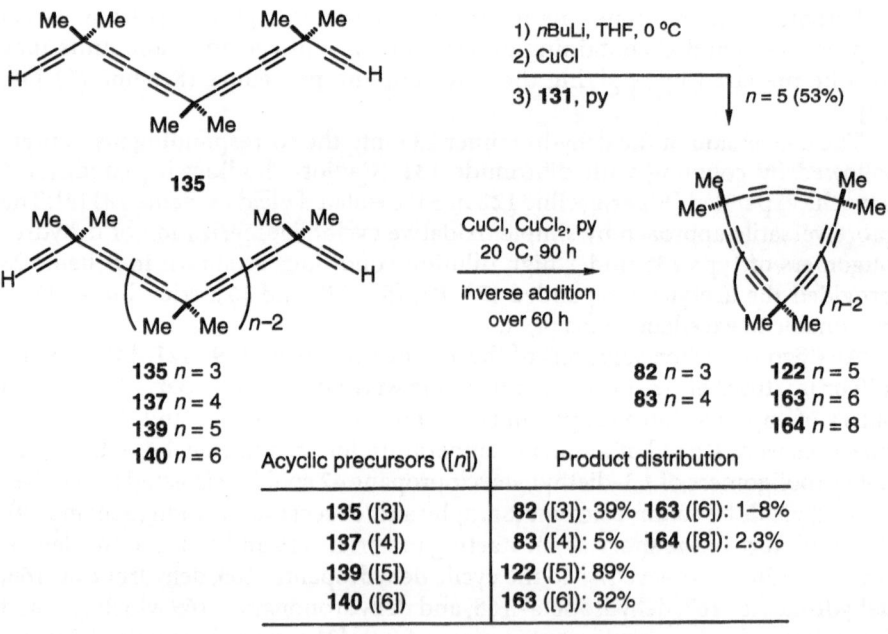

Acyclic precursors ([n])	Product distribution
135 ([3])	**82** ([3]): 39% **163** ([6]): 1–8%
137 ([4])	**83** ([4]): 5% **164** ([8]): 2.3%
139 ([5])	**122** ([5]): 89%
140 ([6])	**163** ([6]): 32%

Scheme 28. One-component synthesis of expanded [n]pericyclines

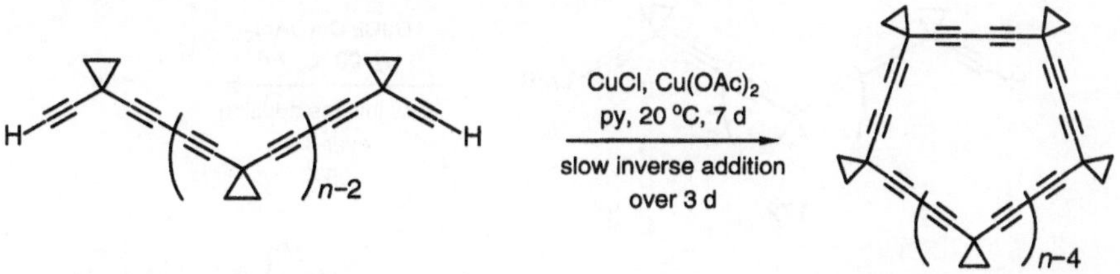

CuCl, Cu(OAc)$_2$
py, 20 °C, 7 d
———————————→
slow inverse addition
over 3 d

119 $n = 2$	**145** $n = 5$
120 $n = 3$	**156** $n = 6$
121 $n = 4$	**159** $n = 12$

165 $n = 5$	**169** $n = 9$
166 $n = 6$	**170** $n = 10$
167 $n = 7$	**171** $n = 12$
168 $n = 8$	

Acyclic precursors ([n])	Product distribution			
119 + 120 ([2] + [3])	**165** ([5]): 35%	**166** ([6]): 20%	**167** ([7]): 13%	**168** ([8]): 3%
	169 ([9]): trace			
120 ([3])	**166** ([6]): 39%	**169** ([9]): 8%		
120 + 121 ([3] + [4])	**166** ([6]): 28%	**167** ([7]): 44%	**168** ([8]): 14%	**169** ([9]): trace
121 ([4])	**168** ([8]): 46%	**171** ([12]): 1%		
145 ([5])	**165** ([5]): 70%	**170** ([10]): 0.2%		
156 ([6])	**166** ([6]): 49%	**171** ([12]): 5%		
159 ([12])	**171** ([12]): 7%			

Scheme 29. One-component synthesis of expanded [n]rotanes

acyclic dehydrotetramer **121** was not observed, but compound **121** readily dimerized to an acyclic dehydrooctamer, and that cyclized reasonably well to the "blown-up" [8]rotane **168** in 46% yield. The acyclic dehydropentamer **145** cyclized rather efficiently to give the "blown-up" [5]rotane **165** in 70% yield. As side products the higher cyclic dehydrooligomers **170** ($n = 10$) and **171** ($n = 12$) were obtained in negligible quantities (0.2 and 1%, respectively) [36, 37].

When the acyclic dehydrohexamer **156** was oxidatively cyclized, an increased yield (49%) of the butadiyne-expanded [6]rotane **166** was observed, but the yield of the expanded [12]rotane **171** with a C$_{60}$ inner ring suffered from the fact that the corresponding acyclic dehydrododecamer **159** slowly decomposed in the syringe pump [48].

The attempted oxidative dimerization of the monosilylated acyclic dehydrotetramer **172** led to the cyclic dehydrooctamer, the "blown-up" [8]rotane **168**, in even better yield (55%) than the analogous cyclodehydrodimerization of the desilylated precursor **121** (46% **168**) [37]. This demonstrates that desilylation does occur under the applied conditions and that at least a trimethylsilyl residue does not provide sufficient protection if long-chain acyclic dehydrooligomers are to be made (Scheme 30). However, cyclodehydrodimerization of the silylated permethyldehydrotrimer **147** gave a lower yield of the permethyl-exp-[6]rotane **173** (37%) than that of its desilylated analogue **148** (49%), and the yield of **173** obtained by oxidative cyclization of the acyclic hexamer **162** was even lower (23%) [48]. This phenomenon must be associated with an unexpected increased

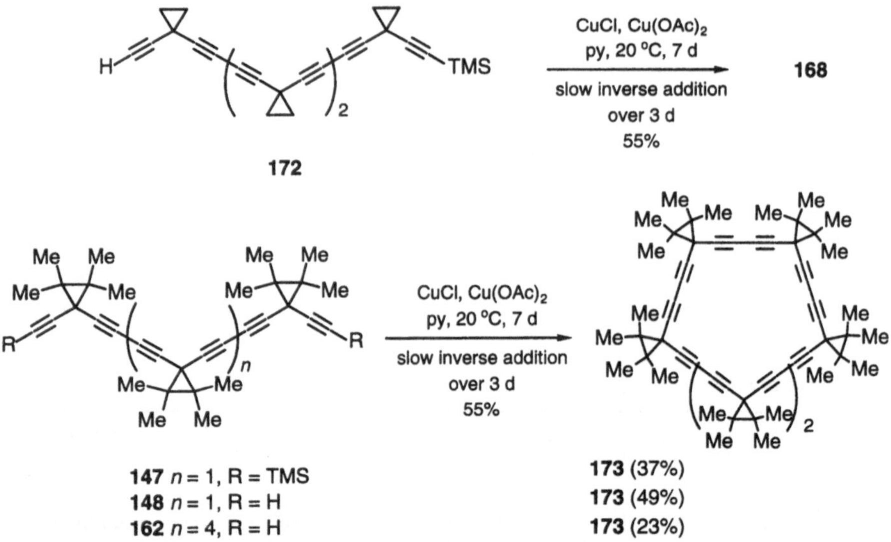

Scheme 30. One- and two-component preparations of permethyl-exp-[6]rotane 173

tendency of the permethylated acyclic higher dehydrooligomers to decompose at room temperature in the syringe pump.

Apparently, the current methodology does not allow the preparation of macrocyclic oligodiacetylenes which would be even more highly strained than the butadiyne-expanded [n]rotanes 165–171. At least, no cyclic dehydro-oligomers could be isolated upon the attempted oxidative coupling of the 7,7-diethynyldispiro[2.0.2.1]heptane dehydrotrimer 151. In the case of dehydrotrimer 150, with only one dispiro[2.0.2.1]heptane unit, one product was isolated in low yield, and its NMR data indicate a cyclic structure of type 174, in which one dispiro[2.0.2.1]heptane moiety has been retained and a second one has undergone ring opening by formal addition of chlorine (Scheme 31) [48].

Upon changing the applied strategy from the completely random approach (cf. Scheme 22) to a block assembly, a complete change of the product distribution is observed in the preparation of the protected expanded pericyclinones of type 89: Only the three macrocycles 123, 125 and 176 were isolated after the oxidative cyclooligomerization of the dehydrodimer 175 (Scheme 32) [39].

All cyclic compounds of types 83–88 have been characterized by their diagnostically simple ^{1}H and ^{13}C-NMR spectra. Some general regularities can be observed, namely, a small but steady upfield shift in the ^{13}C-NMR spectra of the carbon signals of the polyacetylenic macrocycle with increasing ring sizes.

While the molecular masses of expanded [n]pericyclines 82, 83, 122 can easily be determined by GC-MS analysis, higher dehydrocyclooligomers fail to vaporize sufficiently [7]. Fast atom bombardment mass spectometry (FAB-MS) had to be applied for the characterization of the acetal-protected expanded pericyclinones 123–126 and 176 [39]. Attempts to determine the molecular masses of the

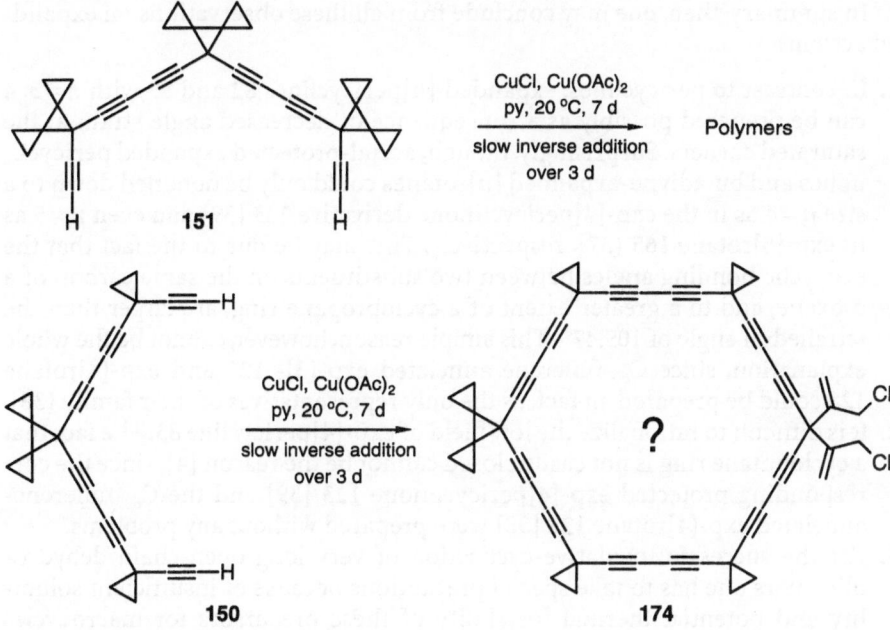

Scheme 31. Attempted synthesis of spirocyclopropanated butadiyne-expanded [*n*]rotanes from the acyclic precursors **150** and **151**

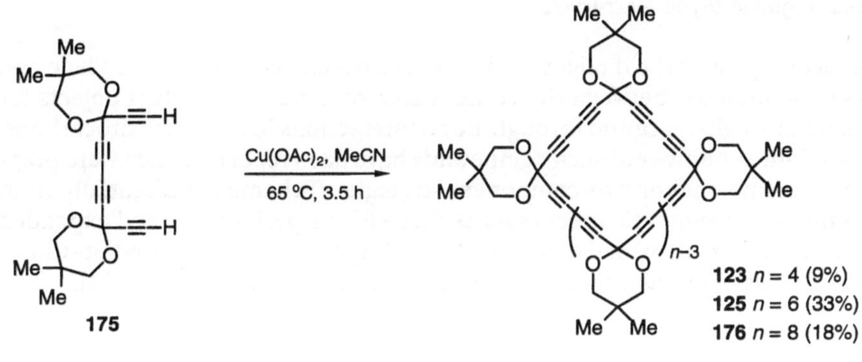

Scheme 32. Multicomponent preparation of protected expanded [*n*]pericyclinones

exploding [*n*]rotanes **165–169** and **173** by mass spectrometry with any of the available routine ionization methods (EI, CI, DCI, FD) failed, as the compounds apparently decomposed irregularly in the inlet system of the mass spectrometer. Several of their molecular masses were successfully determined by vapor pressure osmometry in dichloromethane solution or by the matrix-assisted laser desorption time-of-flight mass spectrometry (MALDI-TOF-MS) method [36, 37]. This method was also used for the C_{60}-annelated exploding [*n*]rotane **127** [38].

In summary, then, one may conclude from all these observations on expanded systems:

1. In contrast to pericyclines, expanded [*n*]pericyclines **82** and **83** with $n = 3, 4$ can be prepared possibly as a consequence of decreased angle strain at the saturated corners. Surprisingly, though, acetal-protected expanded pericyclinones and butadiyne-expanded [*n*]rotanes could only be obtained down to a size $n = 4$ as in the exp-[4]pericyclinone derivative **123** [39] and even $n = 5$ as in exp-[5]rotane **165** [37], respectively. This may be due to the fact that the exocyclic bonding angles between two substituents on the same carbon of a dioxane, and to a greater extent of a cyclopropane ring, are larger than the tetrahedral angle of 109.47°. This simple reason, however, cannot be the whole explanation, since C_{60}-fullerene-annelated exp-[3]- **127** and exp-[4]rotane **128** could be prepared, in fact, as the only representatives of their family [38].
2. It is difficult to rationalize the low yield of exp-[4]pericycline **83**. The fact that a cyclobutane ring is not easily closed cannot be the reason [4], since the corresponding protected exp-[4]pericyclinone **123** [39] and the C_{60}-fullerene-annelated exp-[4]rotane **128** [38] were prepared without any problems.
3. For the successful oxidative cyclization of very long open-chain dehydro-oligomers one has to take special precautions because of insufficient solubility and potential thermal instability of these precursors for macrocycles [48].

3.2
Mixed Oligoyne-Diyne Macrocycles

Homoconjugated hybrid oligoyne-diyne macrocycles containing acetylene and butadiyne moieties between the vertices also constitute interesting objects for probing through-space and through-bond interactions in neutral hydrocarbons. Only a limited number of such compounds have been reported. They were prepared following one- or two-component strategies (Scheme 16) essentially from the same set of synthetic precursors as those for [*n*]pericyclines and expanded [*n*]pericyclines with oxidative acetylene coupling as key steps. The Cadiot–Chodkiewicz cross-coupling procedure has also been applied (Scheme 33) [4].

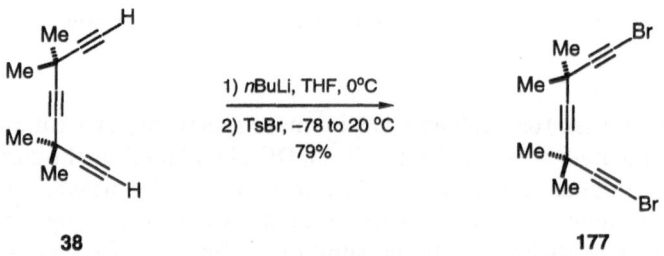

Scheme 33. A two-component approach to the hybrid tetrakisdiyne-monoethyne macrocycle **178**

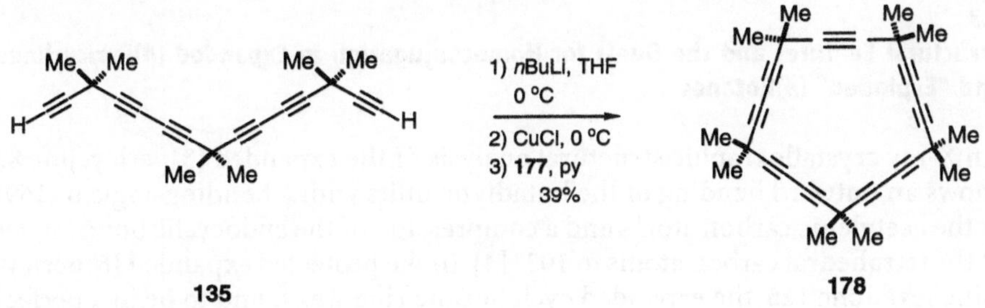

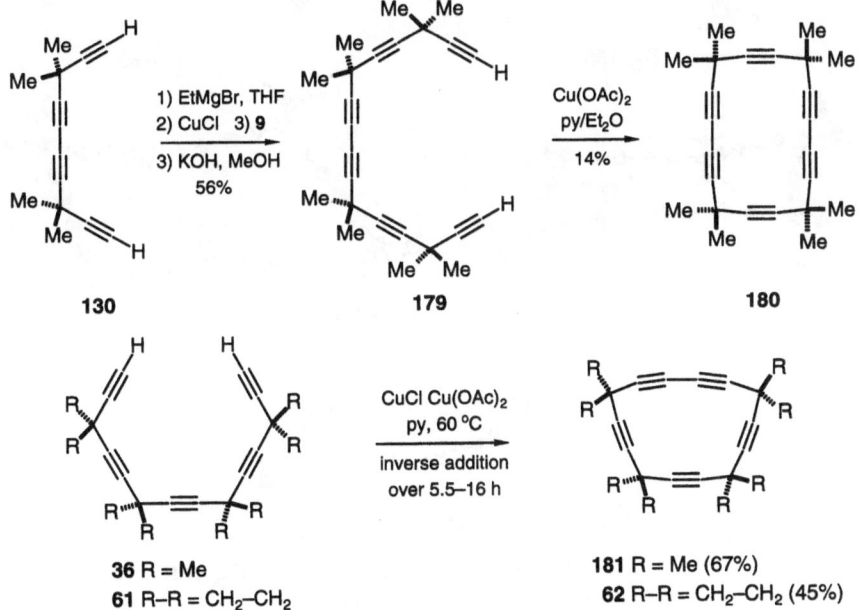

Scheme 33 (continued)

The oxidative coupling, however, normally provides better results, as was demonstrated by the reasonably efficient cyclization of **179** to the homoconjugated 16-membered macrocycle **180** (14% yield) with alternating ethyne and butadiyne units (Scheme 34); the alternative Cadiot–Chodkiewicz coupling of the bisterminal diyne **38** and the dibromotriyne **177** gave only a 2.6% yield of **180** [4].

Apparently, the oxidative acetylene coupling can overcome a drastic strain increase as demonstrated by the highly efficient cyclization of octamethyltetradeca-1,4,7,10,13-pentayne **36** and its perspirocyclopropanated analogue **61** to the corresponding 14-membered macrocycles **181** (67%) and **62** (45%), respectively (Scheme 34) [18].

Scheme 34. One-component syntheses of the hybrid oligoethyne-butadiyne macrocycles **62**, **180** and **181**

3.3
Structural Features and the Quest for Homoconjugation in Expanded [*n*]Pericyclines and "Exploded" [*n*]Rotanes

An X-ray crystallographic structural analysis of the expanded [3]pericycline **82** shows an outward bending of the butadiyne units with a bonding angle of 169° at the acetylenic carbon atoms and a compression of the endocyclic bond angles at the tetrahedral carbon atoms to 103° [4]. In the protected expanded [6]pericy-clinehexanone **125**, the expanded cyclohexane ring was found to be in a perfect chair conformation, and the bond lengths, as well as the angles, were inconspicuous [39].

As indicated above for [*n*]pericyclines, the X-ray crystallographic structural analyses of the exploding [*n*]rotanes **165–168** closely resemble the corresponding [*n*]pericyclines and cycloalkanes with the same number of vertices [36, 37]. Except for the molecules of exp-[5]rotane **165** in crystals obtained from CH_2Cl_2 (D_{5h}-**165** in Fig. 8), all the "blown-up" [*n*]rotanes are not planar. In crystals of exp-[5]rotane obtained from CCl_4 the molecules adopt an envelope conformation (C_s-**165** in Fig. 8). The five-sided macrocycle of **165** with torsional angles of

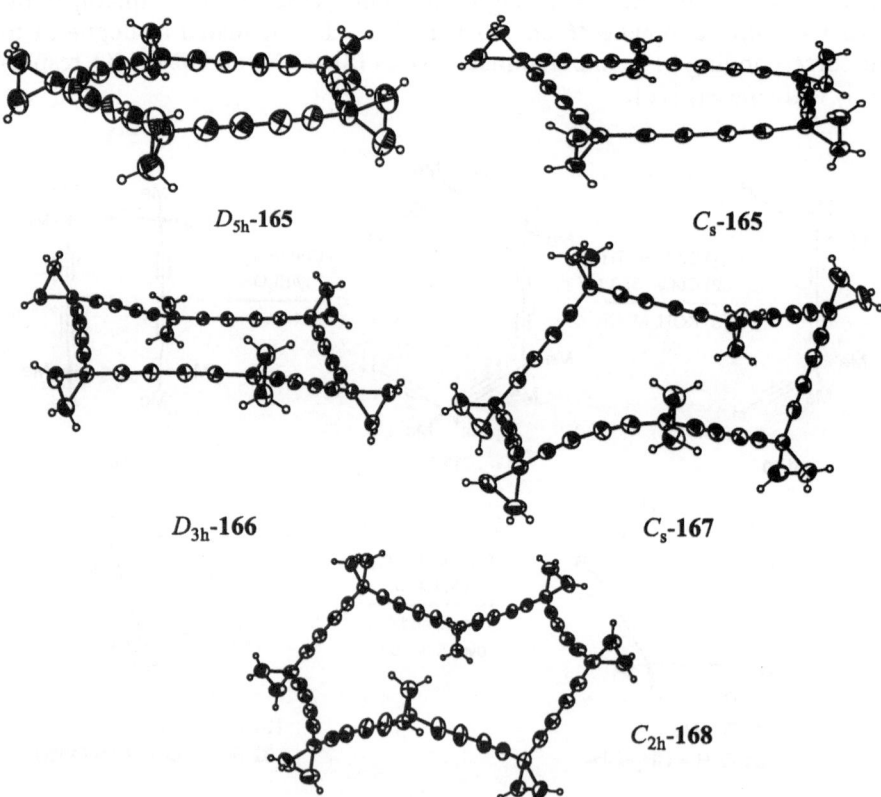

D_{5h}-**165** C_s-**165**

D_{3h}-**166** C_s-**167**

C_{2h}-**168**

Fig. 8. Crystallographic structures of "exploded" [*n*]rotanes **165–168**

−25.0, 17.0, −2.1, −13.9 and 25.2° is, however, less puckered than that in [5]rotane (corresponding angles: −39.6, 27.2, 4.6, 36.7 and −19.8°) [55, 56] and cyclopentane (−40.3, 25.0, 0.0, 40.3 and −25.0°) [57]. This decrease in the puckering angle is caused by the fact that the exocyclic bonding angle between two substituents on the same carbon of a cyclopropane ring is usually significantly larger than the tetrahedral angle of 109.47°. In the case of 165 the average endocyclic angle at the cyclopropane corners is 114.3° in D_{5h}-165 and 114.7° in C_s-165, and, apparently, the small degree of puckering can easily be lifted by crystal packing forces, as the crystals of D_{5h}-165 from CH_2Cl_2 with a planar 25-membered ring demonstrate.

Similarly, the chair of the 30-membered ring of exp-[6]rotane 166 with an average torsional angle of $\varphi = 41.4(4)°$ is less puckered than cyclohexane ($\varphi = 55.1°$) [58] and [6]rotane ($\varphi = 54.6°$) [56].

The cyclic hepta- and octadiacetylenes, exp-[7]rotane 167 and exp-[8]rotane 168, also adopt chair-like conformations in the crystals, and the "blown-up" seven-membered ring in compound 167 has almost the same puckering (torsional angles: 69.0, −90.8, 81.2, −13.9, −45.8, 99.4 and −97.0°) as cycloheptane (torsional angles: 66.0, −89.5, 70.7, 0.0, −70.7, 89.5 and −66.0°) [59]. A crystal structure of cyclooctane itself has never been obtained; the free molecule in the gas phase according to an electron diffraction structural analysis and to molecular mechanics calculations does not adopt a chair-like conformation [60] analogous to that found for exp-[8]rotane 168 (Fig. 8). It is remarkable that the exocyclic bonding angles on the spirocyclopropane moieties in the whole series of "blown-up" [n]rotanes 165 – 168 are retained as closely as possible to the same value as that between the two exocyclic C–C bonds in 1,1-diethynylcyclopropane 47 (115.1°) [31]. This forces the macrocycles 165 and 166 to be less puckered than the five-membered rings in cyclopentane (106.1°) [57] and [5]rotane (105.3°) [55] as well as the six-membered rings in cyclohexane (111.4°) [58] and [6]rotane (111.5°) [56]. The corresponding angle in exp-[7]rotane 167 is only 1.8° larger than the mean value in cycloheptane (115.3 vs 113.5°) [59]. Only in the exp-[8]rotane 168 is this bonding angle 116.2°, and the 1,3-butadiyne units in 168 are bent inward by 4.6°. Apparently, the bonding angle on the spirocyclopropane rings is an energetic compromise between further inward bending of the 1,3-butadiyne units and exocyclic angle deformation on the three-membered ring.

The three-membered rings in these exp-[n]rotanes thus provoke a significant bending of the triple bond angles. The largest bending, corresponding to the smallest bonding angle 172.9(4)° was found for the exp-[7]rotane 167. In one respect both 167 and 168 differ from the smaller members of the family in that the 1,3-butadiyne units in these larger macrocycles are bent inwards rather than outwards. This feature, although to a much lesser extent, was previously observed only for [5]pericyclinosilane 71a [26], while outward bending is quite common [13, 61]. The ethynyl groups on the cyclopropane rings in 165 – 168 apparently exert an electron-withdrawing action, as the proximal and the distal bonds in the three-membered rings are found to be significantly different, as expected for acceptor-substituted cyclopropanes [31, 62]. The observed lengthening of the proximal and shortening of the distal bonds are similar to those for 1,1-diethynylcyclopropane **47** [1.526(1) vs 1.483(1) Å] [31]. The converse electron-donating action of the spirocyclopropane groups on the other hand does not affect the C≡C bond lengths.

An X-ray crystallographic structural analysis of 7,7-diethynyldispiro [2.0.2.1]heptane **109** (Fig. 9) [63] – the subunit of any perspirocyclopropanated expanded [*n*]rotane – reveals that this molecule can essentially be described in terms of a superposition of 1,1-diethynylcyclopropane **47** [31] and dispiro [2.0.2.1]heptane [64] as far as bond lengths are concerned. The exocyclic bonding angle between the ethynyl groups on the dispiroheptane moiety is larger than that in **47** (117.0 vs 115.1°).

The most remarkable structural features of the planar cyclic mixed trisethyne-monobutadiyne pentaacetylenes **181** and **62** are their drastically bowed diyne moieties [18]. The acetylenic carbon atoms deviate from linearity by an average of 11.7° in **181** and even 13.4° in **62**. The internal C–C–C angle is smaller for **181** (103.8°) than for **62** (109.2°). The disparity of the C–C bond lengths in the spirocyclopropane moieties of **62** is analogous to those observed in the expanded [*n*]rotanes **165 – 168**.

As in the case of [*n*]pericyclines, the experimentally determined structural parameters for the protected expanded [6]pericyclinehexaone **125**, as well as those for the "blown-up" [*n*]rotanes **165 – 168**, do not reveal any bond length equalization which might be due to cyclic delocalization. The UV spectra of all the expanded pericyclines of type **84** were virtually all superimposable, except for that for exp-[3]pericycline **82** which showed a large bathochromic shift of 18 nm of its long-wavelength absorption maximum. This was interpreted as a result of cyclic homoconjugation, namely through-space interaction of the in-plane *p*-orbitals [4, 7]. The UV spectra for the macrocyclic "blown-up" [*n*]rotanes **165 – 168** and **171** all look alike and are quite similar to those for the basic chromophore 1,4-dicyclopropylbutadiyne and the acyclic dehydrotrimer **120** in the long-wavelength region, with bathochromic shifts of about 2 nm

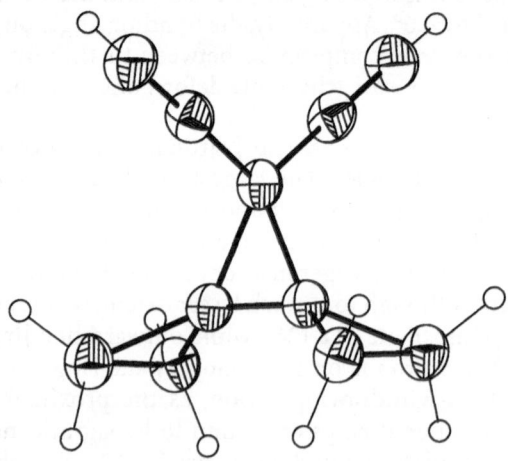

Fig. 9. Crystallographic structure of 7,7-diethynyldispiro[2.0.2.1]heptane **109**

for the absorption maxima of 165–168 and 171. The intense main absorptions in the region 200–220 nm ($\varepsilon = 128,000 - 450,000$), however, are strikingly different. A distinct bathochromic shift is observed on going from 1,4-dicyclopropyl-butadiyne via 165–168 to 171; and, more interestingly, this absorption splits into two bands for macrocycles 165–168 and 171, but with different magnitudes [37]. This must be due to the expected increased homoconjugative interaction between ethynyl units in the "blown-up" [n]rotanes, as corresponding effects are not observed in the UV spectra of the analogous permethylated macrocycles of types 3 and 84 [4, 6, 7]. The UV spectra of expanded [n]rotanes 127 and 128, however, were similar to those of the monomeric diyne 118 except for the increased molar extinction coefficient. Thus for these compounds there is no indication of strong homoconjugation in the hexayne macroring [38].

The hybrid monoyne-tetrakisdiyne macrocycle 178 having $4n + 2$ electrons in both out-of-plane and in-plane π-systems has essentially the same UV spectrum as [5]pericycline 3 which has only $4n$ electrons [4]. The same holds for the hybrid hexayne 180 with two ethyne and two butadiyne units. In the hybrid monoexpanded [4]pericycline 181, the compression of the endocyclic C–C–C bond angles at the saturated carbon atoms enhances the through-space π–π interactions and causes a bathochromic shift of the long-wavelength UV absorbtion to 266 nm, although this shift is only half as large as that seen in the UV spectrum of expanded [3]pericycline 82. In the perspirocyclopropanated pentayne 62 the through-bond π–π interactions are apparently enhanced by the spirocyclopropane rings which cause a shift of the long-wavelength UV absorption maximum to even longer wavelengths ($\lambda_{max} = 273$ nm) [18].

3.4
Thermodynamic Stabilities and Chemical Transformations

The hybrid perspirocyclopropanated monoexpanded [4]pericycline 62, the acetal-protected expanded [n]pericyclinones 123–125, 176 and the "blown-up" [n]rotanes 165–171 are all stable under laboratory conditions for days, but decompose slowly over a period of weeks [18, 37, 39]. The hybrid octamethylhexayne 181 melts cleanly without decomposition at 182–184 °C, while its perspirocyclopropanated analogue 62 had already turned brown at 155 °C and, in one instance, exploded at 192 °C [18]. The most striking feature common to all of the "blown-up" [n]rotanes 165–171 is their shock sensitivity: When struck too hard with a spatula, a pestle or a falling metal ball, even under argon, they go off with a flame and yield a cloud of black soot [36,37]. An attempt to determine the melting point for the macrocyclic decayne 165 led to the complete destruction of a Büchi melting-point apparatus. This unusual behavior of the whole family, which led to them being termed "exploding" [n]rotanes, must be attributed to the additional strain introduced by the spiroannelated three-membered rings; the analogous permethylated macrocyclic oligodiynes of type 84 are by far not as sensitive. Even the permethylated expanded [6]rotane 173 is at least not shock sensitive, although it does not melt without decomposing [4,6,7,48].

In view of the strain energy incorporated in a carbocyclic three-membered ring (28.1 kcal/mol [65]), the butadiyne-expanded [n]rotanes must be true high-energy molecules, as their butadiyne units also have highly unfavorable heats of formation (acetylene itself has $\Delta H_f^\circ = +54.5$ kcal/mol [32]). According to AM1 [66] calculations [37], the six-sided "exploding" [6]rotane 166 has $\Delta H_f^\circ = +773.1$ kcal/mol. Incidentally, the experimentally determined heat of formation for 1,1-diethynylcyclopropane 47 of $\Delta H_f^\circ = +128.7$ kcal/mol [33] and the sixfold value ($128.7 \times 6 = +772.2$ kcal/mol) almost coincide with the calculated value for 166. Compared to that the current "trendsetter" C_{60}-buckminsterfullerene 112 with $\Delta H_f^\circ = +609.6$ kcal/mol is a "low-energy" molecule [67].

A more detailed investigation of the thermal behavior of the "exploding" [n]rotanes by differential scanning calorimetry (DSC) measurements performed in aluminum crucibles with a perforated lid under an argon atmosphere revealed that slow decomposition of exp-[5]rotane 165 has already started at 90 °C and an explosive quantitative decomposition sets on at 150 °C with a release of energy to the extent of $\Delta H_{decomp} = 208$ kcal/mol. Exp-[6]rotane 166 decomposes from 100 °C upwards with a maximum rate at 154 °C and an energy release of $\Delta H_{decomp} = 478$ kcal/mol. The difference between the onset (115 °C) and the maximum-rate decomposition temperature (125–136 °C) in the case of exp-[8]rotane 168 is less pronounced, and ΔH_{decomp} is only 358 kcal/mol. The methylated exp-[6]rotane 173 is thermally less labile and its decomposition is moderated with an onset at 135 °C and a maximum decomposition rate at 194.5 °C with $\Delta H_{decomp} = 285$ kcal/mol. For example, ΔH_{decomp} of the well-known explosive hexogen (1,3,5-trinitro-1,3,5-triazacyclohexane, RDX) determined under similar experimental conditions was only 143 kcal/mol [68]. Applying an evolved gas analysis (EGA) technique to the thermal decomposition of 165 and 166 in different heatable optical cells with rapid-scan Fourier transform infrared (FTIR) spectroscopy monitoring, the formation of only methane, ethylene and acetylene could be detected, and the only gaseous products evolved upon the decomposition of exp-[8]rotane 168 and permethyl-exp-[6]rotane 173 were ethylene and tetramethylethylene, respectively [68]. The composition of the black soot formed in these thermal transformations is of special interest, as the formation of ordered tube- and onion-type carbon layers, along with the evolution of methane and hydrogen gas, was detected upon the analogous explosive thermal decomposition of cyclic oligoynes with aromatic connectors [69]. However, only amorphous carbon with small graphitic areas was found in the explosion products of the butadiyne-expanded [n]rotanes 165–168. Nevertheless, traces of C_{60}-fullerene 112 were detected by MS after the explosion of exp-[6]rotane 166.

The fact that only ethylene and tetramethylethylene are evolved from exp-[8]rotane 168 and permethyl-exp-[6]rotane 173 upon thermal decomposition leads to the conclusion that the spirocyclopropane moieties in these expanded [n]rotanes fragment only externally and leave carbene moieties behind. Indeed, the MALDI-TOF mass spectra of several exp-[n]rotanes show fragment ions with M$^+$ minus 28. Thus, if this fragmentation in an exp-[n]rotane were to continue n times, a cyclic C_n carbon cluster would be left over. So far, however, a fragment ion with $m/z = 480$ corresponding to 182 has not been recorded in the mass spectrum of exp-[8]rotane 168 and it remains to be seen whether a C_{60} cluster 183 will be detected in the mass spectrum of exp-[12]rotane 171 (Scheme 35).

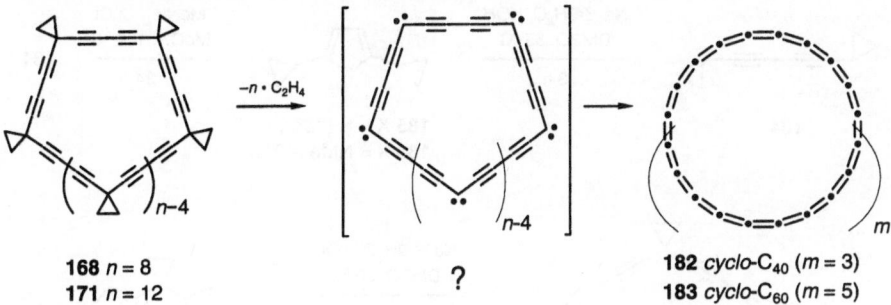

168 $n = 8$
171 $n = 12$

?

182 *cyclo*-C_{40} ($m = 3$)
183 *cyclo*-C_{60} ($m = 5$)

Scheme 35. Potential approaches to *cyclo*-C_{40} **182** and *cyclo*-C_{60} **183** carbon clusters

The MALDI-TOF mass spectra of the C_{60}-fullerene-annelated [3]- and [4]rotanes **127** and **128** also demonstrated that these molecules fragment at the spirocyclopropane units with successive loss of the fullerene moieties. Unfortunately, however, the peaks for *cyclo*-C_{15} and *cyclo*-C_{20} carbon clusters were not observed [38].

The electrochemical oxidation, as well as the reduction of compound **166** in tetrahydrofuran at $-14°C$, proceeds irreversibly, with the reduction potential being remarkably high at -2.81 V [37].

As well as having interesting structural and physical properties, the expanded [n]rotanes should show a rich chemistry due to the high reactivity of their acetylene moieties, as any clean transformation of all their diyne units would lead to novel cyclic structures with alternating spirocyclopropane and various other groups. In view of the known heterocyclizations of 1,3-diynes to thiophenes [70] and pyrroles [71], and the fact that in model experiments 1,4-dicyclopropylbutadiyne **184** and 1,1-bis(trimethylsilylbutadienyl)cyclopropane **187** could be converted to the corresponding thiophene **185**, **188** [72] and N-methylpyrrole **186** [73] derivatives (Scheme 36), the same transformations can be expected for the cyclic oligodiynes **165–168**.

Indeed, upon treatment with sodium sulfide, exp-[5]- **165**, exp-[6]- **166** and exp-[8]rotane **168** were converted to the corresponding crown-type macrocycles **189–191** with alternating thiophene and spirocyclopropane rings in surprisingly good yields (Scheme 36) [72]. An X-ray structural analysis disclosed that the molecules [24](2,5)-cyclopropylthiophene-crown-6 in the crystal [74] adopts a chair-like conformation (Fig. 10) in which the two cyclopropyl groups on each thiophene unit are, in an alternating fashion almost bisected and perpendicular with respect to the thiophene unit so that each cyclopropyl group has a close to optimal orientation for conjugation only with one of its neighboring thiophene moieties. This same feature was also observed for 1,1-di(2-thiophenyl)cyclopropane **188** in the crystalline form (Fig. 10) [63].

With three sulfur atoms above and three below the equatorial plane of the molecule, **190** was expected to be a good ligand for metal ions. However, a dichloromethane solution of **190** did not extract any ions from an aqueous solutions of various metal salts (in total 23 different cations including Al^{3+}, Ba^{2+},

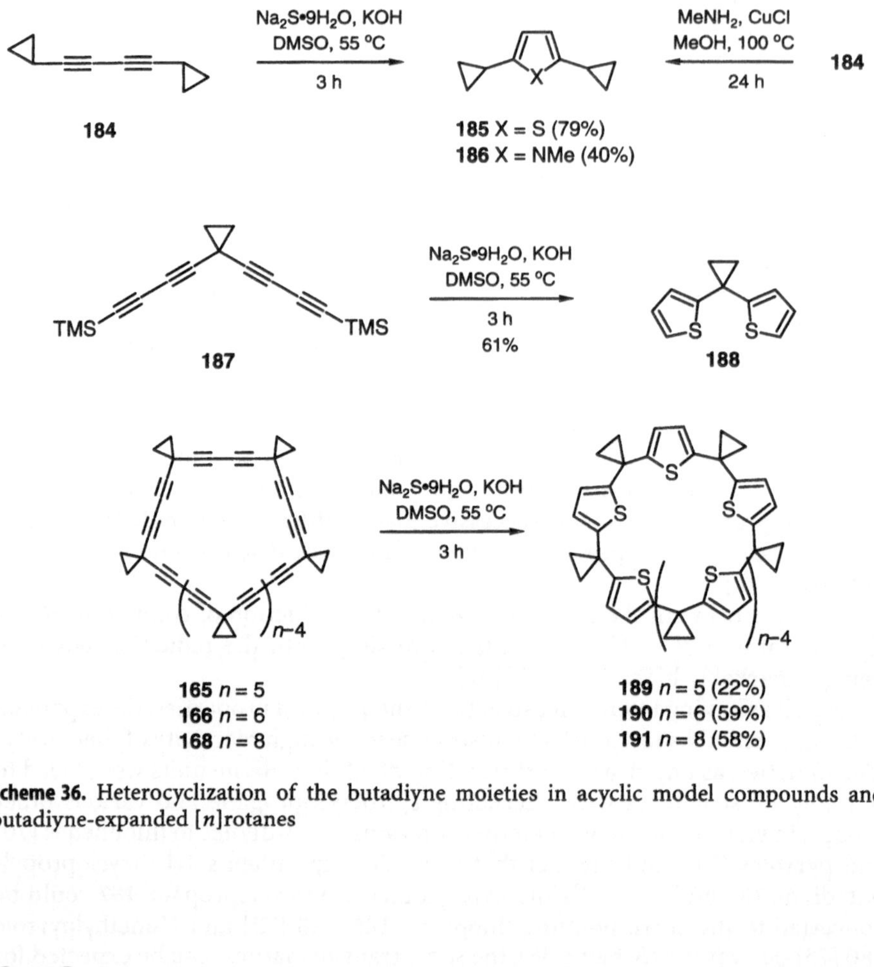

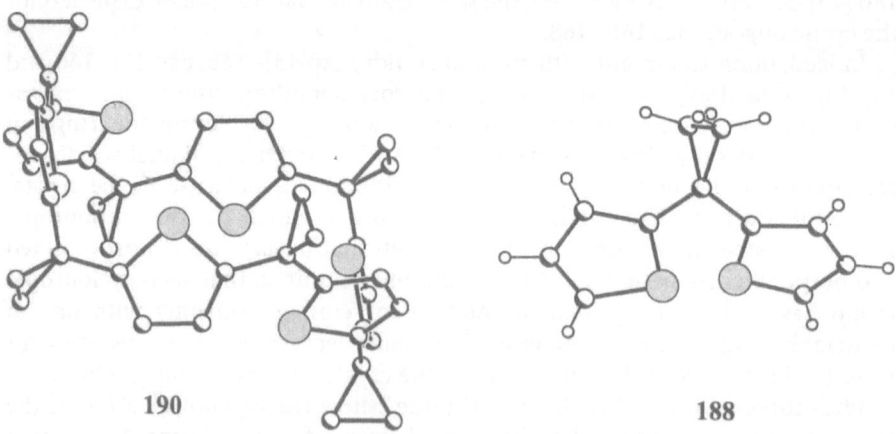

Scheme 36. Heterocyclization of the butadiyne moieties in acyclic model compounds and butadiyne-expanded [*n*]rotanes

Fig. 10. Crystallographic structures of [24](2,5)-cyclopropylthiophene-crown-6 **190** [74] and 1,1-(2-dithiophenyl)cyclopropane **188**

Ca^{2+}, Cr^{3+}, Hg^{2+}, Mg^{2+}, Na^+, Zn^{2+}, Pb^{2+}, Ni^{2+}, Sr^{2+}, etc. [75]), as was also demonstrated for an analogous [24](2,5)-thiophene-crown-6 with dimethylsilyl connectors instead of the spirocyclopropane linkers [76]. This may be due to the relatively small size of the cavity in 190, as the interatomic distances between the sulfur atoms are equal or even smaller than the sum of their van der Waals radii. The hybrid monoexpanded tetraspirocyclopropanated [4]pericycline 62 could not be converted to a cyclic monothiophene derivative; under the conditions applied for 165, 166, and 168 only complete decomposition of 62 was observed. In addition, the exp-[6]rotane 166 did not survive the more drastic reaction conditions required to convert the model dicyclopropylbutadiyne to N-methyl-2,5-dicyclopropylpyrrole 186, so that the pyrrolocrown analogous to 190 [77] was not obtained [48].

The hybrid cyclic pentayne 181 underwent Diels–Alder reaction with the electron-deficient diene tetrachlorothiophene 1,1-dioxide 192, but only at one of the two triple bonds of the 1,3-diyne moiety. This was followed by loss of SO_2 to give the tetrachlorobenzannelated cyclotetradecenetetrayne 193 (Scheme 37) [18].

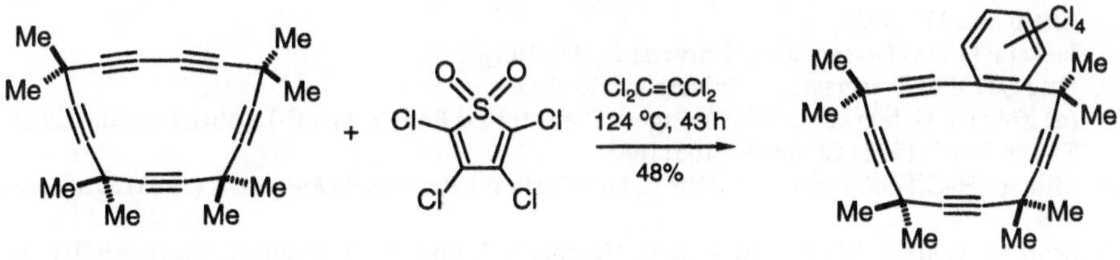

Scheme 37. Thermal cycloaddition of tetrachlorothiophene-1,1-dioxide 192 to the cyclic pentayne 181

4
Perspectives

The correlation of chemical and spectroscopic properties with the structures of molecules in order to ultimately gain predictive power has been one of the major goals of chemistry ever since the earliest concepts of molecular structure were created. The remarkably efficient methods for the syntheses of acyclic and macrocyclic structurally homoconjugated oligoacetylenes and oligodiacetylenes presented here can undoubtedly be applied to generate a large variety of similar compounds. The structure–property relationships which have been uncovered for the known compounds have deepened our understanding of the principles of structural homoconjugation. By applying appropriate physical and chemical methods one can definitely gain more information about physical properties and chemical reactivities of these unusual molecules. The scarce experimental evidence of chemical transformations of such macrocyclic oligoacetylenes and oligodiacetylenes gathered so far is just a hint as to what these types of molecules can do.

5
References

1. (a) McQuilkin RM, Garratt PJ, Sondheimer F (1970) J Am Chem Soc 92:6682; (b) Sondheimer F (1963) Pure Appl Chem 7:363
2. Diederich F, Stang PJ (1998) Metal-catalyzed cross-coupling reactions. Wiley-VCH, Weinheim
3. Stang PJ, Diederich F (1995) Modern acetylene chemistry. VCH, Weinheim
4. Scott LT, Cooney MJ (1995) Macrocyclic homoconjugated polyacetylenes. In: Stang PJ, Diederich F (eds) Modern acetylene chemistry. VCH, Weinheim, p 321
5. Scott LT, DeCicco GJ, Hyun JL, Reinhardt G (1983) J Am Chem Soc 105:7760
6. Scott LT, DeCicco GJ, Hyun JL, Reinhardt G (1985) J Am Chem Soc 107:6546
7. Scott LT, Cooney MJ, Johnels D (1990) J Am Chem Soc 112:4054
8. Roth WR, Staemmler V, Neumann M, Schmuck C (1995) Liebigs Ann 1061
9. de Meijere A (1979) Angew Chem Int Ed Engl 18:809
10. Wiberg KB (1997) In: de Meijere A (ed) Houben-Weyl, Thieme, Stuttgart, vol E17a, p 1
11 (a) Eckert-Maksic M, Gleiter R, Zefirov NS, Kozhushkov SI, Kuznetsova TS (1991) Chem Ber 124:371; (b) Gleiter R, Spanget-Larsen J (1992) In: Halton B (ed) Advances in strain in organic chemistry. vol 2, JAI Press, London, p 143
12 de Meijere A, Kozhushkov SI, Scott LT, Cooney MJ, Gleiter R unpublished results
13. de Meijere A, Jaekel F, Simon A, Borrmann H, Köhler J, Johnels D, Scott LT (1991) J Am Chem Soc 113:3935
14. Jaekel F (1988) Dissertation, Universität Hamburg
15. Otte C (1994) Dissertation, Universität Göttingen
16. (a) Köbrich G, Merkel D (1970) Angew Chem Int Ed Engl 9:243; (b) Köbrich G, Merkel D, Thiem K-W (1972) Chem Ber 105:1683
17. Militzer H-C, Schömenauer S, Otte C, Puls C, Hain J, Bräse S, de Meijere A (1993) Synthesis 998
18. Scott LT, Cooney MJ, Otte C, Puls C, Haumann T, Boese R, Carroll PJ, Smith AB, III, de Meijere A (1994) J Am Chem Soc 116:10275
19. Scott LT, Unno M (1990) J Am Chem Soc 112:7823
20. Voronkov MG, Yarosh OG, Zhilitskaya LV, Albanov AI, Vitkovskii VYu (1991) Metalloorg Khim 4:368, (1991) Chem Abstr 115:29446d
21. Voronkov MG, Pavlov SF (1973) J Gen Chem USSR (Engl Transl) 43:1397, (1973) Chem Abstr 79:66448q
22. Hengge E, Baumegger A (1989) J Organomet Chem 369:C39
23. Bortolin R, Parbhoo B, Brown SSD (1988) J Chem Soc Chem Commun 1079
24. Sakurai H, Eriyama Y, Hosomi A, Nakadaira Y, Kabuto C (1984) Chem Lett 595
25. Houk KN, Scott LT, Rondan NG, Spellmeyer DC, Reinhardt G, Hyun JL, DeCicco GJ, Weiss R, Chen MHM, Bass LS, Clardy J, Jørgensen FS, Eaton TA, Sarkozi V, Petit CM, Ng L, Jordan KD (1985) J Am Chem Soc 107:6556
26. Baumegger A, Hengge E, Gamper S, Hardtweck E, Janoschek R (1991) Monatsh Chem 122:661
27. Scott LT, Cooney MJ, Rogers DW, Dejroongruang K (1988) J Am Chem Soc 110:7244
28. Dewar MJS, Holloway MK (1984) J Chem Soc Chem Commun 1188
29. Schaad LJ, Hess BA Jr, Scott LT (1993) J Phys Org Chem 6:316
30. Jiao H, van Eikema Hommes NJR, Schleyer PvR, de Meijere A (1996) J Org Chem 61:2826
31. Haumann T, Boese R, Kozhushkov SI, Rauch K, de Meijere A (1997) Liebigs Ann/Recueil 2047
32. Domalski ES, Hearing ED (1988) J Phys Chem Ref Data 17:1637
33. Kuznetsova TS (1991) Habilitation Thesis, Moscow State University
34. Gleiter R, Schäfer W, Sakurai H (1985) J Am Chem Soc 107:3046
35. Verevkin SP, Beckhaus H-D, Rüchardt C, Haag R, Kozhushkov SI, Zywietz T, de Meijere A, Jiao H, Schleyer PvR (1998) J Am Chem Soc 120:in press
36. de Meijere A, Kozhushkov SI, Puls C, Haumann T, Boese R, Cooney MJ, Scott LT (1994) Angew Chem Int Ed Engl 33:869

37. de Meijere A, Kozhushkov SI, Haumann T, Boese R, Puls C, Cooney MJ, Scott LT (1995) Chem Eur J 1:124
38. (a) Isaacs L, Seiler P, Diederich F (1995) Angew Chem Int Ed Engl 34:1466; (b) Isaacs L, Diederich F, Haldimann RF (1997) Helv Chim Acta 80:317
39. Brake M, Enkelmann V, Bunz UHF (1996) J Org Chem 61:1190
40. Diederich F (1995) Oligoacetylenes. In: Stang PJ, Diederich F (eds) Modern acetylene chemistry. VCH, Weinheim, p 443
41. Glaser C (1869) Chem Ber 2:422
42. Behr OM, Eglinton G, Galbraith AR, Raphael RA (1960) J Chem Soc 3614
43. Hay AS (1962) J Org Chem 27:3320
44. Cadiot P, Chodkiewicz W (1969) Couplings of acetylenes. In: Viehe HG (ed) Chemistry of acetylenes. chap 9, Dekker, New York, p 630
45. Roth WR, Kowalczik U, Maier G, Reisenauer HP, Sustmann R, Müller W (1987) Angew Chem Int Ed Engl 26:1285
46. Herberich GE, Bauer E, Hengesbach J, Kölle U, Huttner G, Lorenz H (1977) Chem Ber 110:760
47. Liese T, de Meijere A (1986) Chem Ber 119:2995
48. Kozhushkov SI, de Meijere A unpublished results
49. Zefirov NS, Kozhushkov SI, Kuznetsova TS, Gleiter R, Eckert-Maksic M (1986) J Org Chem USSR (Engl Transl) 22:95
50. de Meijere A, Kozhushkov SI, Spaeth T, Zefirov NS (1993) J Org Chem 58:502
51. (a) Hauptmann H (1975) Tetrahedron Lett 1931; (b) Hauptmann H (1976) Tetrahedron 32:1293; (c) An Y-Z, Rubin Y, Schaller C, McElvany SW (1994) J Org Chem 59:2927
52. Anderson HL, Faust R, Rubin Y, Diederich F (1994) Angew Chem Int Ed Engl 33:1366
53. Berscheid R, Vögtle F (1992) Synthesis 58
54. O'Krongly D, Denmeade SR, Chiang MY, Breslow R (1985) J Am Chem Soc 107:5544
55. Boese R, Haumann T, Belov VN, de Meijere A unpublished results.
56. Prangé T, Pascard C, de Meijere A, Behrens U, Barnier J-P, Conia J-M (1980) Nouv J Chim 4:321
57. (a) Saebø S, Cordell FR, Boggs JE (1983) J Mol Struct Theochem 104:221; (b) Adams WJ, Geise HJ, Bartell LS (1970) J Am Chem Soc 92:5013; (c) Allinger NL, Geise HJ, Pyckhout W, Paquette LA, Gallucci JC (1989) J Am Chem Soc 111:1106 and references cited therein
58. Kahn R, Fourme R, André D, Renaud M (1973) Acta Cryst B 29:131
59. Dillen J, Geise HJ (1979) J Chem Phys 70:425
60. (a) Dorofeeva OV, Mastryukov VS, Allinger NL, Almenningen A (1985) J Phys Chem 89:252; (b) Siam K, Dorofeeva OV, Mastryukov VS, Ewbank JD, Allinger NL, Schäfer L (1988) J Mol Struct Theochem 164:93 and references cited therein
61. (a) Guo L, Bradshaw JD, Tessier CA, Youngs WJ (1994) J Chem Soc Chem Commun 243 and references cited therein; (b) Zhou Q, Carroll PJ, Swager TM (1994) J Org Chem 59:1294; (c) Anthony J, Knobler CB, Diederich F (1993) Angew Chem Int Ed Engl 32:406
62. (a) Hoffmann R (1970) Tetrahedron Lett: 2907; (b) Günther H (1970) Tetrahedron Lett: 5173
63. Kozhushkov SI, Haumann T, Boese R, de Meijere A unpublished results
64. Boese R, Haumann T, Jemmis ED, Kiran B, Kozhushkov SI, de Meijere A (1996) Liebigs Ann 913
65. Schleyer PvR, Williams JE, Blanchard KR (1970) J Am Chem Soc 92:2377
66. Dewar MJS, Zoebich EG, Healy EF, Stewart JJP (1985) J Am Chem Soc 107:3902
67. Beckhaus H-D, Verevkin S, Rüchardt C, Diederich F, Thilgen C, ter Meer H-U, Mohn H, Müller W (1994) Angew Chem Int Ed Engl 33:996
68. (a) Löbbecke S, Pfeil A, de Meijere A (1997) Pyrolisis of high energetic diacetylenes 28th Int Ann Conf ICT, p 111; (b) Löbbecke S, Pfeil A (1998) Thermochimica Acta in press
69. Boese R, Matzger AJ, Vollhardt KPC (1997) J Am Chem Soc 119:2052
70. For rewiews see (a) Shostakovskii MF, Bogdanova AV (1974) The chemistry of diacetylenes. Wiley, New York, p 96; (b) Nakayama J, Konishi T, Hoshino M (1988) Heterocycles 27:1731; (c) Gronowitz S (1985) In: Gronowitz S (ed) Heterocyclic compounds. Wiley, New York, vol 44, part 1, p 11

71. Schulte KE, Reich J, Walker H (1965) Chem Ber 98:98
72. Kozhushkov SI, Haumann T, Boese R, Knieriem B, Scheib S, Bäuerle P, de Meijere A (1995) Angew Chem Int Ed Engl 34:781
73. Hellwig J (1996) Diplomarbeit, Universität Göttingen
74. Since analogous macrocyclic arrays with pyrrole rings have been termed calix[n]pyrroles (cf. Gale PA, Sessler JL, Král V, Lynch V (1996) J Am Chem Soc 118:5140) it would be appropriate to rename these thiophene-crowns as calix[n]thiophenes
75. König B, Kozhushkov SI, de Meijere A unpublished results
76. König B, Rödel M, Bubenitschek P, Jones PG, Thondorf I (1995) J Org Chem 60:7406
77. This elusive compound should be termed perspirocyclopropanecalix[6]pyrrol: cf. [74]

Cyclic and Linear Acetylenic Molecular Scaffolding

François Diederich* · Luca Gobbi

Laboratorium für Organische Chemie, Universitätstrasse 16, ETH Zürich CH-8092 Zürich, Switzerland. * E-mail: diederich@org.chem.ethz.ch

During the past decade, the construction and investigation of expanded acetylenic π-chromophores has become a central area of chemical research. It has been fueled by the availability of new synthetic methods, in particular Pd(0)-catalyzed cross-coupling reactions, the discovery of the antitumor activity of a series of natural compounds possessing reactive enediyne π-chromophores, and the need for new nanoscale molecular and polymeric materials that exhibit unusual electronic and optical functions and properties. In this review, synthetic approaches to the cyclo[n]carbons (cyclo-C_n), n-membered monocyclic rings of sp-hybridized C-atoms with unique electronic structures resulting from two perpendicular systems of conjugated π-orbitals – one in-plane and one out-of-plane – are presented. In the following sections, the syntheses and properties of perethynylated molecules, constituting a versatile "molecular construction kit" for acetylenic molecular scaffolding, are discussed. Examples of such compounds are perethynylated annulenes, radialenes, olefins and cumulenes, or transition metal complexes. The article concludes by outlining advances ion novel acetylenic polymers, such as poly(triacetylene), the third linearly conjugated polymer with a non-aromatic, all-carbon backbone.

Keywords: Acetylene chemistry, Cross-coupling reactions, Cyclo[n]carbons, Expanded radialenes, Molecular scaffolding, Nanostructures, Perethynylated chromophores, Poly(triacetylene), Tetraethynylethene.

Topics in Current Chemistry, Vol. 201
© Springer-Verlag Berlin Heidelberg 1999

1
Introduction

Acetylenic molecular scaffolding has turned into a major area of research over the past decade, fueled by several different yet simultaneous developments. The mass-spectrometric observation of buckminsterfullerene (C_{60}) and the proposal that it has a stable soccer-ball-type structure in 1985 [1], followed in 1990 by the isolation of this first molecular allotrope of carbon in bulk quantities [2], led chemists to search for other stable molecular and polymeric – non-fullerenic – forms of carbon, which it might be possible to synthesize and isolate. Many of the all-carbon molecules, as well as one-, two- and three-dimensional networks of carbon that have been proposed based on theoretical calculations, or targeted by synthesis, are acetylenic in nature; this challenging research area has been reviewed on other occasions [3–5]. At the same time, the natural enediyne anti-tumor antibiotics were discovered and the synthesis of analogs was vigorously pursued in order to establish structure-activity relationships [6]. In yet another avenue of research, chemists became increasingly interested in advanced materials, the electronic, optical and mechanical properties of which are tunable by synthesis [7]. Many of these carbon-rich designer materials are constructed from acetylenic building blocks, and several of them will be the subject of this review. However, all these efforts would not have been possible without the simultaneous and vigorous synergistic development in recent years of novel synthetic methodology. Preparative acetylene chemistry [8] has seen an amazing renaissance, and acetylenic covalent assembly and scaffolding in particular have benefited from the discovery of novel, metal-catalyzed, cross-coupling reactions for C–C bond formation [9].

In view of these broad developments, the scope of this review needs to be limited. We do not intend to discuss the elegant early work of Scott and co-workers on "exploded cycloalkanes", in which –C≡C– ("[n]pericyclynes") or –C≡C–C≡C– fragments are inserted between each pair of adjacent sp^3-hybridized C-atoms in n-membered cycloalkanes [10]. Furthermore, phenyl-acetylenic scaffolding, which has prospered so well in the hands of Moore and co-workers in particular, and has culminated in the construction of some of the largest structurally defined dendrimers known [11] will not be considered here. Phenyl-acetylenic scaffolding [12] has provided some of the longest molecular disperse conjugated rods [13] as well as large planar macrocycles with multinanometer dimensions [14]. Phenylacetylenic scaffolding, together with recent developments in the field of benzoannelated dehydroannulenes, is covered by M. Haley elsewhere in this volume [15]. The ongoing efforts aimed at the total synthesis of endohedral metal ion complexes of C_{60} *via* intermediate acetylenic spheroidal transition metal complexes in the laboratories of Y. Rubin at UCLA has recently been reviewed [16]. Terminal transition metal complexes have also found application in the stabilization of some of the longest carbyne $-(C{\equiv}C)_n-$ fragments known [17–19]. A chapter by U. Bunz on transition-metal stabilized carbon-rich acyclic and cyclic scaffolds is included in this monograph [20]. Intriguing acetylenic molecular architecture has been reported from the laboratories of A. Vasella at ETH Zürich, who combined acetylene with carbohydrate chemistry and prepared multinanometer-long linear "acetylenosaccharides" [21], molecular rods constructed from alternating alkyne and carbohydrate moieties, as well as cyclic derivatives [22] that display receptor properties [22b].

This article summarizes efforts undertaken towards the synthesis of the cyclo[n]carbons, the first molecular carbon allotropes for which a rational preparative access has been worked out. Subsequently, a diversity of perethynylated molecules will be reviewed; together, they compose a large molecular construction kit for acetylenic molecular scaffolding in one, two and three dimensions. Finally, progress in the construction and properties of oligomers and polymers with a poly(triacetylene) backbone, the third linearly conjugated, non-aromatic all-carbon backbone, will be reviewed.

2
The Cyclo[*n*]carbons

2.1
Theoretical Considerations

In 1987, before macroscopic quantities of the fullerenes became available [2], one of us (FD), together with his graduate student Y. Rubin, initiated a research program aimed at preparing molecular carbon allotropes from stable, well characterized precursors. The first target compound in this program was *cyclo*-C_{18}, a member of the cyclo[n]carbon family [23]. The cyclo[n]carbons (1, Fig. 1) are defined as *n*-membered monocyclic rings of sp-hybridized C-atoms, with unique electronic structures resulting from two different orthogonal systems of conjugated π-orbitals, one in-plane and one out-of-plane. Since the two cyclo-conjugated π-systems in *cyclo*-C_{18} each possess $(4n + 2)$ π-electrons, the planar compound might display particular stability due to a double Hückel aromaticity. The term double aromaticity was first employed in 1979 by Schleyer and co-workers to explain the particular stability of the 3,5-dehydrophenyl cation [24, 25]; later on, Schleyer et al. also demonstrated that the D_{6h}-cyclo[6]carbon is the most favorable form of the small all-carbon molecule C_6 according to the geometric, energetic and magnetic criteria of aromaticity [26]. Double – in-plane and out-of-plane – aromaticity has also been advanced to explain the stability of cyclocarbon ions such as C_{11}^+, which are formed on laser vaporization of graphite

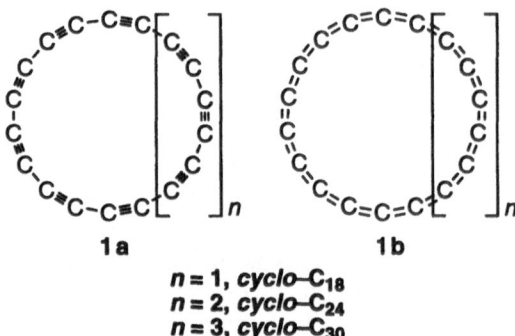

1a   **1b**

n = 1, *cyclo*–C₁₈
n = 2, *cyclo*–C₂₄
n = 3, *cyclo*–C₃₀

Fig. 1. Cyclo[*n*]carbons in their polyyne and their cumulenic forms

[27] or by dehalogenation of suitable arene precursors [28], and are presumed to be intermediates in the formation of fullerenes in flames or carbon plasmas [29–32].

The structures of small non-fullerenic C_n-compounds and cyclo[*n*]carbon structures ($n = 4 - 18$) have been subject to intense theoretical work, which has been comprehensively reviewed by Houk and co-workers [33]. Different theoretical predictions of the electronic structure had been made for *cyclo*-C_{18}, the structure of which was first proposed by Hoffmann [34] in 1966 and which was calculated at that time to display special Hückel-aromatic stabilization as a result of the two orthogonal $(4n + 2)$ π-electron systems. Self-consistent field (SCF) calculations with a 3–21G or larger basis set by Houk and co-workers [23] predicted that the cyclic acetylenic D_{9h}-structure **1a** (Fig. 1) with alternating bond lengths represents the ground-state geometry. However, optimization by Almlöf and co-workers at the level of Møller-Plesset second-order perturbation theory (MP2) with inclusion of valence electron correlations [35], as well as density functional theory calculations by Lüthi and co-workers [36], favored the cumulenic D_{18h}-structure **1b** as the most stable planar monocyclic structure. The most recent calculations by Plattner and Houk, by both a corrected density functional theory approach as well as by the RHF/6–31G* ab initio approximation, now provide strong evidence that *cyclo*-C_{18} is a flat circular polyyne of C_{9h}-symmetry, corresponding to **1a** [33, 37]. These calculations also revealed that the $(4n + 2)$ π-electron cyclocarbons tend towards greater bond localization with increasing ring size. Accordingly, *cyclo*-C_{10} is predicted to prefer a cumulenic D_{5h} structure [38], *cyclo*-C_{14} is borderline, and the energies of its cumulenic D_{7h} and polyynic C_{7h} structures are nearly degenerate, whereas the larger *cyclo*-C_{18} is a polyyne [39]. It is clear that this fascinating theoretical controversy over the past years can only be definitively solved with the synthesis and characterization of *cyclo*-C_{18}.

The choice of *cyclo*-C_{18} as a prime synthetic target was based on the evaluation of its chemical stability and on the availability of a suitable synthetic sequence [23]. In structure **1a**, the bond angle at each sp-hybridized C atom is distorted from the ideal value, 180°, to ≈160°. This distortion requires about

4 kcal mol^{-1}, leading to a predicted angle strain of 72 kcal mol^{-1}. Several cyclic alkynes with similar, or even much smaller, bond angles at sp-hybridized C atoms have been isolated as stable molecules at room temperature [40]. Therefore, we expected that *cyclo*-C$_{18}$ could be isolated under ambient conditions. Several synthetic routes towards the preparation of *cyclo*-C$_{18}$ that have been pursued by us [41] and by others will be presented below.

2.2
The Transition-Metal Route to Cyclocarbons

Although the preparation of bulk quantities of free *cyclo*-C$_{18}$ remains elusive, a transition metal complex could be prepared and characterized by X-ray crystallography [41, 42]. Complexation of an alkyne with [Co$_2$(CO)$_8$] leads to a reduction of the C≡C–C angles in the formed (μ-acetylene)dicobalt hexacarbonyl complex to values around 140° [43, 44]. Therefore, dicobalt hexacarbonyl fragments have been used as protecting groups to allow geometrically disfavored cyclization reactions by bending an alkyne moiety [45, 46]. The free alkyne can usually be liberated from its complex *via* oxidation [47], alkyne-ligand exchange [48], or flash vacuum pyrolysis [49].

On the way to *cyclo*-C$_{18}$, 1,6-bis(triisopropylsilyl)-1,3,5-hexatriyne (2) [50] was subjected to reaction with [Co$_2$(CO)$_8$], followed by ligand exchange with the bridging bis(diphenyphosphino)methane (dppm) ligand, whereupon smooth deprotection yielded the stable dark-red dicobalt complex 3 (Scheme 1). Oxida-

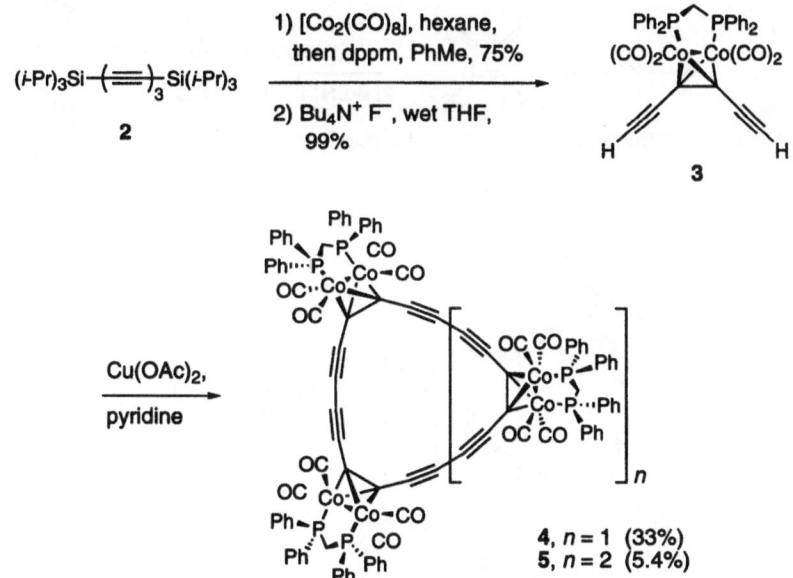

Scheme 1. Synthesis of the cobalt complexes of *cyclo*-C$_{18}$ (**4**) and *cyclo*-C$_{24}$ (**5**) [41 a, 42]. (*Dppm* bis(diphenylphosphino)methane)

tive coupling under conditions of high dilution subsequently afforded the very stable cobalt complexes of *cyclo*-C_{18} (**4**) and *cyclo*-C_{24} (**5**), both as shiny black needles. The X-ray crystal structure of **4** revealed considerable angle bending of the three buta-1,3-diynediyl moieties within the nearly planar C_{18} ring. With values as low as 161°, several of the C≡C–C angles in **4** approach the degree of bending expected for *cyclo*-C_{18}. The UV/Vis spectra of **4** and **5** provided evidence for the presence of partial cyclic π-electron delocalization in the central all-carbon rings. Such conjugation across transition metal-complexed alkynes had been previously predicted by theory [51]. Unfortunately, all attempts to generate the free cyclocarbons by decomplexation of **4** and **5** failed. The dppm ligands generate particularly stable complexes, and the bulky phosphine residues hinder the attack of a reagent at the metal center. Alternatively, it cannot be excluded that *cyclo*-C_{18} or its partially deprotected precursors decompose under the conditions of the various deprotection methods [47–49] that have been applied.

2.3
The Carbon Oxide Route to Cyclocarbons

An attractive approach to cyclocarbon molecules required the preparation of the higher carbon oxides **6–8** followed by either thermal or photochemical CO-elimination (Scheme 2) [41, 52]. The direct synthesis of **6–8** by oxidative macrocyclization of **9** was not possible, since all attempts to prepare the latter

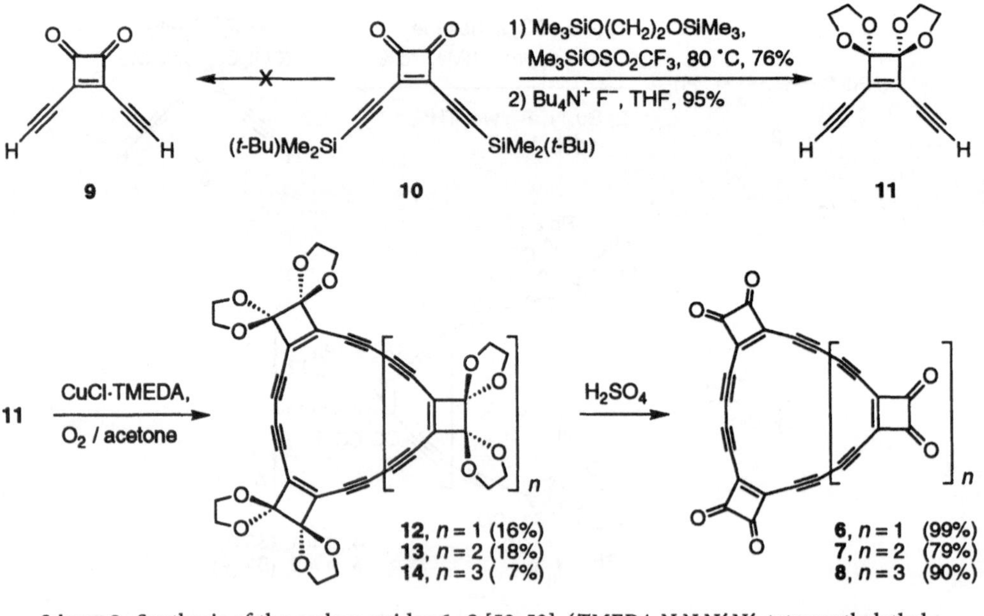

Scheme 2. Synthesis of the carbon oxides **6–8** [52, 53]. (*TMEDA N,N,N',N'*- tetramethylethylenediamine)

compound by deprotection of **10** or other silyl-protected 3,4-dialkynylcyclobut-3-ene-1,2-diones [53] failed. Instead, the diacetal **11** was formed from **10** under unusually forcing conditions, followed by desilylation. Hay-coupling [54] of **11** subsequently yielded the three cyclobuteno-fused dehydroannulenes **12–14** in good overall yield.

The parent dehydroannulenes **15** and **16** (Fig. 2) had been previously prepared by Sondheimer and co-workers as highly unstable compounds that exploded upon heating [55, 56]. In sharp contrast, the three macrocycles **12–14** are stable in moderate temperature ranges and when exposed to air. This extra stabilization of the cyclobutene-fused derivatives is a result of the enhanced rigidity of the ring skeleton due to annelation, which makes the bending and out-of-plane distortions that are normally required to reach reaction transition states energetically more difficult [57]. The analysis of the UV/Vis and ^{1}H NMR spectra showed that trimeric **12** possesses a planar, rigid, diatropic [18]annulene perimeter, while tetrameric **13** has a planar, rigid, paratropic [24]annulene perimeter. While the parent trimeric compound **15** is also planar, tetrameric **16** is a nonplanar, conformationally flexible chromophore which, according to AM1 and MM2 calculations – prefers a cyclooctatetraene-type conformation [57]. These calculations showed that the peculiar stereochemistry of the 1,2-diethynylcyclobut-1-ene moiety defines the difference in geometry between the two tetrameric macrocycles **13** and **16**. With its large $\equiv$C–C$=$C bond angle of 136.3° this unit is accommodated in a nearly strain-free way into a planar tetrameric **13** ($\equiv$C–C$=$C bond angle of 135.8°), whereas the *cis*-1,2-diethynylethene moiety, with an internal $\equiv$C–C$=$C bond angle of 125.4°, prefers incorporation into a nonplanar, cyclooctatetraene-type structure of **16** ($\equiv$C–C$=$C bond angle of 125.3°

This structural analysis was later confirmed by Komatsu and co-workers who prepared two new series of dehydroannulenes fused with bicyclo-[2.2.2]octene moieties [58]. In the series of compounds **17–20** (Fig. 2), they showed by X-ray crystallography that trimeric **18** has a planar π-electron peri-

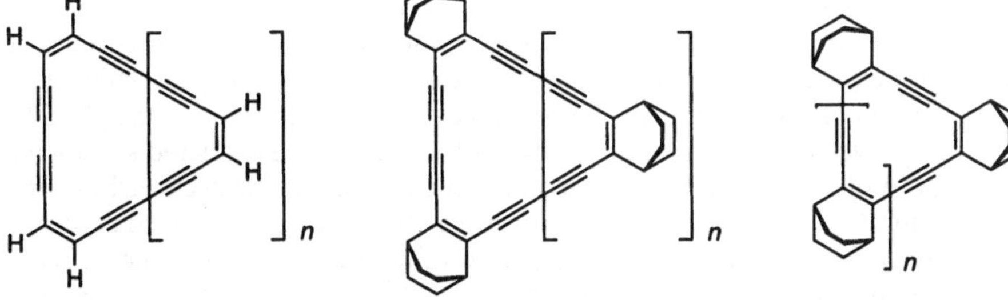

15, *n* = 1
16, *n* = 2

17, *n* = 0
18, *n* = 1
19, *n* = 2
20, *n* = 3

21, *n* = 1
22, *n* = 2

Fig. 2. Dehydroannulenes prepared by Sondheimer and co-workers (**15**, **16** [55, 56] and Komatsu and co-workers (**17–22**, [58])

meter whereas PM3 calculations suggested that tetrameric **19**, like the parent **16**, prefers a nonplanar, cyclooctatetraene-type structure. Such a tub-like structure was actually found by X-ray crystallography for the tetrameric derivative in the series of smaller fused dehydroannulenes **21–22** (Fig. 2). With a calculated $\equiv$C–C$=$C bond angle of 126.3° (PM3), the 2,3-diethynylbicyclo[2.2.2]octene moiety, like the *cis*-diethynylethene moiety in **16**, fits well geometrically into the nonplanar, tub-like structures of tetramers **19** ($\equiv$C–C$=$C bond angle of 126.0°) and **22** ($\equiv$C–C$=$C bond angles of 126.7(3) and 125.9(2)°, X-ray), respectively.

Removal of the acetal protecting groups in **12–14**, to give the carbon oxides **6–8**, proved to be exceptionally difficult, succeeding only when the compounds were dissolved in concentrated sulfuric acid [52]. The rapidly formed carbon oxides were subsequently extracted into 1,2-dichloroethane that had been treated with powdered $CaCO_3$ to remove residual traces of acid. The X-ray crystal structure of orange-yellow trimeric **6** showed a perfectly planar annulene perimeter with considerable angle strain in the three buta-1,3-diynediyl moieties [52]. The carbon oxides **6–8** are extremely sensitive to nucleophiles – including water – which induce rapid polymerization. They also undergo rapid Diels-Alder reactions with 1,3-dienes, such as furan, leading to strained derivatives that subsequently polymerize. They are stable at room temperature but explode above 80 °C.

The results of low-temperature matrix-isolation studies with **6** [41a] are quite consistent with the photochemical formation of *cyclo*-C_{18} *via* 1,2-diketene intermediates [59] and subsequent loss of six CO molecules. When **6** was irradiated at $\lambda > 338$ nm in a glass of 1,2-dichloroethane at 15 K, the strong cyclobut-3-ene-1,2-dione C$=$O band at 1792 cm^{-1} in the FT-IR spectrum disappeared quickly and a strong new band at 2115 cm^{-1} appeared, which was assigned to 1,2-diketene substructures. Irradiation at $\lambda > 280$ nm led to a gradual decrease in the intensity of the ketene absorption at 2115 cm^{-1} and to the appearance of a weak new band at 2138 cm^{-1} which was assigned to the CO molecules extruded photochemically from the 1,2-diketene intermediates. Attempts to isolate *cyclo*-C_{18} preparatively by flash vacuum pyrolysis of **6** or low-temperature photolysis of **6** in 2-methyltetrahydrofuran in NMR tubes at liquid-nitrogen temperature have not been successful.

Results of Fourier-transform mass spectrometric (FT-MS) investigations of the carbon oxides **6–8** conclusively demonstrated that one reaction pathway to the formation of fullerenes is the coalescence of large cyclocarbon ions [29, 52]. In the negative-ion mass spectra of the carbon oxides, the expected successive loss of CO molecules from the precursor anions, producing the cyclocarbon ions C_{18}^-, C_{24}^- and C_{30}^-, was observed. In the positive-ion mode, gas-phase coalescence reactions of the cyclocarbon ions produced fullerene ions. Ion-molecule reactions starting from the cyclocarbon cations C_{18}^+ (formed by laser desorption of **6**) and C_{24}^+ (formed from **7**) led through distinct intermediates ($C_{18}^+ \rightarrow C_{36}^+ \rightarrow C_{54}^+ \rightarrow C_{72}^+ \rightarrow C_{70}^+ + C_2$ and $C_{24}^+ \rightarrow C_{48}^+ \rightarrow C_{72}^+ \rightarrow C_{70}^+ + C_2$, respectively) to fullerene C_{70} as the major product ion. These reactions are remarkably selective, since the formation of the C_{60}^+ ion is not observed. In contrast, the ion-molecule dimerization reaction of *cyclo*-C_{30} (produced from carbon oxide **8**) led selectively to the buckminsterfullerene cation C_{60}^+. These experiments, together with others [30–32, 60], provide strong support for the chain-to-ring-to-sphere mechanism for the Krätschmer-Huffmann fullerene production process starting from graphite [2].

In unrelated work, matrix-assisted laser-desorption-ionization time-of-flight mass spectrometry (MALDI-TOF MS) provided evidence for the formation of ions corresponding to mono-fullerene adducts of the cyclocarbons *cyclo*-C_{15} and *cyclo*-C_{20}, respectively [61]. The solubilized derivatives of C_{195} (**23**) and C_{260} (**24**), two members of a new class of fullerene-acetylene hybrid C-allotropes with the general formula $C_{n(60+5)}$, underwent rapid retro-addition of two and three fullerene moieties, respectively, to yield fragment ions **25⁺** and **26⁺** (Scheme 3). These findings suggest that further investigations by FT-MS may permit the observation of the free *cyclo*-C_{15} and *cyclo*-C_{20} ions, from **23** and **24**, respectively, and allow study of their gas-phase ion-molecule coalescence reactions.

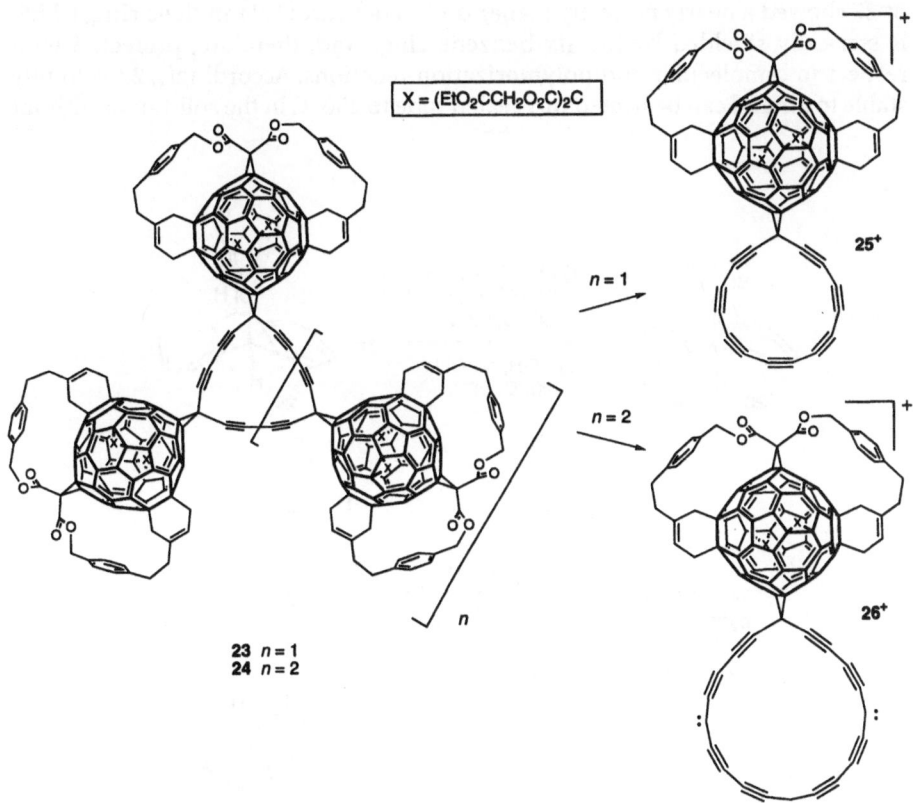

Scheme 3. *Retro*-addition of **23** and **24** under MALDI-TOF MS conditions generates the ions of the mono-fullerene adducts of *cyclo*-C_{15} (**25⁺**) and *cyclo*-C_{20} (**26⁺**), respectively [61]

2.4
Cycloreversion Routes to Cyclocarbons

2.4.1
Cyclo-C₁₈ by Retro-Diels-Alder Reaction

Compound **27** was prepared as a direct precursor to *cyclo*-C_{18}, since it should lose three anthracene molecules in a retro-Diels-Alder reaction under thermal conditions (Scheme 4) [23, 41a]. In the synthesis of **27**, Pd(0)-catalyzed alkynylation [62] of **28** [41a, 63], followed by deprotection, gave the *cis*-enediyne **29** as very unstable crystals which, in one case, exploded spontaneously. Oxidative coupling of **29** under Eglinton-Glaser conditions [64] afforded cyclic trimeric **27** as the only identifiable product, in 25% yield. The absence of dimeric, tetrameric, or pentameric products can be explained by steric matching between monomer **29** and macrocyclic oligomer, as discussed in Section 2.3. The internal ≡C–C=C bond angle, α, in **29** (Scheme 4) is calculated as 125° (MM2) and 127° (AM1) – values close to the average value of 123.3° found for the corresponding bond angles in the X-ray crystal structure of **27** [65]. The X-ray crystal structure of **27** showed a nearly perfectly planar dodecadehydro[18]annulene ring, which is sterically shielded by the six benzene rings and, therefore, protected with respect to bimolecular and polymerization reactions. Accordingly, **27** is highly stable to air and can be heated in a sealed tube to 250 °C in the solid state without

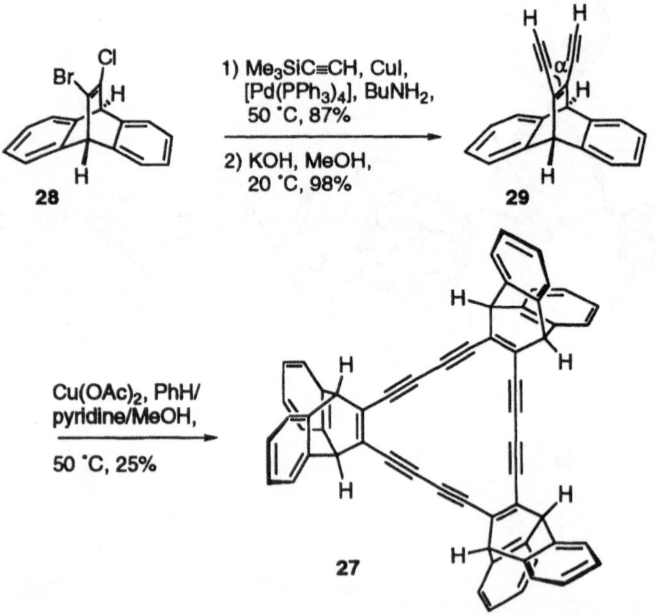

Scheme 4. Synthesis of dodecadehydro[18]annulene **27** for the gas-phase preparation of C_{18} by retro-Diels-Alder reaction [23, 41a]

appreciable decomposition. Both ^{1}H-NMR and electronic absorption spectroscopic data suggest that the dehydro[18]annulene perimeter in 27 is aromatic.

Starting from 27, *cyclo*-C_{18} was prepared in the gas phase by laser flash heating and the neutral product, formed by stepwise elimination of three anthracene molecules in retro-Diels-Alder reactions, was detected by resonant two-photon-ionization time-of-flight mass spectrometry [23]. However, all attempts to prepare macroscopic quantities of the cyclocarbon by flash vacuum pyrolysis using solvent-assisted sublimation [50] only afforded anthracene and polymeric material.

2.4.2
Cyclocarbons by [2 + 2] Cycloreversion

Elegant work by Tobe et al. provided additional strong evidence that cycloreversion reactions of well-defined precursor molecules may be the methods of choice for generation of bulk quantities of cyclocarbons in the future [66 – 68]. Eglinton-Glaser coupling [64] of the dialkynylated [4.3.2]propellatriene 30 yielded 31 as the major product (29%) in addition to a small amount of tetramer 32, both as mixtures of diastereoisomers (Scheme 5) [66]. Upon changing the oxidative coupling conditions by employing Cu(OAc)$_2$ and CuCl in dry pyridine [10b, 69], an approximately 1 : 1 mixture of 31 and 32 (mixtures of diastereoisomers) was obtained in ca. 56% yield. In analogy to the cyclobutene-fused dehydroannulenes 12 and 13 (Section 2.3, Scheme 2), the electronic absorption- and H NMR spectra supported the presence of a diatropic, planar dodecadehydro[18]annulene perimeter in 31 and a paratropic, planar hexadecadehydro[24]annulene perimeter in 32. In another study [68a], Pd(0)-catalyzed cross-coupling of the dichloro[4.3.2]propellatriene 33 with 2-methylbut-3-yn-2-ol afforded 34 which was subjected under phase-transfer conditions to deprotection and subsequent Pd(0)-catalyzed cross-coupling to afford the dehydroannulenes 35 (0.9%), 36 (11%), and 37 (0.9%) as mixtures of diastereoisomers (Scheme 5). By a longer, stepwise construction route, 35 was also selectively prepared in higher yield. Again, cyclobutene-annelation enforces planar macrocyclic π-electron perimeters in the paratropic dehydro[12]annulene 35 and dehydro[16]annulene 36, whereas the dodecadehydro[20]annulene perimeter in 37 is presumably nonplanar.

Mass spectrometric studies were undertaken to explore the conversion of the dehydroannulenes 31, 32 and 35 – 37 into cyclocarbon ions by [2 + 2] cycloreversion [66, 67]. The only peaks displayed in the positive ion-mode laser-desorption time-of-flight (LD-TOF) mass spectra of 31 and 32 were due to the indan fragment ($C_9H_{10}^+$). In contrast, the negative ion-mode spectrum showed peaks resulting from the successive loss of indan fragments by cycloreversion, leading ultimately to the cyclocarbon ions C_{18}^- and C_{24}^-, respectively, as parent ions. Similarly, the LD-TOF mass spectra of 35 – 37 displayed the same characteristic fragmentation pattern, resulting from the successive loss of indan fragments and ultimately leading to the cyclocarbon ions: C_{12}^-, C_{16}^-, and C_{18}^-, respectively.

The photochemical [2+2] cycloreversion [70] of the C_{18}-precursor 31 was studied at 0 °C in furan as the solvent (Scheme 6) [66, 67]. Irradiation with a

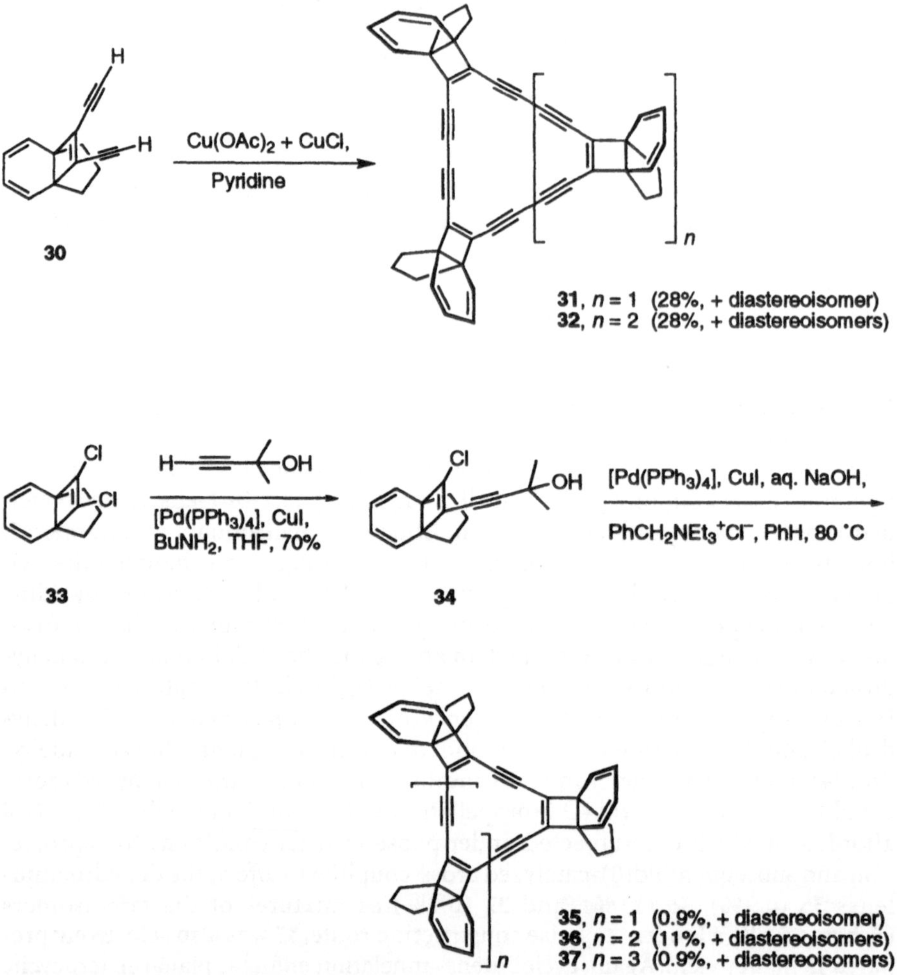

31, *n* = 1 (28%, + diastereoisomer)
32, *n* = 2 (28%, + diastereoisomers)

35, *n* = 1 (0.9%, + diastereoisomer)
36, *n* = 2 (11%, + diastereoisomers)
37, *n* = 3 (0.9%, + diastereoisomers)

Scheme 5. Synthesis of the dehydroannulenes 31/32 and 35–37 for the preparation of cyclo-carbons by [2 + 2] cycloreversion [66, 68a]

low-pressure Hg lamp yielded the three oxanorbornadiene-fused dodecadehy-dro[18]annulenes **38–40** as diastereoisomeric mixtures in respectable yields (**38**:15%; **39**:27% and **40**:11%) beside some polymeric material. A closer examination of this interesting conversion suggested that **38** was formed first and was then converted to **39** and, ultimately, to **40**. Clearly, these products must result from efficient [2+2] cycloreversions under loss of indan fragments, followed by regioselective [4+2] Diels-Alder cycloaddition of furan at exactly the same positions that were previously occupied by the propellane moieties. Semiempirical AM1 calculations explained this regioselectivity of the cycloaddition nicely, by showing that the new alkyne bond formed by loss of one indan from **31** is the most strained and introduces the greatest distortion from lineari-

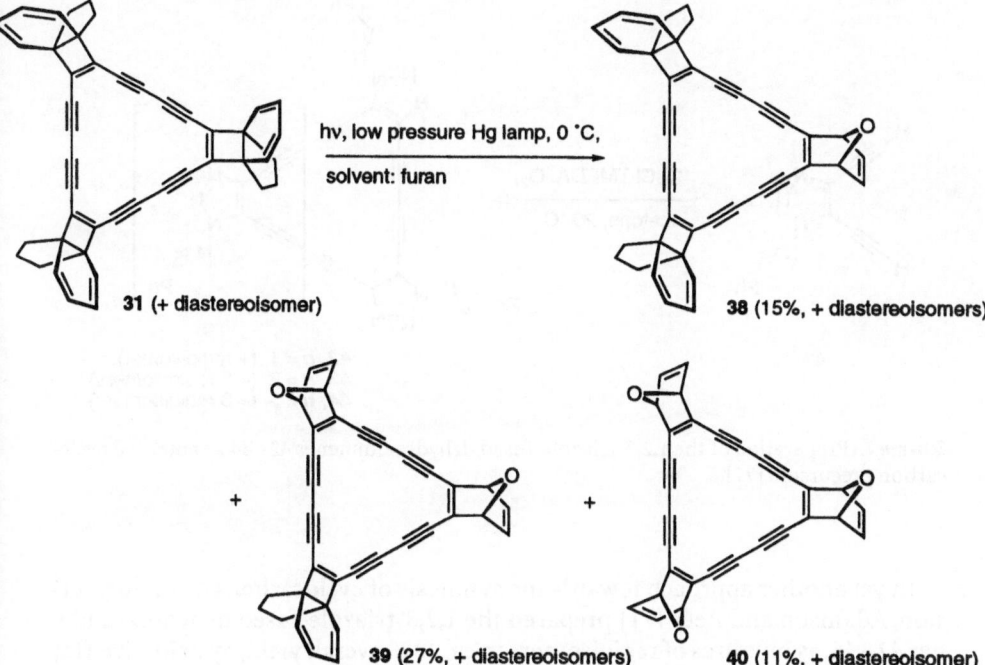

31 (+ diastereoisomer)

hv, low pressure Hg lamp, 0 °C,

solvent: furan

38 (15%, + diastereoisomers)

+

39 (27%, + diastereoisomers)

+

40 (11%, + diastereoisomer)

Scheme 6. Photochemical [2+2] cycloreversion of the dodecadehydro[18]annulene in furan as the solvent, followed by [4+2] Diels-Alder cycloaddition, gives the oxanorbornadiene adducts **38–40** [66]

ty into the macrocyclic perimeter. This study does not necessarily imply direct formation of *cyclo*-C_{18} by photolysis of **31**; however there is good probability that the same reaction in an appropriate inert, non-reacting matrix might ultimately lead to isolable amounts of the cyclocarbon.

In another study, the dehydroannulenes **31, 32**, and **35–37** were laser-ablated by 355 nm photons to generate the bare cyclocarbon skeletons by cycloreversion [68b]. The mono-anions produced were first analyzed by mass spectrometry and then subjected to a second laser pulse at 266 nm to obtain the photoelectron (PE) spectra for the cyclocarbon anions C_n^- ($n = 12, 16, 18, 20$, and 24). These spectra were compared to those obtained by using graphite as the target for laser ablation. The comparative analysis of the two types of PE spectra suggested that laser ablation of the dehydroannulene precursors led to monocyclic C_{12}^- which had isomerized to a large extent to linear C_{12}^-, whereas $C_{16}^-, C_{18}^-, C_{20}^-$ and C_{24}^- were formed as the monocyclic cyclocarbon species. Dimers and trimers of the monocyclic oligocarbon anions were also formed during laser ablation, and the PE spectra suggested that the monocyclic C_{16}^- anions, generated from the dehydroannulene precursor **36**, had undergone spontaneous coalescence, to produce structures that are not monocyclic but possibly cap- or cage-like [30c]. The nature of these structures is under further investigation.

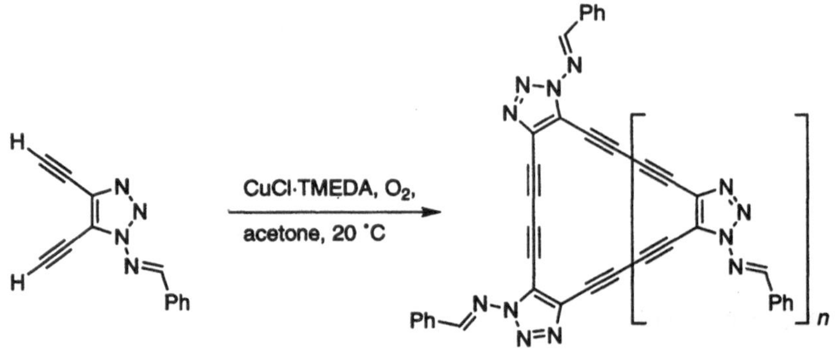

Scheme 7. Preparation of the 1,2,3-triazole-fused dehydroannulenes **42 – 44** as potential cyclo-carbon precursors [71]

In yet another approach towards the synthesis of cyclocarbons by cyclorever-sion, Adamson and Rees [71] prepared the 1,2,3-triazole-fused dehydroannule-nes **42 – 44**, as mixtures of regioisomers in ca. 30 % overall yield, by oxidative Hay coupling of the protected 4,5-diethynyl-1,2,3-triazole **41** (Scheme 7). No investi-gations have yet been reported on the thermal or mass spectrometric [3+2] cycloreversions of **42 – 44**, with loss of the triazole moieties and ultimate forma-tion of the cyclocarbons C_{18}, C_{24}, and C_{30}, respectively.

3
Tetraethynylethene (TEE) Molecular Scaffolding

During the past decade, scientists have become increasingly interested in the preparation of acetylenic molecular and polymeric carbon allotropes and car-bon-rich nanometer-sized compounds that display desirable characteristics, such as unusual structures, high stability, and superior electronic and nonlinear optical properties [4]. For the construction of these materials of fundamental and technological interest at the interface between chemistry and materials science, they developed a large molecular construction kit of peralkynylated building blocks. Figure 3 shows several of these molecules that are already in hand (**45 – 66**) [72 – 92] as well as several that remain elusive (**67 – 69**) [3, 93, 94]. Derivatives of tetraethynylethene (**54**, TEE, 3,4-diethynylhex-3-ene-1,5-diyne) are particularly useful building blocks in this modular chemistry [95]. Today, TEEs of nearly any desired functionalization and silyl-protection are available for the construction of nanometer-sized functional molecular objects. The earlier work by the Diederich group and by others [81, 96, 97] on the synthesis of TEEs has been recently reviewed [41b, 98]; therefore, this article is more concerned with several of the unusual properties displayed by the advanced materials which result from two-dimensional TEE molecular scaffolding [99].

A) Known Modules for Two-dimensional Scaffolding

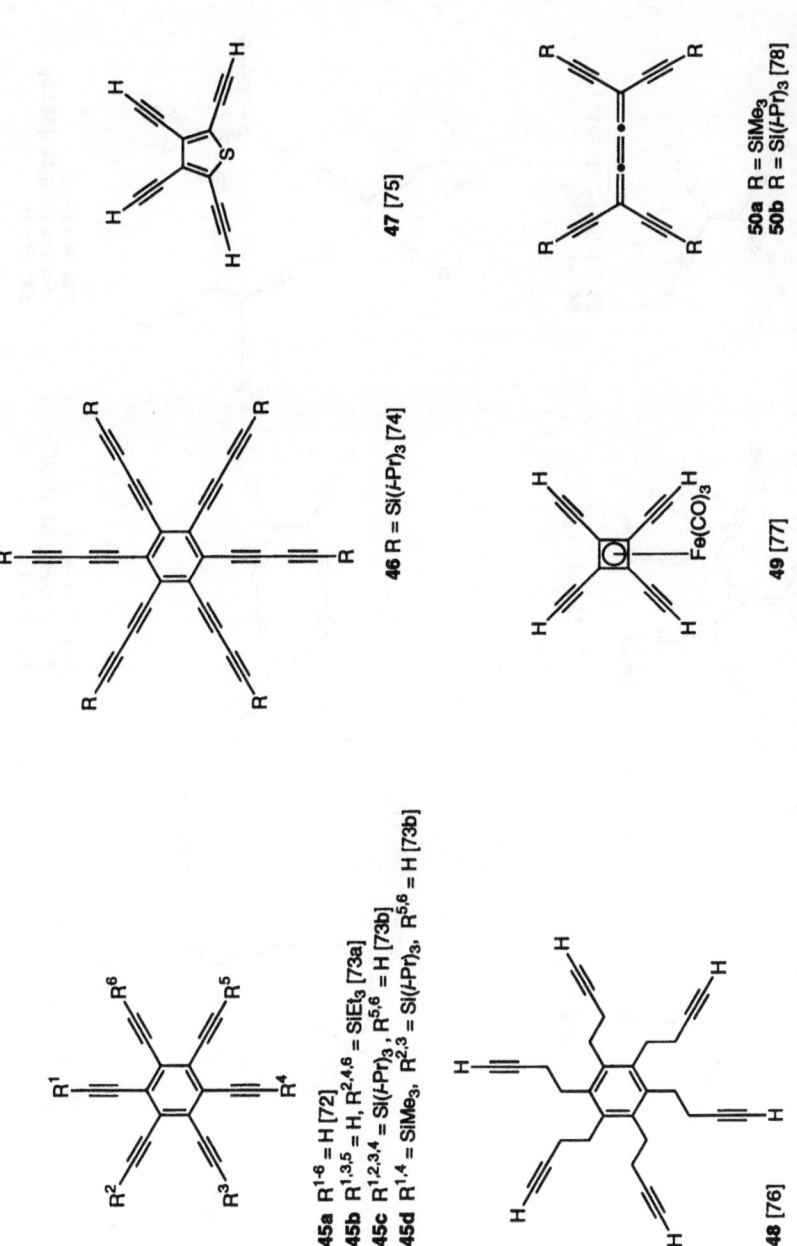

45a R$^{1-6}$ = H [72]
45b R^{1,3,5} = H, R^{2,4,6} = SiEt$_3$ [73a]
45c R^{1,2,3,4} = Si(i-Pr)$_3$, R^{5,6} = H [73b]
45d R^{1,4} = SiMe$_3$, R^{2,3} = Si(i-Pr)$_3$, R^{5,6} = H [73b]

46 R = Si(i-Pr)$_3$ [74]

47 [75]

48 [76]

49 [77]

50a R = SiMe$_3$ [78]
50b R = Si(i-Pr)$_3$ [78]

Fig. 3. Known and hitherto elusive perethynylated building blocks for two- and three-dimensional acetylenic scaffolding. The only compounds shown are those with either free or silyl-protected ethynyl groups

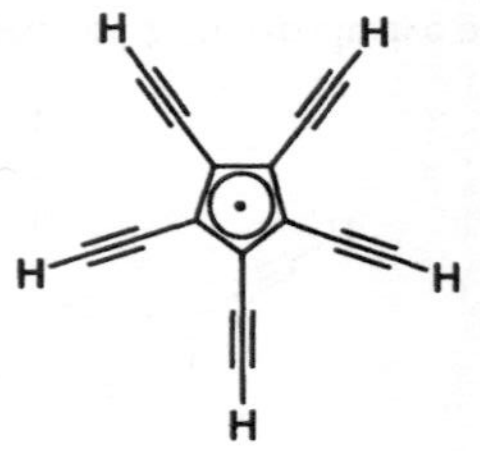

53 R = SiMe$_3$ [81,82]
54 R = H [82]

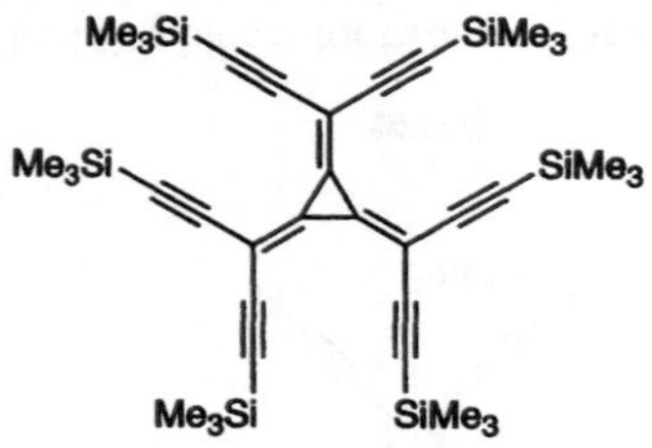

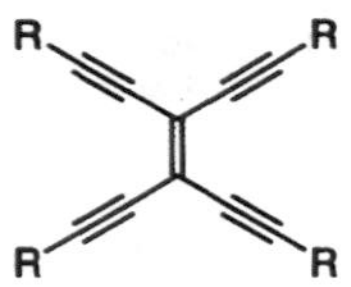

52 [80]

51 [79]

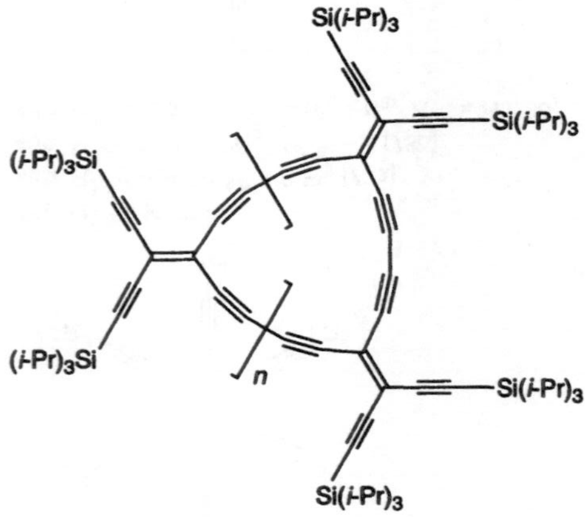

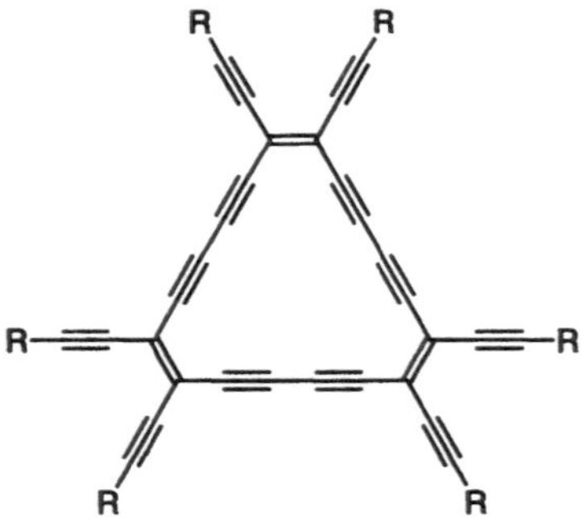

59a R = Me$_3$Si
59b R = (i-Pr)$_3$Si [84,85]
59c R = H

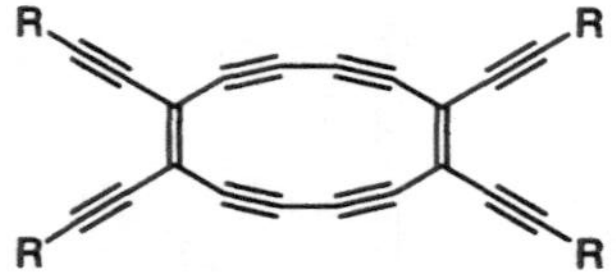

58a R = Me$_3$Si
58b R = (i-Pr)$_3$Si [84,85]
58c R = H

55 n = 2
56 n = 3 [83,84]
57 n = 4

Fig. 3 (continued)

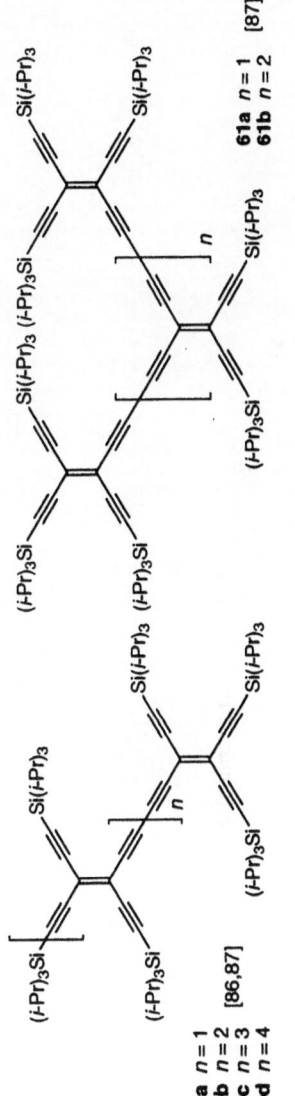

60a *n* = 1
60b *n* = 2 [86,87]
60c *n* = 3
60d *n* = 4

61a *n* = 1 [87]
61b *n* = 2

B) Known Modules for Three-dimensional Scaffolding

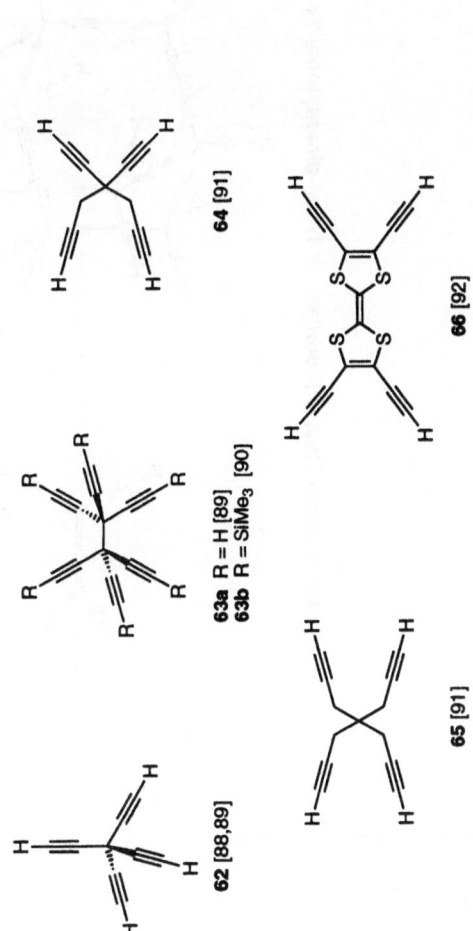

62 [88,89]

63a R = H [89]
63b R = SiMe₃ [90]

64 [91]

65 [91]

66 [92]

Fig. 3 (continued)

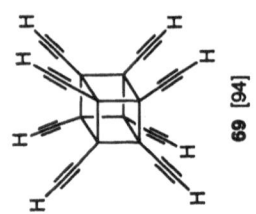

C) Some Hitherto Elusive Modules for Three-dimensional Scaffolding

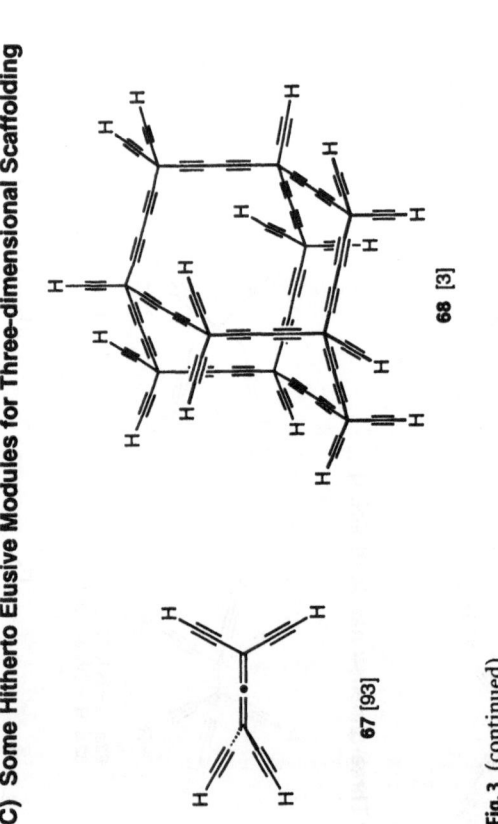

Fig. 3 (continued)

3.1
Macrocyclic TEE Architecture

3.1.1
Perethynylated Dehydroannulenes

The per(silylethynyl)ated octadehydro[12]annulenes **58a, b** and the dodecade-hydro[18]annulenes **59a, b** were prepared by oxidative Hay macrocyclization of the corresponding *cis*-bis(trialkylsilyl)-protected TEEs [84, 85]. X-ray crystal structures of (i-Pr)$_3$Si-protected **58b** and Me$_3$Si-protected **59b** showed that both annulene perimeters are perfectly planar. Considerable strain in the 12-membered ring of **58b** is expressed by the strong bend of the two buta-1,3-diynediyl moieties, with C≡C–C angles as low as 164.5° and by the reduction of the inner C–C=C bond angles to 116.6° and 118.5°. Comparable C–C=C angles in unstrained TEEs adopt values between 120° and 125° [82,100]. In contrast, the buta-1,3-diynediyl moieties in the [18]annulene perimeter of **59b** are practically linear and the inner C–C=C bond angles correspond to those seen in acyclic, strain-free TEEs.

A comparative evaluation of the electronic absorption and ^{1}H NMR spectra of **59a, b** and other dodecadehydro[18]annulenes (**12** [57], **15** [55], **18** [58], **27** [23], **31** [66], and **70** [84] Fig. 4) with planar conjugated ($4n+2$) π-electron perimeters showed that all of these compounds are diatropic. In agreement with their aromaticity, the HOMO-LUMO gaps of these compounds are large, with yellow-colored **59a, b** displaying an optical end absorption around 480 nm and a HOMO-LUMO gap of 2.57 eV in pentane. In contrast, the HOMO-LUMO gaps of the octadehydro[12]annulenes **17** [58], **58a, b**, **71** [84], and **72** [101], and also of the hexadehydro[12]annulenes **21** [58], **35** [68a], and **73** [102] are much smaller, in keeping the antiaromatic character of these $4n$ π-electron systems. Thus, the purple-colored perethynylated derivatives **58a, b**, in pentane, display an optical end absorption around 660 nm and a HOMO-LUMO gap of 1.87 eV. For several of these compounds, a pronounced paratropicity was clearly revealed by ^{1}H NMR spectroscopy. On the other hand, **74** [64, 103], **75** [104] and other benzo[12] annulenes [105] do not exhibit paramagnetic ring currents as a result of the benzoannelation, and display optical properties that differ entirely from those of the antiaromatic systems. Thus, compound **74** is pale yellow and its electronic end absorption occurs at 450 nm, corresponding to a HOMO-LUMO gap of 2.74 eV.

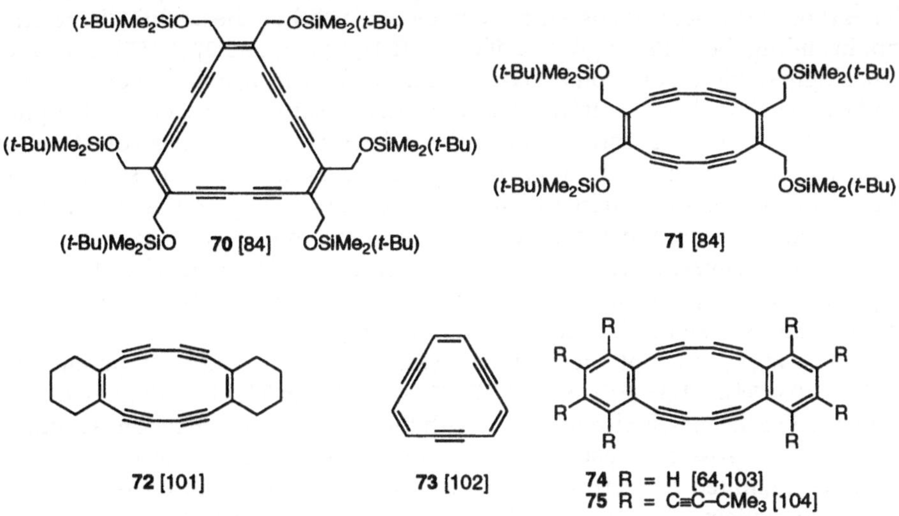

Fig. 4. Diatropic (70), paratropic (71–73), and benzoannelated dehydroannulenes (74, 75)

In agreement with these findings, octadehydro[12]annulene **58b** underwent two stepwise one-electron reductions ($E° = -0.99$ and -1.46 V vs Fc/Fc$^+$ (ferrocene/ferricinium couple) in THF) more readily than the dodecadehydro[18]annulene **59a** ($E° = -1.12$ and -1.52 V) [84]. This redox behavior is best explained by the formation of an aromatic $(4n+2)$ π-electron dianion from antiaromatic **58b**, whereas **59a** loses its aromaticity upon reduction. The strain-free 18-π-electron system showed, higher stability than the strained 12-π-electron system, as expected. Nevertheless, **58b** was found to be quite stable. At room temperature, dilute pentane solutions remained unchanged over weeks and crystals were stable to air and light for months. However, if a solution of **58b** was concentrated without crystallization, significant decomposition occurred. The remarkable stability of **58b** [m.p. 200 °C (dec.)] clearly originates from the way the individual molecules are arranged in the crystal. An examination of the crystal lattice of **58b** showed that the delicate annulenic chromophores are offset in the stacking, so that they are completely surrounded by the bulky, inert (i-Pr)$_3$Si groups. This insulating matrix-type effect of (i-Pr)$_3$Si groups has been observed in many X-ray crystal structures of TEEs, and seems to represent a more general mode of stabilization of these compounds in the solid state [84, 85, 87, 100a, 106, 107].

The Me$_3$Si protecting groups in **58a** and **59a** could be removed by treatment with sodium tetraborate (borax) in MeOH/THF, yielding **58c** and **59c**, respectively, as very unstable compounds [84]. Any attempts to obtain characterizable two-dimensional all-carbon network structures [3, 4] by oxidative polymerization of **59c** have failed.

3.1.2
Perethynylated Expanded Radialenes

Compounds **55–57** (Fig. 3) are the first representatives of the expanded radialenes (Fig. 5) [83, 84] and possess nanometer-sized carbon sheets with diameters, not including the (i-Pr)$_3$Si groups, of ca. 17 (**55**), 19 (**56**) and 22 Å (**57**). They are amazingly stable, readily soluble compounds, with melting points above 220 °C, and can be viewed as persilylated C$_{40}$, C$_{50}$ and C$_{60}$ isomers, respectively. Despite their high degree of unsaturation, all three derivatives are yellow colored and their end-absorption in the UV/Vis spectra occurs at nearly the same wavelength, below 500 nm. Apparently, macrocyclic cross-conjugation is quite inefficient, and π-electron delocalization in all three macrocycles extends only through the longest linearly conjugated fragment, which is equal to the conjugated dodeca-3,9-diene-1,5,7,11-tetrayne π-backbone in **60a** (Fig. 3). Compounds **55–57** exhibit 2, 3 and 4 one-electron reduction waves, respectively [84]. The appearance of the first reversible reduction at similar potentials: [$E° = -1.08$ (**55**), -1.35 (**56**) and -1.27 (**57**) vs Fc/Fc$^+$ in THF], in all three of the expanded radialenes, is also in accord with the fact that π-conjugation is limited by inefficient cross-conjugation, and possibly – to some extent – by nonplanarity as well. Limited π-electron delocalization, due to inefficient cross-conjugation, also explains the unusual UV/Vis spectra of the expanded dendralenes **61a/b** (Figs. 3 and 5) [87]. The spectra of both compounds are nearly identical to

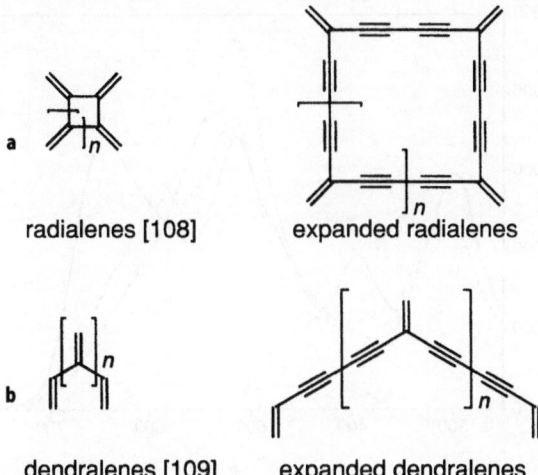

a radialenes [108]   expanded radialenes

b dendralenes [109]   expanded dendralenes

Fig. 5. Insertion of buta-1,3-diynediyl fragments (**a**) into radialenes [108] produces carbon-rich expanded radialenes and (**b**) into dendralenes [109] produces expanded dendralenes

that of **60 a** with regard to absorption wavelength maxima, end absorption, and vibrational fine structure. As in the case of the expanded radialenes **55 – 57**, the π-backbone of **60 a** is the longest linearly conjugated fragment in **61 a, b**.

The physical properties of the expanded radialenes were greatly enhanced upon donor functionalization, leading to the stable derivatives **76 – 78** with fully planar conjugated π-chromophores [110]. These compounds exhibit large third-order nonlinear optical coefficients, can be reversibly reduced or oxidized, and

$(C_{12}H_{25})_2N$

$(C_{12}H_{25})_2N$

$-N(C_{12}H_{25})_2$

76 $n = 1$
77 $n = 2$ [110]
78 $n = 3$

$-N(C_{12}H_{25})_2$

$(C_{12}H_{25})_2N$

$(C_{12}H_{25})_2N$

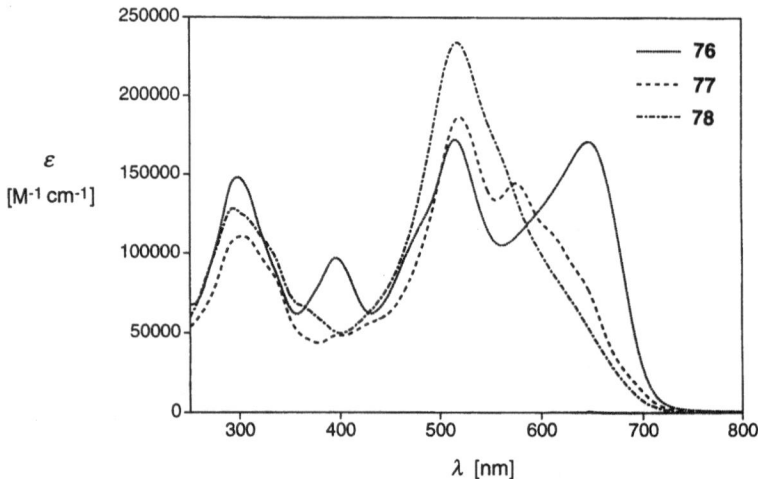

Fig. 6. Electronic absorption spectra of donor-functionalized expanded radialenes **76–78** in CHCl$_3$ at 298 K

form Langmuir monolayers at the air-water interface. Of particular interest is the electronic absorption spectrum of trimeric **76**, which displays a strong low-energy absorption band in the Vis region with an exceptionally large molar extinction coefficient ($\varepsilon = 171{,}000$ dm^3 cm^{-1} mol^{-1} at $\lambda_{max} = 646$ nm; Fig. 6). The origin of this remarkable band is not yet well understood. The high stability and solubility of the large carbon cores in the expanded radialenes raises great hope that much larger acetylenic carbon surfaces can be prepared and characterized in the future.

3.2
Molecular Wires: Oligomers and Polymers with the Poly(triacetylene) (PTA) Backbone

Molecular conducting [111] or photonic [112] wires, as well as molecular wires isolated in various polymeric, inorganic, or dendritic matrices [113], are seen as the ultra-miniaturized molecule-sized electronic components for molecular electronics devices [13, 114, 115]. The design and construction of conjugated polymers and corresponding oligomers of precise length and constitution [17, 116] increasingly takes advantage of high-level computational approaches, which predict with confidence, key properties such as band gap and nonlinear optical response [117].

Oxidative polymerization of *trans*-bis-deprotected **79** under Hay coupling conditions [54] yielded, after end-capping with phenylacetylene, the high-melting and readily soluble oligomers **80a–e** with the poly(triacetylene) back-bone [87, 106] (Scheme 8). Poly(triacetylene)s [PTAs, $-(C{\equiv}C-CR{=}CR-C{\equiv}C)_n-$] are the third class of linearly conjugated polymers with a non-aromatic all-carbon backbone in the progression which starts with polyacetylene [PA,

$-(CR=CR)_n-]$ [118] and poly(diacetylene) [PDA, $-(C\equiv C-CR=CR)_n-]$ [119], and ultimately leads to carbyne $[-(C\equiv C)_n-]$ [17-19]. Compounds **80a-e** extend up to ≈ 50 Å in length (**80e**). They undergo facile one-electron reduction, the number of reversible reduction steps being equal to the number of TEE moieties in each molecular rod [87, 106]. Thus, the first reduction of **80a** occurs at $E(= -1.57$ V (vs Fc/Fc$^+$, in THF + 0.1 M Bu$_4$NPF$_6$), whereas the first reduction of **80e** is much more facile, occurring at -1.07 V. The highly colored oligomers are amazingly stable to air, and can be stored on the laboratory bench for months without decomposition. Correspondingly, no oxidation of these rods was observed in THF below 1.0 V (vs Fc/Fc$^+$) in electrochemical studies.

When the oxidative polymerization of **79** was carried out in 1,2-dichlorobenzene at 65 °C in the presence of 3,5-bis(*tert*-butyl)phenylacetylene as end-capping reagent, an air-stable longer-chain PTA (**81**) was obtained (Scheme 8); its number-averaged molecular weight was determined to be $M_n=9600$ (degree of oligomerization $X_n=22$) [120]. Deep red-brown **81** was soluble in hot chloroform and in 1,2-dichlorobenzene above 65 °C and showed an optical gap of $E_g=2.0$ eV, i.e. in the range of values measured for poly(diacetylenes) (PDAs) [119b]. It can be reversibly reduced at the remarkably low potential of $E(=-0.65$ V (vs Fc/Fc$^+$). The materials properties of the PTA polymers were further enhanced by the introduction of lateral p-C$_6$H$_4$N(C$_{12}$H$_{25}$)$_2$ donor groups [110]. Thus, polymer **82** (Scheme 8, $M_n=16800$, $X_n=17$) undergoes electrochemically one reversible $1-e^-$ reduction and one reversible $1-e^-$ oxidation step, and the optical gap ($E_g=1.6$ eV (770 nm)) is substantially reduced as compared to **81**.

PTA oligomers and polymers were also constructed starting from 1,2-diethynylethenes ((*E*)-hex-3-ene-1,5-diynes, DEEs) [110, 120, 121]. In terms of stability and redox behavior, the properties of oligomers **83a-f** (Scheme 9) resemble those of **80a-e**. Their enhanced solubilities, however, allowed estimation of the effective conjugation length [122] of PTA polymers [123] for the first time. The effective conjugation length indicates the number of repeat units in a conjugated polymer that is required to furnish size-independent properties: redox, optical, etc. The chain-length-dependent electronic absorption spectra and the third-order nonlinear optical properties of oligomers **83a-f** were compared with those of the longer-chain polymers **84a, b** (Scheme 9). The comparison yielded convergence of the linear and nonlinear optical properties in the range of 7 to 10 monomer units, corresponding to 21 to 30 conjugated double and triple bonds.

Poly(triacetylene) oligomers, with backbones as in **83a-f** but with lateral dendritic side chains, have recently been prepared as insulated molecular wires [113e, 124]. In these tubular macromolecules, such as **85a, b** (Scheme 9), the insulating layer, created by the dendritic, Fréchet-type [125, 126] wedges, protects and stabilizes the central conjugated backbone but does not alter its electronic characteristics. UV/Vis measurements indicate that there is no loss of π-electron conjugation along the PTA backbone in the higher generation compounds, despite their distortion away from planarity due to the bulky dendritic wedges.

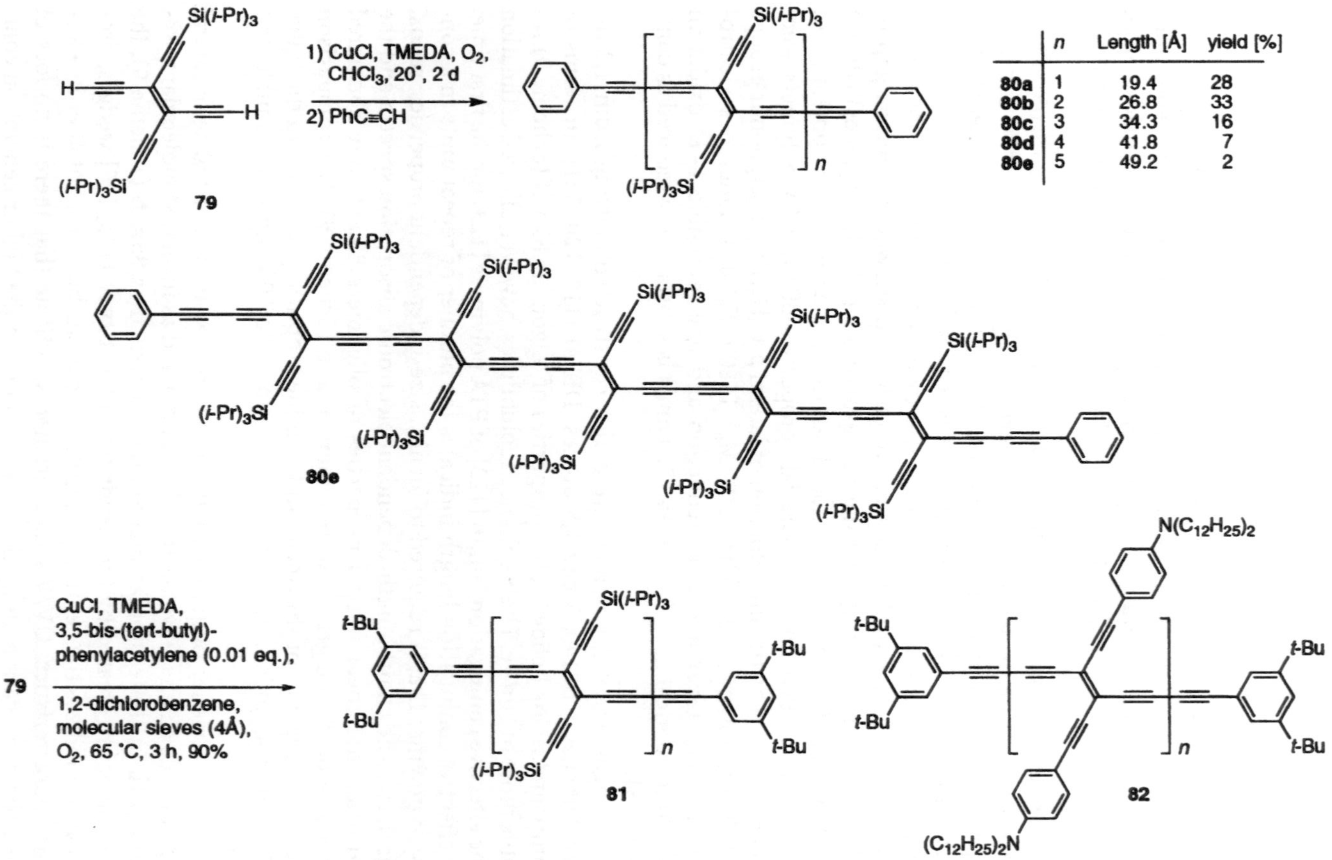

	n	Length [Å]	yield [%]
80a	1	19.4	28
80b	2	26.8	33
80c	3	34.3	16
80d	4	41.8	7
80e	5	49.2	2

Scheme 8. Oligomers and polymers with a poly(triacetylene) (PTA) backbone that are derived from tetraethynylethenes (TEEs)

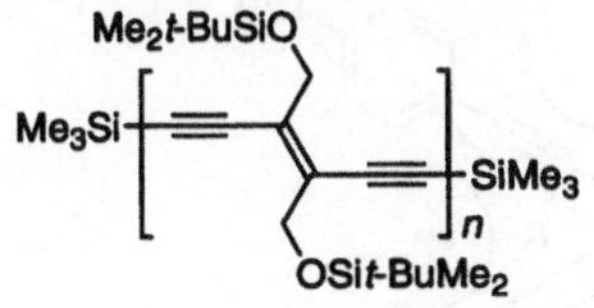

	n	l [Å]
83a	1	9.6
83b	2	16.9
83c	3	24.2
83d	4	31.5
83e	5	38.8
83f	6	46.1

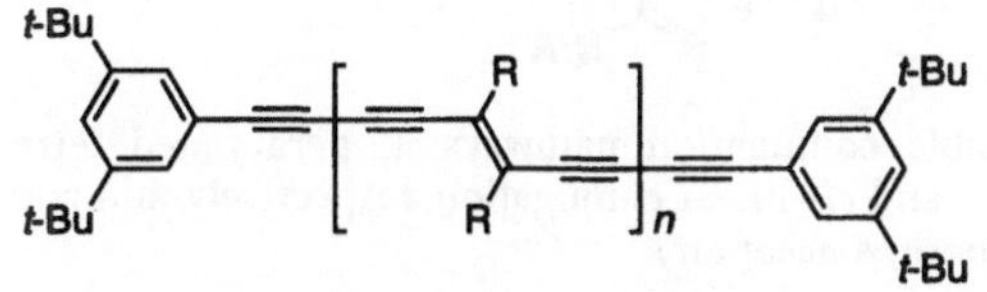

	R	X_n	M_n
84a	CH$_2$OSi-t-BuMe$_2$	31	11300
84b	CH$_2$OSi-t-BuMe$_2$	22	8000

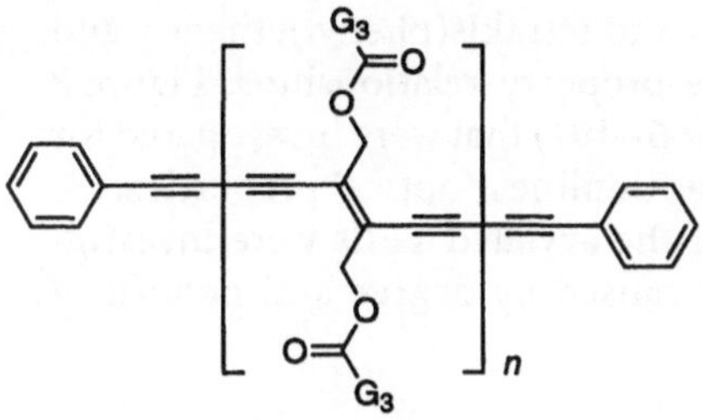

85a n = 1
85b n = 2

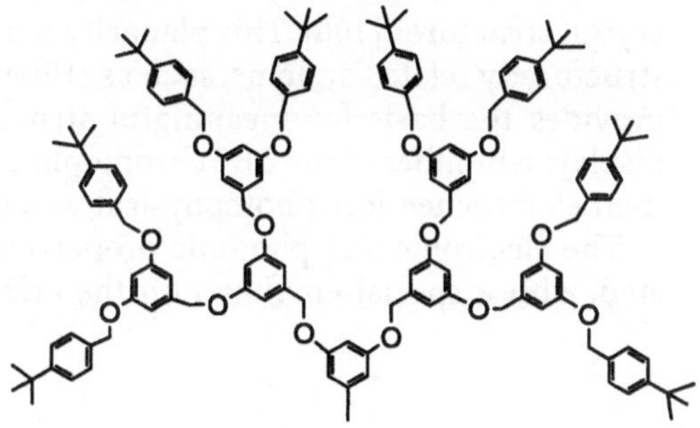

G$_3$

Scheme 9. Oligomers and polymers with a PTA backbone that are derived from 1,2- diethynylethenes (DEEs)

3.3
Donor-Acceptor Substituted Tetraethynylethenes

3.3.1
Preparation, Physical Properties, and Photochemical trans → cis Isomerization

A library of more than 50 different arylated tetraethynylethenes (TEEs) and 1,2-diethynylethenes (DEEs) were prepared [100] for systematic investigation of nonlinear optical properties in two-dimensionally conjugated molecules [127]. In these chromophores, a total of six conjugation pathways in two dimensions are effective: two linear *cis*- and two linear *trans*-conjugation paths as well as two *geminal* cross-conjugation paths (Fig. 7). High-yielding synthetic entries into this class of compounds take remove advantage of Pd(0)-catalyzed cross-coupling reactions [9a, 98, 100]. Photochemical *trans → cis* isomerizations provided elegant access to some of these derivatives as illustrated by the syntheses of **86** (Scheme 10) and **87** (Scheme 11) [98, 100]. The planarity of the π-conjugated carbon frames in many of these chromophores was revealed by their X-ray

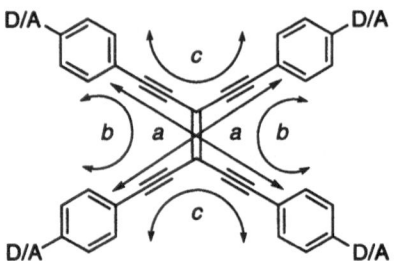

Fig. 7. Schematic representation of possible conjugation pathways in perarylated tetra-ethynylethenes. Paths *a* and *b* depict *trans*- and *cis*-linear conjugation, respectively, and path *c* depicts geminal cross-conjugation. (*D* donor, *A* acceptor)

crystal structures [100]. This planarity is a unique feature, as compared to other structurally related systems, such as stilbenes and tetrakis(phenyl)ethenes, and provides the basis for meaningful structure-property relationships. Figure 8 displays a number of the TEE chromophores (**86–101**) that were investigated for their electrochemical, photophysical, as well as nonlinear optical properties.

The electronic and photonic properties of the arylated TEEs were investigated, with a special emphasis on the effects caused by degree and pattern of

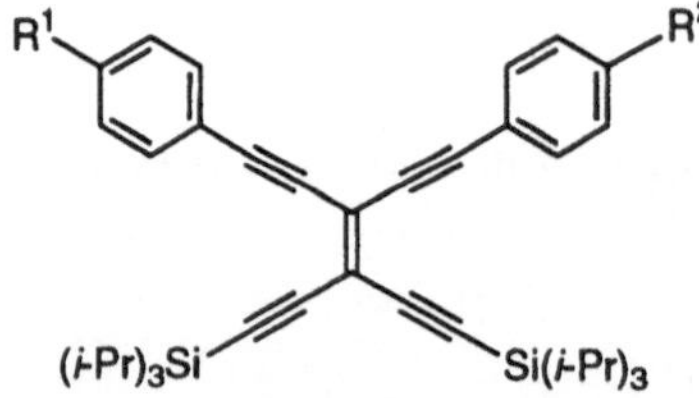

88 $R^1 = R^2 = NMe_2$ (310)
89 $R^1 = NMe_2, R^2 = NO_2$ (59, 280)

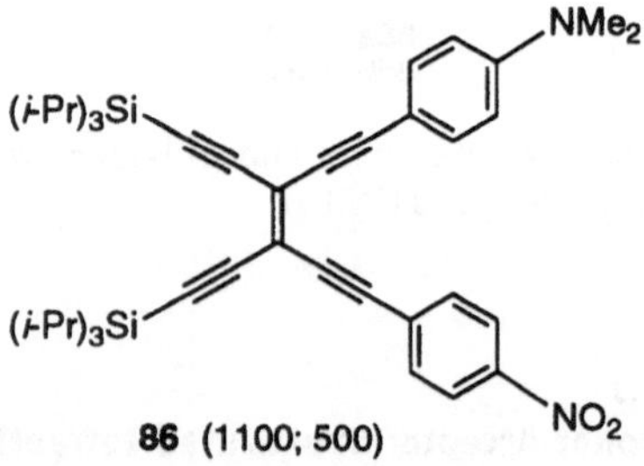

86 (1100; 500)

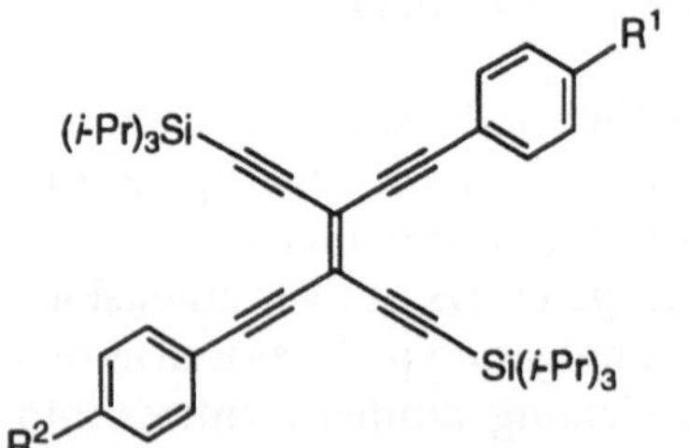

90 $R^1 = R^2 = NMe_2$ (420)
91 $R^1 = R^2 = NO_2$ (230)
92 $R^1 = NMe_2, R^2 = NO_2$ (1000; 790)

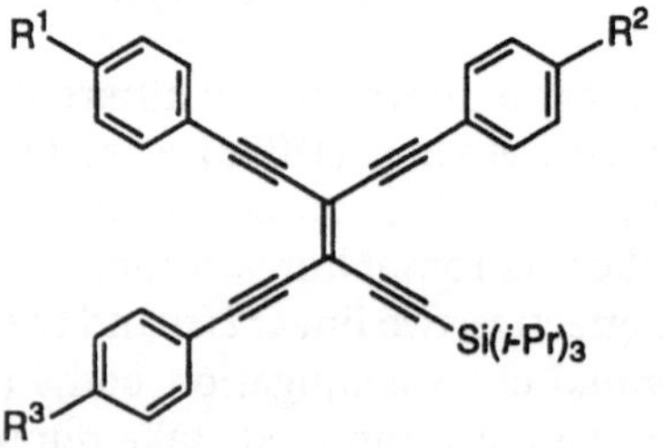

93 $R^1 = R^2 = NMe_2, R^3 = NO_2$ (1500; 1110)
94 $R^1 = R^2 = OMe, R^3 = NO_2$ (240)

Fig. 8

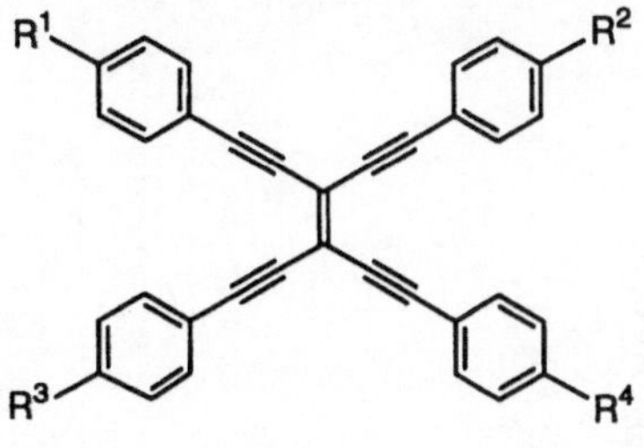

95 $R^1 = R^2 = NMe_2$, $R^3 = R^4 = NO_2$ (3200; 2000)
96 $R^1 = R^4 = NO_2$, $R^2 = R^3 = NMe_2$ (n.o.; 890)
87 $R^1 = R^3 = NMe_2$, $R^2 = R^4 = NO_2$ (n.o.; 1590)

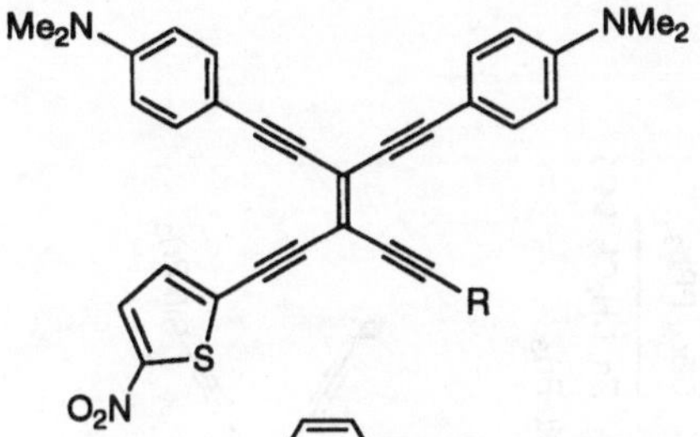

97 $R = $ ⟨benzene⟩$-NO_2$ (n.o.; 2550)

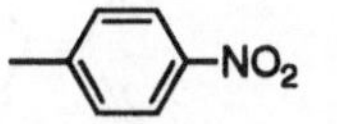

98 $R = $ ⟨thiophene⟩$-NO_2$ (n.o.; 2200)

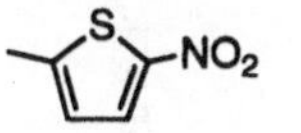

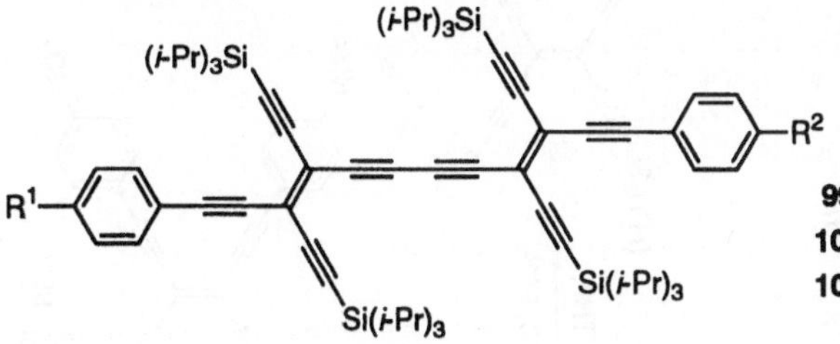

99 $R^1 = R^2 = NMe_2$ (1900)
100 $R^1 = R^2 = NO_2$ (530)
101 $R^1 = NO_2$, $R^2 = NMe_2$ (3600)

Fig. 8. Examples of some of the donor-acceptor substituted TEEs prepared for the exploration of structure-property relationships in the second- and third-order nonlinear optical effects of fully two-dimensionally-conjugated chromophores. For all compounds, the second hyperpolarizability γ [10^{-36} esu], measured by third harmonic generation experiments in CHCl$_3$ solution at a laser frequency of either $\lambda = 1.9$ or 2.1 (second value if shown) μm is given in parentheses. *n.o.* not obtained

donor/acceptor substitution. The large majority of these compounds are thermally and environmentally stable molecules [100a]. As solids, they can be stored for months in air at room temperature, and many decompose only upon heating to temperatures above 200 °C. Intramolecular donor-acceptor interactions, as evidenced by a long-wavelength charge-transfer band, are considerably more effective in TEEs **86** and **92**, which have *cis* and *trans* linearly conjugated electronic pathways between donor and acceptor, than in **89**, with a geminal, cross-conjugated electronic pathway. UV/Vis spectroscopy revealed a steady bathochromic shift of the longest wavelength absorption band (λ_{max}) as the number of donor-acceptor conjugation paths increased from bis-arylated-, to tris-arylated-, and to tetrakis-arylated TEEs. Many of these molecules fluoresce strongly. Electronic emission spectroscopic investigations demonstrated a considerable solvent dependence of the fluorescence of donor-acceptor substituted TEEs such as **89** and **92**, which is in agreement with the presence of highly polarized excited states in these molecules.

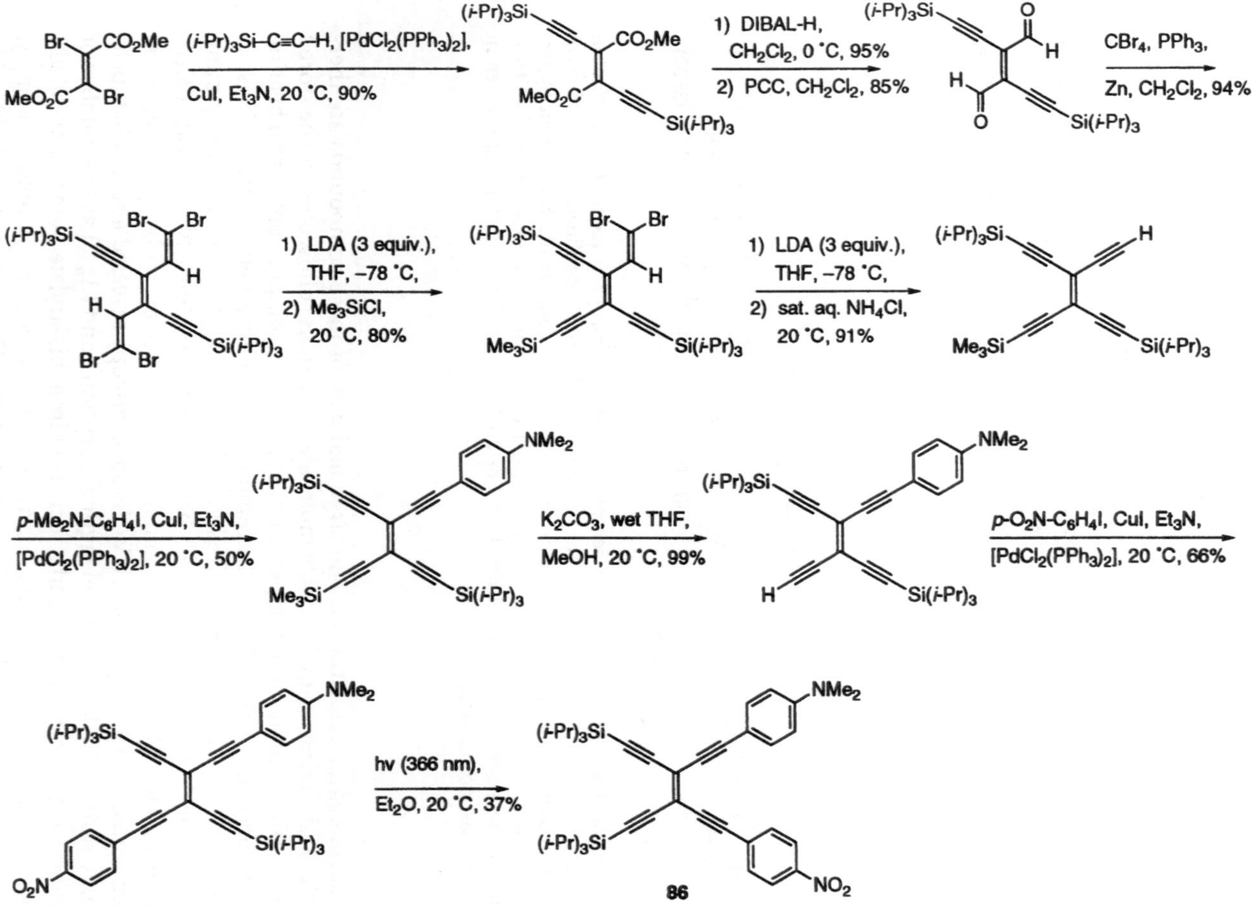

Scheme 10. Synthesis of the *cis*-donor-acceptor-substituted TEE **86** [100]. *DIBAL-H* diisobutylaluminum hydride, *PCC* pyridinium chlorochromate, *LDA* lithium diisopropylamide

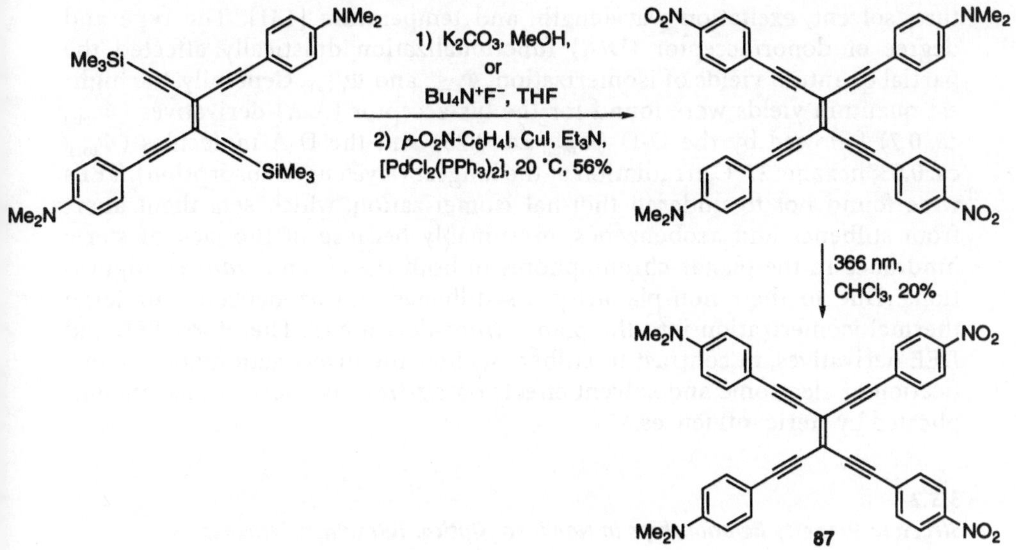

Scheme 11. Synthesis of the tetrakis-arylated TEE 87 by photochemical isomerization [100]

Electrochemical analysis and *ab initio* calculations were employed in a systematic study to determine the ability of the conjugated carbon core to promote electronic communication between the pendant donor/acceptor functionalities [128]. It was found experimentally that multiple *p*-nitrophenyl redox centers present on the TEE core apparently behave independently from one another in electrochemical reduction steps. Thus, the first reversible one-electron reduction of one *p*-nitrophenyl ring in **86, 89, 92, 93,** and **94** occurred in CH_2Cl_2 (+ 0.1 M Bu_4NPF_6) at the same potential (around –1.37 V vs Fc/Fc$^+$) as the two first reduction steps in **91, 95,** and **96** [129]. Although these findings suggested at first that the electrochemically generated charges are localized on individual *p*-nitrophenyl rings, high level computing clearly revealed charge delocalization; the charges are effectively conveyed by the alkyne moieties into the entire π-conjugated carbon framework. In the dianion, this delocalization actually imparts enough single bond character to the central TEE double bond to allow rotation and *cis-trans* isomerization. The computationally predicted isomerization was indeed realized subsequently in spectroelectrochemical studies with a bis (*p*-nitrophenyl)-substituted DEE [128].

Photochemical *trans → cis* isomerization, which provided elegant and facile access to a variety of TEEs (see Schemes 10 and 11), was comprehensively investigated to elucidate the structural and electronic parameters as well as the environmental conditions controlling the reaction. Similar studies had previously been conducted on stilbenes and azobenzenes [130]. For a total of seven TEEs and DEEs, including **90–92** and an analog of **96** (with $(n\text{-}C_{12}H_{25})_2N$ instead of Me_2N groups), the quantum yields of the *trans → cis* and *cis → trans* isomerization steps were investigated as a function of structure, solvent, excitation wavelength, and temperature [131]. The type and degree of donor/acceptor (D/A) functionalization drastically affected the partial quantum yields of isomerization, $\Phi_{t \to c}$ and $\Phi_{c \to t}$. Generally, the highest quantum yields were found for the bis-acceptor (A-A) derivatives (Φ_{total} ca. 0.7) followed by the D-D (Φ_{total} ca. 0.25) and the D-A molecules (Φ_{total} ca. 0.03; hexane, 27 °C, irradiation at the longest wavelength absorption). TEEs were found not to undergo thermal isomerization, which sets them apart from stilbenes and azobenzenes, presumably because of the lack of steric hindrance in the planar chromophores in both the cis and trans configurations. Due to their non-planarity, cis-stilbenes and azobenzenes undergo thermal isomerization into the planar trans-derivatives. Therefore, TEE and DEE derivatives, in contrast to stilbenes, allow the investigation and quantification of electronic and solvent effects on cis-trans isomerizations, uncomplicated by steric influences.

3.3.2
Structure-Property Relationships in Nonlinear Optical Tetraethynylethenes

The second hyperpolarizability, γ, that describes molecular third-order nonlinear optical effects [132], was measured for over 35 TEE and DEE derivatives in third harmonic generation (THG) experiments in $CHCl_3$ solutions (Fig. 8). The experiments, at a laser frequency of either $\lambda = 1.9$ or 2.1 μm were performed by rotating the samples parallel to the polarization, to generate Maker-fringe interference patterns that were subsequently analyzed [133]. Since all compounds in solution are planar, differences in γ cannot be due to nonplanarity-induced deconjugation. Furthermore, the absorption of the compounds at the third harmonic is either zero or negligible, so that the measured γ-values do not originate from resonance effects. These are important criteria for meaningful structure-function relationships, since corrections for large resonance contributions to γ are either not straightforward or not possible. A series of fundamental conclusions was reached [134, 135]:

1. Donor substitution of TEEs gives higher γ-values than acceptor substitution (see **90** vs **91** or **99** vs **100**, Fig. 8). This correlation might be specific for the TEEs, which are strongly electron-accepting residues, as revealed by electrochemical studies [128, 136], since donor-acceptor interactions lead to enhanced γ-values (see below). The magnitude of γ increases with the donor strength (see **93** vs **94**, Fig. 8).
2. Acentricity greatly enhances the γ-value (see **92** vs **91** and **90** or **101** vs **99** and **100**, Fig. 8). Such a trend had been predicted for certain ranges of compounds by theory [137]; however when the first hyperpolarizability, β, which determines second-order nonlinear optical properties, is maximized, γ is predicted to be zero [138].
3. Molecules with donor and acceptor in trans- and cis-configurations give much higher nonlinearities, due to favorable linear donor-acceptor conjugation, than those with substituents at the geminal position, where only the weaker cross-conjugation is effective (see **89** vs **86** and **92**, Fig. 8).

4. In contrast to other findings [139], a substantial increase in γ is observed upon extending the conjugation length (see **99–101** vs **90–92**, Fig. 8). However, this increase in γ goes along with a considerable increase in molecular mass and, therefore, only a moderate enhancement of the macroscopic $\chi^{(3)}$, which expresses the third-order nonlinearity per unit volume, is observed.

5. Full two-dimensional conjugation, with as many as six conjugation paths (see Fig. 7) leads to very large γ-values. The value measured at 1.9 μm for **95** is the largest in the series of tetrakis-arylated derivatives, and full dimensional conjugation is nearly as effective in increasing the third-order nonlinearity as extension of one specific conjugation path (see **95** vs **101**, Fig. 8). Comparison of the macroscopic $\chi^{(3)}$-values, which take the higher mass of **101** into account, demonstrate an even larger advantage of the two-dimensionally fully-conjugated system **95** over **101**. Resonance effects were observed at 1.9 μm in the series of two-dimensionally fully conjugated systems **87** and **95–98**. Therefore, these chromophores were also measured at 2.1 μm, at which laser wavelength resonance effects are very small and could be corrected for; the largest γ-values were clearly recorded for the compounds of highest acentricity (see **95** vs **96** or **97** vs **98**, Fig. 8). They are among the largest nonresonant γ-values measured for small molecules of this size.

These structure-function relationships provide extremely useful guidance for the future rational design of molecules and polymers with even higher optical nonlinearities. For non-centrosymmetric molecules such as **95**, very high first hyperpolarizabilities β that determine the second-order nonlinear optical properties were also measured [140].

4
Conclusions

The state of research on the two classes of acetylenic compounds described in this article, the cyclo[*n*]carbons and tetraethynylethene derivatives, differs drastically. The synthesis of bulk quantities of a cyclocarbon remains a fascinating challenge in view of the expected instability of these compounds. These compounds would represent a fourth allotropic form of carbon, in addition to diamond, graphite, and the fullerenes. The full spectral characterization of macroscopic quantities of *cyclo*-C$_n$ should provide a unique experimental calibration for the power of theoretical predictions dealing with the electronic and structural properties of conjugated π-chromophores of substantial size and number of heavy atoms. We believe that access to bulk cyclocarbon quantities will eventually be accomplished by controlled thermal or photochemical cycloreversion reactions of structurally defined, stable precursor molecules similar to those described in this review.

In contrast, the synthesis of tetraethynylethene (TEE, C$_{10}$H$_4$) was described in 1991 and, since then, a rich variety of cyclic and acyclic molecular scaffolds incorporating this carbon-rich molecule as a construction module have been prepared. The majority of these compounds, such as the expanded radialenes or the oligomers and polymers of the poly(triacetylene) type, are highly stable and soluble and display a variety of interesting electronic and optical properties. In addition to the compounds that have been explicitly discussed in this review, TEE molecular scaffolding has generated liquid crystalline PTA oligomers [121], solid state charge-transfer complexes with electrostatically controlled layered structures [141], and organometallic derivatives with TEE units as η^1-ligands coordinating to Pt centers, in which metal-to-ligand charge-transfer leads to electronic delocalization over the entire planar π-system [142]. It is clear today that only the tip of a large iceberg of structural diversity in TEE chemistry and its technological perspectives has been explored [143].

The link between cyclo[n]carbons and tetraethynylethene is the occurrence of both structural motifs as repeat units in fascinating two-dimensional all-carbon networks [3, 4]. The development of viable preparative approaches toward these elusive acetylenic networks represents one of the true challenges for synthesis at the turn of the millennium.

5
References

1. Kroto HW, Heath JR, O'Brien SC, Curl RF, Smalley RE (1985) Nature (Lond) 318:162
2. Krätschmer W, Lamb LD, Fostiropoulos K, Huffman DR (1990) Nature (Lond) 347: 354
3. Diederich F, Rubin Y (1992) Angew Chem Int Ed Engl 31:1101
4. Diederich F (1994) Nature (Lond) 369:199
5. For recent proposals on novel carbon modifications, see: (a) Tyutyulkov N, Dietz F, Müllen K, Baumgarten M (1997) Chem Phys Lett 272:111; (b) Rajca A, Safronov A, Rajca S, Ross CR, II, Stezowski JJ (1996) J Am Chem Soc 118:7272; (c) Schulman JM, Disch RL (1996) J Am Chem Soc 118:8470; (d) Kusner RB, Lahti PM, Lillya CP (1995) Chem Phys Lett 241:522; (e) Bucknum MJ, Hoffmann R (1994) J Am Chem Soc 116:11456; (f) Visscher GT, Bianconi PA (1994) J Am Chem Soc 116:1805; g) see also: Tykwinski RR, Diederich F, Gramlich V, Seiler P (1996) Helv Chim Acta 79:634 and references cited therein
6. (a) Nicolaou KC, Dai WM (1991) Angew Chem Int Ed Engl 30:1387; (b) Nicolaou KC, Smith AL (1995) The Enediyne Antibiotics. In: Stang PJ, Diederich F (eds) Modern Acetylene Chemistry. VCH, Weinheim, p 203; (c) Nicolaou KC (1993) Angew Chem Int Ed Engl 32:1377; (d) Lhermitte H, Grierson DS (1996) Contemporary Organic Synthesis 3:41 and 93
7. (a) Nalwa HS, Miyata S (1997) Nonlinear Optics of Organic Molecules and Polymers, CRC Press, Boca Raton; (b) Nalwa HS (1997) Handbook of Organic Conductive Molecules and Polymers, Vols. 1–4, Wiley, Chichester
8. Stang PJ, Diederich F (1995) Modern Acetylene Chemistry, VCH, Weinheim
9. (a) Diederich F, Stang PJ (1998) Metal-catalyzed Cross-coupling Reactions, Wiley-VCH, Weinheim; see also: (b) Wright ME, Porsch MJ, Buckley C, Cochran BB (1997) J Am Chem Soc 119:8393; (c) Liu Q, Burton DJ (1997) Tetrahedron Lett 38:4371; (d) Rossi R, Carpita A, Bellina F (1995) Org Prep Proced Int 27:127
10. (a) Scott LT, Cooney MJ (1995) Macrocyclic Homoconjugated Polyacetylenes. In: Stang PJ, Diederich F (eds) Modern Acetylene Chemistry. VCH, Weinheim, p 321; (b) Scott LT, Cooney MJ, Otte C, Puls C, Haumann T, Boese R, Carroll PJ, Smith AB III, Meijere A de (1994) J Am Chem Soc 116:10275; (c) Brake M, Enkelmann V, Bunz UHF (1996) J Org Chem 61:1190
11. (a) Moore JS (1997) Acc Chem Res 30:402; (b) Young JK, Moore JS (1995) Acetylenes in Nanostructures. In: Stang PJ, Diederich F (eds) Modern Acetylene Chemistry. VCH, Weinheim, p 415

12. (a) Bunz UHF (1994) Angew Chem Int Ed Engl 33:1073; (b) Gleiter R, Kratz D (1993) Angew Chem Int Ed Engl 32:842
13. (a) Jones L II, Schumm JS, Tour JM (1997) J Org Chem 62:1388; (b) Lavastre O, Ollivier L, Dixneuf PH, Sibandhit S (1996) Tetrahedron 52:5495
14. (a) Morrison DL, Höger S (1996) Chem Commun 2313; (b) Kawase T, Ueda N, Darabi HR, Oda M (1996) Angew Chem Int Ed Engl 35:1556; (c) Bradshaw JD, Solooki D, Tessier CA, Youngs WJ (1994) J Am Chem Soc 116:3177
15. (a) Haley MM, Pak JJ, Brand SC, this volume pp 81; (b) see also: Haley MH, Bell ML, English JJ, Johnson CA, Weakley TJR (1997) J Am Chem Soc 119:2956; (c) Boese R, Matzger AJ, Vollhardt KPC (1997) J Am Chem Soc 119:2052
16. Rubin Y (1997) Chem Eur J 3:1009
17. Bartik T, Bartik B, Brady M, Dembinski R, Gladysz JA (1996) Angew Chem Int Ed Engl 35:414
18. For carbyne, see: (a) Kavan L (1997) Chem Rev 97:3061; (b) Kudryavtsev YP, Heimann, RB, Evsyukov SE (1996) J Mater Sci 31:5557
19. Lagow RJ, Kampa JJ, Wei HC, Battle SL, Genge JW, Laude DA, Harper CJ, Bau R, Stevens RC, Haw JF, Munson E (1995) Science 267:362
20. (a) Bunz UHF, this volume pp 131; (b) Bunz UHF (1997) Synlett 1117
21. (a) Alzeer J, Vasella A (1995) Helv Chim Acta 78:1219; (b) Cai C, Vasella A (1996) Helv Chim Acta 79:255
22. (a) Bürli R, Vasella A (1997) Helv Chim Acta 80:1027; (b) Bürli R, Vasella A (1997) Angew Chem Int Ed Engl 36:1852; (c) Bürli R, Vasella A (1997) Helv Chim Acta 80:2215
23. Diederich F, Rubin Y, Knobler CB, Whetten RL, Schriver KE, Houk KN, Li Y (1989) Science 245:1088
24. Chandrasekhar J, Jemmis ED, Schleyer PvR (1979) Tetrahedron Lett 3707
25. Schleyer PvR, Jiao H, Glukhovtsev MN, Chandrasekhar J, Kraka E (1994) J Am Chem Soc 116:10129
26. Minkin VI, Glukhoytsev MN, Simkin BY (1994) Aromaticity and Antiaromaticity, Electronic and Structural Aspects. Wiley, New York
27. (a) Weltner W Jr, Van Zee RJ (1989) Chem Rev 89:1713; (b) Parent DC, McElvany SW (1989) J Am Chem Soc 111:2393
28. (a) Sun J, Grützmacher HF (1997) Eur Mass Spectrom 3:121; (b) Sun J, Grützmacher HF, Lifshitz C (1994) Int J Mass Spectrom Ion Processes 138:49; (c) Sun J, Grützmacher HF, Lifshitz C (1994) J Phys Chem 98:4536
29. McElvany SW, Ross MM, Goroff NS, Diederich F (1993) Science 259:1594
30. (a) Helden G von, Kemper PR, Gotts NG, Bowers MT (1993) Science 259:1300; (b) Helden G von, Gotts NG, Bowers MT (1993) J Am Chem Soc 115:4363; (c) Gotts NG, Helden G von, Bowers MT (1995) Int J Mass Spectrom Ion Processes 149/150:217
31. (a) Hunter J, Fye J, Jarrold MF (1993) Science 260:784; (b) Hunter JM, Fye JL, Roskamp EJ, Jarrold MF (1994) J Phys Chem 98:1810
32 (a) Goroff N (1996) Acc Chem Res 29:77; (b) Crane, JD (1995) Computers Chem 19:1; (c) Babic D, Trinajstic N (1996) J Mol Struct 376:507
33. Plattner, DA, Li Y, Houk KN (1995) Modern Computational and Theoretical Aspects of Acetylene Chemistry. In: Stang PJ, Diederich F (eds) Modern Acetylene Chemistry. VCH, Weinheim, p 1
34. Hoffmann R (1966) Tetrahedron 22:521
35. Parasuk V, Almlöf J, Feyereisen MW (1991) J Am Chem Soc 113:1049
36. Hutter J, Lüthi HP, Diederich F (1994) J Am Chem Soc 116:750
37. Plattner DA, Houk KN (1995) J Am Chem Soc 117:4405
38. (a) Liang C, Schaefer HF III (1990) J Chem Phys 93:8844; (b) Watts JD, Bartlett RJ (1992) Chem Phys Lett 190:19
39. See also: Zahradník R, Hobza P, Burcl R, Hess BA Jr, Radziszewski JG (1994) J Mol Struct (Theochem) 313:335

40. (a) Krebs A, Wilke J (1983) Top Curr Chem 109:189; (b) Meier H (1972) Synthesis 235; (c) Gleiter R, Schäfer W (1990) Acc Chem Res 23:369; (d) Gleiter R, Merger R (1995) Cyclic Alkynes: Preparation and Properties. In: Stang PJ, Diederich F (eds) Modern Acetylene Chemistry, VCH, Weinheim p 285
41. (a) Diederich F, Rubin Y, Chapman OL, Goroff NS (1994) Helv Chim Acta 77:1441; (b) Diederich F (1995) Oligoacetylenes. In: Stang PJ, Diederich F (eds) Modern Acetylene Chemistry, VCH, Weinheim p 443
42. Rubin Y, Knobler CB, Diederich F (1990) J Am Chem Soc 112:4966
43. Dickson RS, Fraser PJ (1974) Adv Organomet Chem 12:323
44. Melikyan GG, Nicholas KM (1995) The Chemistry of Metal-Alkyne Complexes. In: Stang PJ, Diederich F (eds) Modern Acetylene Chemistry, VCH, Weinheim p 99
45. (a) Magnus P, Carter PA (1988) J Am Chem Soc 110:1626; (b) Schreiber SL, Sammakia T, Crowe WE (1986) J Am Chem Soc 108:3128; (c) Magnus P, Miknis GF, Press NJ, Grandjean D, Taylor GM, Harling J (1997) J. Am Chem Soc 119:6739
46. Haley MM, Langsdorf BL (1997) Chem Commun 1121
47. (a) Petersen H, Meier H (1980) Nouv J Chim 4:687; (b) Shvo Y, Hazum E (1974) J Chem Soc Chem Commun 336
48. Cetini G, Gambino O, Rossetti R, Sappa E (1967) J Organomet Chem 8:149
49. Allison NT, Fritch JR, Vollhardt KPC, Walborsky EC (1983) J Am Chem Soc 105:1384
50. Rubin Y, Lin SS, Knobler CB, Anthony J, Boldi AM, Diederich F (1991) J Am Chem Soc 113:6943
51. Sautet P, Eisenstein O, Canadell E (1987) New J Chem 11:797
52. Rubin Y, Kahr M, Knobler CB, Diederich F, Wilkins CL (1991) J Am Chem Soc 113:495
53. (a) Rubin Y, Diederich F (1989) J Am Chem Soc 111:6870; (b) Rubin Y, Knobler CB, Diederich F (1990) J Am Chem Soc 112:1607
54. Hay AS (1962) J Org Chem 27:3320
55. Okamura WH, Sondheimer F (1967) J Am Chem Soc 89:5991; (b) Figeys HP, Gelbcke M (1970) Tetrahedron Lett 5139
56. McQuilkin RM, Garratt PJ, Sondheimer F (1970) J Am Chem Soc 92:6682
57. Li Y, Rubin Y, Diederich F, Houk KN (1990) J Am Chem Soc 112:1618
58. Nishinaga T, Kawamura T, Komatsu K (1997) J Org Chem 62:5354
59. Allen AD, Colomvakos JD, Diederich F, Egle I, Hao X, Liu R, Lusztyk J, Ma J, McAllister MA, Rubin Y, Sung K, Tidwell TT, Wagner BD (1997) J Am Chem Soc 119:12125
60. (a) Grösser T, Hirsch A (1993) Angew Chem Int Ed Engl 32:1340; (b) Schermann G, Grösser T, Hampel F, Hirsch A (1997) Chem Eur J 3:1105
61. (a) Isaacs L, Seiler P, Diederich F (1995) Angew Chem Int Ed Engl 34:1466; (b) Isaacs L, Diederich F, Haldimann RF (1997) Helv Chim Acta 80:317
62. Sonogashira K (1998) Cross-coupling Reactions to sp Carbon Atoms. In: Diederich F, Stang PJ (eds) Metal-catalyzed Cross-coupling Reactions. Wiley-VCH, Weinheim, p 203
63. Hart H, Shamouilian S, Takehira Y (1981) J Org Chem 46:4427
64. Behr OM, Eglinton G, Galbraith AR, Raphael RE (1960) J Chem Soc 3614
65. For further discussion of the propensity of cis-enediynes to undergo oxidative macrocyclization and of the macroring-size preferences, see: (a) ref. 20b); (b) Altmann M, Friedrich J, Beer F, Reuter R, Enkelmann V, Bunz UHF (1997) J Am Chem Soc 119:1472
66. Tobe Y, Fujii T, Matsumoto H, Naemura K, Achiba Y, Wakabayashi T (1996) J Am Chem Soc 118:2758
67. Tobe Y, Fujii T, Matsumoto H, Naemura K (1996) Pure Appl Chem 68:239
68. (a) Tobe Y, Matsumoto H, Naemura K, Achiba Y, Wakabayashi T (1996) Angew Chem Int Ed Engl 35:1800; (b) Wakabayashi T, Kohno M, Achiba Y, Shiromaru H, Momose T, Shida T, Naemura K, Tobe Y (1997) J Chem Phys 107:4783
69. O'Krongly D, Denmeade SR, Chiang MY, Breslow R (1985) J Am Chem Soc 107:5544
70. Tobe Y, Fujii T, Naemura K (1994) J Org Chem 59:1236
71. Adamson GA, Rees CW (1996) J Chem Soc Perkin Trans 1 1535
72. Diercks R, Armstrong JC, Boese R, Vollhardt KPC (1986) Angew Chem Int Ed Engl 25:268

73. (a) Anthony JE, Khan SI, Rubin Y (1997) Tetrahedron Lett 38:3499; (b) Tobe Y, Kubota K, Naemura K (1997) J Org Chem 62:3430
74. Boese R, Green JR, Mittendorf J, Mohler DL, Vollhardt KPC (1992) Angew Chem Int Ed Engl 31:1643
75. Neenan TX, Whitesides GM (1988) J Org Chem 53:2489
76. Marx HW, Moulines F, Wagner T, Astruc D (1996) Angew Chem Int Ed Engl 35:1701
77. (a) Bunz UHF, Enkelmann V (1993) Angew Chem Int Ed Engl 32:1653; (b) Bunz UHF, Enkelmann V, Räder J (1993) Organometallics 12:4745
78. Loon JD van, Seiler P, Diederich F (1993) Angew Chem Int Ed Engl 32:1187
79. Jux N, Holczer K, Rubin Y (1996) Angew Chem Int Ed Engl 35:1986
80. Lange T, Gramlich V, Amrein W, Diederich F, Gross M, Boudon C, Gisselbrecht JP (1995) Angew Chem Int Ed Engl 34:805
81. (a) Hauptmann H (1976) Tetrahedron 32:1293; (b) Hauptmann H (1975) Angew Chem Int Ed Engl 14:498; (c) Hauptmann H (1975) Tetrahedron Lett 1931
82. (a) Rubin Y, Knobler CB, Diederich F (1991) Angew Chem Int Ed Engl 30:698; (b) Anthony J, Boldi AM, Rubin Y, Hobi M, Gramlich V, Knobler CB, Seiler P, Diederich F (1995) Helv Chim Acta 78:13
83. Boldi AM, Diederich F (1994) Angew Chem Int Ed Engl 33:468
84. Anthony J, Boldi AM, Boudon C, Gisselbrecht JP, Gross M, Seiler P, Knobler CB, Diederich F (1995) Helv Chim Acta 78:797
85. Anthony J, Knobler CB, Diederich F (1993) Angew Chem Int Ed Engl 32:406
86. Boldi AM, Anthony J, Knobler CB, Diederich F (1992) Angew Chem Int Ed Engl 31:1240
87. Boldi AM, Anthony J, Gramlich V, Knobler CB, Boudon C, Gisselbrecht JP, Gross M, Diederich F (1995) Helv Chim Acta 78:779
88. Feldman KS, Kraebel CM, Parvez M (1993) J Am Chem Soc 115:3846
89. a) Feldman KS, Weinreb CK, Youngs WJ, Bradshaw JD (1994) J Am Chem Soc 116:9019; (b) Feldman KS, Mareska DA, Weinreb CK, Chasmawala M (1996) Chem Commun 865
90. Alberts AH, Wynberg H (1988) J Chem Soc Chem Commun 748
91. Bunz U, Vollhardt KPC, Ho JS (1992) Angew Chem Int Ed Engl 31:1648
92. Solooki D, Parker TC, Khan SI, Rubin Y (1998) Tetrahedron Lett 39:1327
93. (a) Loon JD van, Seiler P, Diederich F (1993) Angew Chem Int Ed Engl 32:1706; (b) Lange T, Loon JD van, Tykwinski RR, Schreiber M, Diederich F (1996) Synthesis 537
94. Eaton PE, Galoppini E, Gilardi R (1994) J Am Chem Soc 116:7588
95. Michl J (ed) (1997) Modular Chemistry. Kluwer, Dordrecht
96. Hori Y, Noda K, Kobayashi S, Taniguchi H (1969) Tetrahedron Lett 3563
97. Hopf H, Kreuzer M, Jones PG (1991) Chem Ber 124:1471
98. Tykwinski RR, Diederich F (1997) Liebigs Ann/Recueil 649
99. Diederich F (1997) Tetraethynylethenes: Versatile Carbon-rich Building Blocks for Two-dimensional Acetylenic Scaffolding. In: Michl J (ed) Modular Chemistry. Kluver, Dordrecht, 1997, p 17
100. a) Tykwinski RR, Schreiber M, Carlon RP, Diederich F, Gramlich V (1996) Helv Chim Acta 79:2249; (b) Tykwinski RR, Schreiber M, Gramlich V, Seiler P, Diederich F (1996) Adv Mater 8:226
101. Pilling GM, Sondheimer F (1971) J Am Chem Soc 93:1970
102. (a) Untch KG, Wysocki DC (1966) J Am Chem Soc 88:2608; (b) Sondheimer F, Wolovsky R, Garratt PJ, Calder IC (1966) J Am Chem Soc 88:2610
103. (a) Grant WK, Speakman JC (1959) Proc Chem Soc Lond 231; (b) Huynh C, Linstrumelle G (1988) Tetrahedron 44:6337; (c) Zhou Q, Carroll PJ, Swager TM (1994) J Org Chem 59:1294
104. Tovar JD, Jux N, Jarrosson T, Khan SI, Rubin Y (1997) J Org Chem 62:3432
105. Staab HA, Bader R (1970) Chem Ber 103:1157
106. Anthony J, Boudon C, Diederich F, Gisselbrecht JP, Gramlich V, Gross M, Hobi M, Seiler P (1994) Angew Chem Int Ed Engl 33:763
107. Nierengarten JF, Schreiber M, Diederich F, Gramlich V (1996) New J Chem 20:1273
108. Hopf H, Maas G (1992) Angew Chem Int Ed Engl 31:931
109. Hopf H (1984) Angew Chem Int Ed Engl 23:948

110. Schreiber M, Tykwinski RR, Diederich F, Spreiter R, Gubler U, Bosshard Ch, Poberaj I, Günter P, Boudon C, Gisselbrecht JP, Gross M, Jonas U, Ringsdorf H (1997) Adv Mater 9:339

111. (a) Bumm LA, Arnold JJ, Cygan MT, Dunbar TD, Burgin TP, Jones II L, Allara DL, Tour JM, Weiss PS (1996) Science 271:1705; (b) Jones II L, Pearson DL, Schumm JS, Tour JM (1996) Pure Appl Chem 68:145

112. Wagner RW, Lindsey JS (1994) J Am Chem Soc 116:9759

113. (a) Zhou Q, Swager TM (1995) J Am Chem Soc 117:12593; (b)Wu CG, Bein T (1994) Science 266:1013; (c) Constable EC (1995) Macromol Symp 98:503; (d) Golden JH, DiSalvo FJ, Fréchet JMJ, Silcox J, Thomas M, Elman J (1996) Science 273:782; (e) Karakaya B, Claussen W, Gessler K, Saenger W, Schlüter AD (1997) J Am Chem Soc 119:3296

114. Ward MD (1996) Chem Ind (Lond) August 5:568

115. (a) Pearson DL, Tour JM (1997) J Org Chem 62:1376; (b) Bunz UHF (1996) Angew Chem Int Ed Engl 35:969

116. (a) Tour JM (1996) Chem Rev 96:537; (b) Bäuerle P, Götz G, Synowczyk A, Heinze J (1996) Liebigs Ann. 279; (c) Baumgarten M, Müllen K (1994) Top Curr Chem 169:3; (d) Grosshenny V, Harriman A, Ziessel R (1995) Angew Chem Int Ed Engl 34:1100; (e) Roncali J (1997) Chem Rev 97:173

117. Brédas JL (1995) Adv Mater 7:263

118. Heeger AJ, McDiarmid AG (1980) Conducting Organic Polymers: Doped Polyacetylene. In: Alcacer L (ed) The Physics and Chemistry of Low-Dimensional Solids. Reidel, Dordrecht, p 353

119. (a) Wegner G (1977) Pure Appl Chem 49:443; (b) Baughman RH, Brédas JL, Chance RR, Elsenbaumer RL, Shacklette LW (1982) Chem Rev 82:209

120. Schreiber M, Anthony J, Diederich F, Spahr ME, Nesper R, Hubrich M, Bommeli F, Degiorgi L, Wachter P, Kaatz P, Bosshard C, Günter P, Colussi M, Suter UW, Boudon C, Gisselbrecht JP, Gross M (1994) Adv Mater 6:786

121. Nierengarten JF, Guillon D, Heinrich B, Nicoud JF (1997) Chem Commun 1233

122. (a) Meier H, Stalmach U, Kolshorn H (1997) Acta Polym 48:379; (b) Jenekhe SA (1990) Macromolecules 23:2848; (c) Grimme J, Kreyenschmidt M, Uckert F, Müllen K, Scherf U (1995) Adv Mater. 7:292; (d) Brédas JL, Silbey R, Boudreaux DS, Chance RR (1983) J Am Chem Soc 105:6555; (e) Guay J, Kasai P, Diaz A, Wu R, Tour JM, Dao LH (1992) Chem Mater 4:1097

123. Martin RE, Gubler U, Boudon C, Gramlich V, Bosshard C, Gisselbrecht JP, Günter P, Gross M, Diederich F (1997) Chem Eur J 3:1505

124. Schenning APHJ, Martin RE, Ito M, Diederich F, Boudon C, Gisselbrecht JP, Gross M (1998) Chem Commun 1013

125. Hawker C, Fréchet JMJ (1990) J Chem Soc Chem Commun 1010

126 Wooley KL, Hawker CJ, Fréchet JMJ (1991) J Am Chem Soc 113:4252

127. (a) Kondo K, Yasuda S, Sakaguchi T, Miya M (1995) J Chem Soc Chem Commun 55; (b) Nalwa HS, Watanabe T, Miyata S (1995) Adv Mater 7:754

128. Hilger A, Gisselbrecht JP, Tykwinski RR, Boudon C, Schreiber M, Martin RE, Lüthi HP, Gross M, Diederich F (1997) J Am Chem Soc 119:2069

129. For similar findings, see: (a) Wolf MO, Fox HH, Fox MA (1996) J Org Chem 61:287; (b) Zhou Q, Swager TM (1995) J Org Chem 60:7096

130. (a) Meier H (1992) Angew Chem Int Ed Engl 31:1399; (b) Görner H, Kuhn HJ (1995) cis-trans Photoisomerization of Stilbenes and Stilbene-Like Molecules. In: Neckers DC, Volman DH, Bünau G von (eds) Advances in Photochemistry, vol 19. Wiley-Interscience, New York, p 1; (c) Schulte-Frohlinde D, Görner H (1979) Pure Appl Chem 51:279; (d) Waldeck DH (1991) Chem Rev 91:415; (e) Malkin S, Fischer E (1962) J Phys Chem 66:2482; (f) Sanchez AM, Rossi RH de (1996) J Org Chem 61:3446

131. Martin RE, Bartek J, Diederich F, Tykwinski RR, Meister EC, Hilger A, Lüthi HP (1998) J Chem Soc Perkin Trans 2 233

132. Bosshard Ch, Sutter K, Prêtre Ph, Hulliger J, Flörsheimer M, Kaatz P, Günter P (1995) Organic Nonlinear Optical Materials, Gordon and Breach Science, Basel 1995

133. Kajzar F, Messier J (1985) Phys Rev A Gen Phys 32:2352
134. Bosshard Ch, Spreiter R, Günter P, Tykwinski RR, Schreiber M, Diederich F (1996) Adv Mater 8:231
135. Tykwinski RR, Gubler U, Martin RE, Diederich F, Bosshard Ch, Günter P (1998) J Phys Chem B 102:4451
136. Boudon C, Gisselbrecht JP, Gross M, Anthony J, Boldi AM, Faust R, Lange T, Philp D, Loon JD van, Diederich F (1995) J Electroanal Chem 394:187
137. (a) Meyers F, Brédas JL (1993) Theoretical Investigation of the Static Third-order Polarizabilities in Push-Push, Pull-Pull, and Push-Pull Polyenes. In: Ashwell GJ, Bloor D (eds) Organic Materials for Non-Linear Optics III. RSC Special Publication No. 137. RSC, London, p 1; (b) Puccetti G, Blanchard-Desce M, Ledoux I, Lehn JM, Zyss J (1993) J Phys Chem 97:9385; (c) Oudar JL, Chemla DS, Batifol E (1977) J Chem Phys 67:1626; (d) Oudar JL, Chemla DS (1977) J Chem Phys 66:2664; (e) Garito AF, Heflin JR, Wong KY, Zamani-Khamiri O (1989) Enhancement of Non-linear Optical Properties of Conjugated Linear Chains through Lowered Symmetry. In: Hann RA, Bloor D (eds) Organic Materials for Non-Linear Optics: RSC Special Publication No. 69. RSC, London, 1989, p 16; (f) Dirk CW, Cheng LT, Kuzyk MG (1992) Int J Quant Chem 43:27
138. Marder SR, Gorman CB, Meyers F, Perry JW, Bourhill G, Brédas JL, Pierce BM (1994) Science 265:632
139. Cheng LT, Tam W, Marder SR, Stiegmann AE, Rikken G, Spangler CW (1991) J Phys Chem 95:10643
140. Spreiter R, Bosshard Ch, Knöpfle G, Günter P, Tykwinski RR, Schreiber M, Diederich F (1998) J Phys Chem B 102:29
141. (a) Diederich F, Philp D, Seiler P (1994) J Chem Soc Chem Commun 205; (b) Philp D, Gramlich V, Seiler P, Diederich F (1995) J Chem Soc Perkin Trans 2 875; (c) Taniguchi H, Hayashi K, Nishioka K, Hori Y, Shiro M, Kitamura T (1994) Chem Lett 1921
142. (a) Diederich F, Faust R, Gramlich V, Seiler P (1994) J Chem Soc Chem Commun 2045; (b) Faust R, Diederich F, Gramlich V, Seiler P (1995) Chem Eur J 1:111
143. Computational studies on TEEs: (a) Ma B, Xie Y, Schaefer HF III (1992) Chem Phys Lett 191:521; (b) Ma B, Sulzbach HM, Xie Y, Schaefer HF III (1994) J Am Chem Soc 116:3529

Macrocyclic Oligo(phenylacetylenes) and Oligo(phenyldiacetylenes)

Michael M. Haley · Joshua J. Pak · Stephen C. Brand

Department of Chemistry, University of Oregon, Eugene, OR 97403 – 1253, USA.
E-mail: haley@oregon.uoregon.edu

The following is a comprehensive survey of the chemistry of macrocycles comprised entirely of phenyl and acetylenic moieties. Although over four decades old, this area of research has come into its own just in the last few years. Widespread interest in the field has been spurred by recent discoveries utilizing these compounds as ligands for organometallic chemistry, hosts for binding guest molecules, models of synthetic carbon allotropes, and precursors to fullerenes and other carbon-rich materials. This review will discuss the preparation of a tremendous variety of novel structures and detail the development of versatile synthetic methods for macrocycle construction.

Keywords: Annulenes, Carbon-rich compounds, Macrocycles, Phenylacetylenes, Phenyldiacetylenes.

Abbreviations

aq	aqueous
BPO	benzoyl peroxide
dba	dibenzylideneacetone
DCC	1,3-Dicyclohexylcarbodiimide
DEAD	Diethyl azodicarboxylate
DIBALH	Diisobutylaluminum hydride
dppm	1,3-Bis(diphenylphosphino)methane
dppp	1,3-bis(diphenylphosphino)propane
Dec	Decyl
DPTS	4-(N,N-Dimethylamino)pyridinium p-toluenesulfonate
DSC	Differential Scanning Calorimetry
HEB	Hexaethynylbenzene
HMPA	hexamethylphosphoric triamide
LHMDS	lithium hexamethyldisilazane
MALDI	Matrix Assisted Laser Desorption Ionization
Oct	Octyl
ODCB	o-dichlorobenzene
PAM	Phenylacetylene macrocycle
PDM	Phenyldiacetylene macrocycle
TEM	Transmission Electron Microscopy
TFA	trifluoroacetic acid
thexyl	1,1,2-trimethylpropyl
TIPS	triisopropylsilyl
TIPSA	Triisopropylsilylacetylene
TMEDA	N,N,N',N' – tetramethylethylenediamine
TMS	trimethylsilyl
TMSA	Trimethylsilylacetylene
TOF-MS	Time of flight – mass spectroscopy

1
Introduction

Over the last decade, the chemistry of the carbon-carbon triple bond has experienced a vigorous resurgence [1]. Whereas construction of alkyne-containing systems had previously been a laborious process, the advent of new synthetic methodology based on organotransition metal complexes has revolutionized the field [2]. Specifically, palladium-catalyzed cross-coupling reactions between alkyne sp-carbon atoms and sp^2-carbon atoms of arenes and alkenes have allowed for rapid assembly of relatively complex structures [3]. In particular, the preparation of alkyne-rich macrocycles, the subject of this report, has benefited enormously from these recent advances. For the purpose of this review, we limit the discussion to cyclic systems which contain benzene and acetylene moieties only, henceforth referred to as phenylacetylene and phenyldiacetylene macrocycles (PAMs and PDMs, respectively). Not only have a wide

variety of new PAM and PDM topologies now become accessible due to organotransition metal chemistry, but also the number of steps to known macrocycles has been shortened and overall yields dramatically increased in some
cases (vide infra) [4, 5]. This report will focus on the new synthetic methods for
preparation of PAMs and PDMs, as well as on some of their subsequent chemical transformations.

2
Historical Perspectives

Prior to the mid-1980s, the number of phenylacetylene and phenyldiacetylene
macrocycles was exceedingly limited. The very first example claimed in the literature was *ortho*-substituted PDM 1 [6]. Beginning in the late 1950s, Eglinton
et al. published a series of articles describing novel carbocyclic products obtained through the oxidative coupling of terminal acetylene moieties [7]. In
particular, these researchers found that gentle heating of both aliphatic and aromatic diynes under high dilution conditions with excess anhydrous cupric acetate in a pyridine-methanol-diethyl ether solvent mixture produced good to
excellent yields of diacetylenic (intramolecular product) and tetraacetylenic
(intermolecular product) macrocycles. Application of the new experimental
conditions to o-diethynylbenzene (2), prepared from known o-divinylbenzene
(Scheme 1), eventually led to the isolation of an unstable canary-yellow solid.
Initially, on the basis of strain considerations and preliminary X-ray data, they
believed this substance to be molecule 1, the macrocyclic trimer of 2; however,
subsequent experimental and analytical studies pointed toward formation of
the strained cyclodimer 3 [8].

Although mass spectrometry is the most obvious method of discerning between molecules 1 and 3, this particular technique was unavailable to Eglinton
and coworkers at that time; thus, the mass of the material was determined by two
less utilized methods:

(1) the "thermistor drop technique" suggested that the molecular weight of the
yellow solid was quite close to 248.28 g mol^{-1}, a value associated with the
$C_{20}H_8$ formula of dimer 3; and

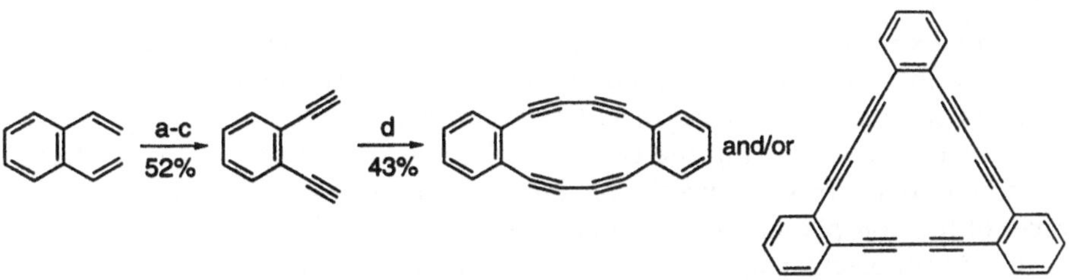

Scheme 1. a) Br_2, CCl_4; b) *t*-BuOK, *t*-BuOH, dioxane; c) *t*-BuOK, PhH; d) $Cu(OAc)_2$, py, MeOH, Et_2O

(2) the application of X-ray techniques gave the dimensions of the unit cell, which were then used in conjunction with an experimentally derived value of the density to calculate a molecular weight representing 99.8% of the theoretical value for the dimer [8].

Cyclodimer 3 proved to be somewhat difficult to manipulate, thus contributing to the complexity of its characterization. The "bowed" diacetylenic linkages revealed in the X-ray data impart surprising physical characteristics to the molecule. The energy-rich hydrocarbon was sufficiently strained that it decomposed explosively upon grinding (i.e. preparing a Nujol mull) or when heated above 80°C. At room temperature, crystals blackened within a few days and apparently auto-polymerized, even when stored under vacuum in the dark. Only dilute solutions of 3 in benzene or pyridine were fairly stable over time, especially when stored cold under an inert atmosphere.

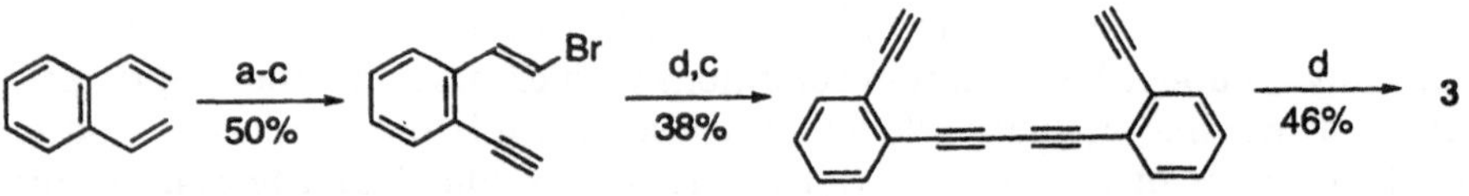

Scheme 2. a) Br_2, CCl_4; b) t-BuOK, t-BuOH, dioxane, 0°C; c) t-BuOK, t-BuOH, dioxane, 100°C; d) $Cu(OAc)_2$, py, MeOH

Definitive evidence for dimer formation was provided by an intramolecular synthesis of 3 (Scheme 2) [9]. On the basis of this successful route, the authors concluded that a stepwise mechanism was plausible for the formation of 3 from o-diethynylbenzene. Furthermore, they speculated that the final product was formed from a transitory copper complex, in which the two acetylene moieties undergoing dimerization were held in physical proximity by a single copper atom. Interestingly, intermolecular coupling of the 1,4-bis(o-ethynyl-phenyl)butadiyne intermediate to produce a cyclic octayne did not occur to any significant degree [9]. Even when the cyclooligomerization of o-diethynyl-benzene was run in more concentrated solutions, Eglinton and coworkers were unable to find evidence of trimeric or tetrameric carbocycles [8]. Instead, they noted a decreased yield of 3 and an increased amount of insoluble polymers and copper salts.

Syntheses of the first *ortho*-PAM were independently reported by Eglinton and Staab in 1966 using two different strategies. The intermolecular approach used by Eglinton et al. was Stephens-Castro coupling [10] of o-iodoethynylben-zene (Scheme 3). Refluxing the corresponding copper acetylide in anhydrous pyridine under nitrogen gave the planar cyclotrimer 4 in 26% yield as well as traces of the non-planar tetramer 5 [11].

The intramolecular approach of Staab and Graf, shown in Scheme 4, precluded formation of 5, but was considerably more involved [12]. The cyclic dienyne 6 was afforded by Wittig reaction of o-phthaldialdehyde with the corresponding bis(ylide) derived from tolane. Bromination of 6 and subsequent treat-

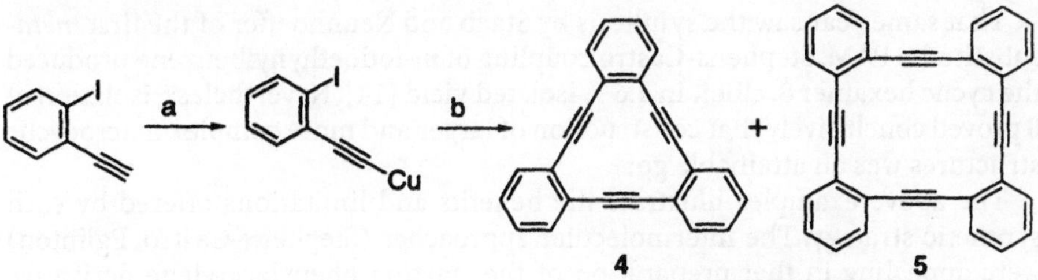

Scheme 3. a) $CuSO_4$, $NH_2OH \cdot HCl$, NH_4OH, EtOH; b) py, reflux

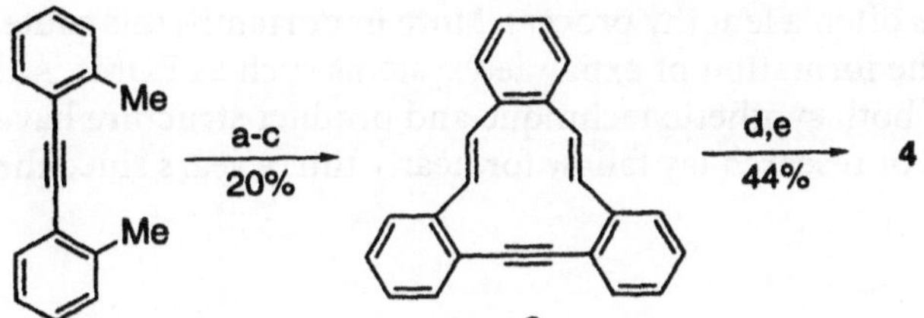

Scheme 4. a) Br_2, CCl_4, NBS, BPO; b) PPh_3, PhH; c) i] PhLi, THF ii] phthaldialdehyde, THF; d) Br_2, CCl_4; e) t-BuOK, THF

ment with excess KO-t-Bu gave 4 in 9% overall yield. X-ray crystallography of the macrocycle showed 4 to be planar and essentially strain-free [12c].

The formal "cyclodimer" of o-iodoethynylbenzene, diyne (7), was prepared in 1974 by Sondheimer's group using analogous bromination/dehydrobromination chemistry [13]. The highly strained molecule was comparatively stable, decomposing around 110°C on attempted melting. Slow decomposition of the solid was observed after 2 days, when unprotected from light and air.

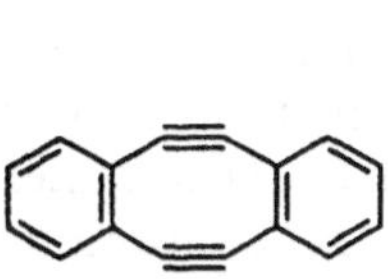

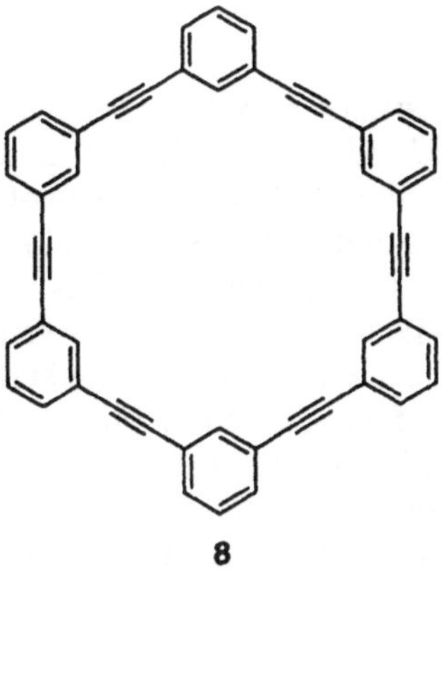

That same year saw the synthesis by Staab and Neunhoeffer of the first *meta*-substituted PAM. Stephens-Castro coupling of *m*-iodoethynylbenzene produced the cyclic hexamer **8**, albeit in 4.6% isolated yield [14]. Nevertheless, isolation of **8** proved conclusively that construction of larger and more complex macrocyclic structures was an attainable goal.

The above examples illustrate the benefits and limitations offered by each synthetic strategy. The intermolecular approaches (Stephens-Castro, Eglinton) were appealing in that preparation of the starting phenylacetylene derivative was generally straightforward; however, yields of a given macrocycle were often low and the desired material had to be isolated from a mixture of products. On the other hand, intramolecular approaches (halogenation/dehydrohalogenation) usually produced a single product that could be isolated and purified with relative ease. The drawback of this approach was that construction of the cyclic structure was often a lengthy process. More importantly, this route could not be utilized for the formation of expanded systems such as PDMs. Although subtle variations of both synthetic technique and product structure have been reported, this field of research lay fallow for nearly thirty years since the preparation of **3**.

3
Phenylacetylenes

In 1980 Sonogashira reported a convenient synthesis of ethynylarenes – the Pd-catalyzed cross-coupling of bromo- or iodoarenes with trimethylsilylacetylene followed by protiodesilylation in basic solution [15]. Prior to this discovery, formation of terminal acetylenes required manipulation of a preformed, two-carbon side chain via methods that include halogenation/dehydrohalogenation of vinyl- and acetylarenes, dehalogenation of β,β-dihaloalkenes, and the Vilsmeier procedure [14]. With the ready availability of trialkylsilylacetylenes, the two-step Sonogashira sequence has become *the* cornerstone reaction for the construction of virtually all ethynylated arenes used in PAM and PDM synthesis (vide infra).

3.1
Ortho

Although other methods have been reported to give PAM **4** [16], the preferred route involves cyclooligomerization of an appropriate *o*-haloethynylbenzene. Youngs et al. have improved the Stephens-Castro method shown in Scheme 3 such that **4** is now isolated in 47% yield along with 8% of tetramer **5** and traces of a hexamer [17]. The Youngs group has utilized this cyclization procedure to prepare methoxy-PAM (**9**) [18] and the related thiophene derivative (**10**) [19].

Vollhardt et al. recently reported the synthesis of hexaethynylated PAM **11** using a combined Sonogashira/Stephens-Castro approach [20]. The requisite compound for cyclization, molecule **12**, was prepared in stepwise fashion from 1,2,3,4-tetrabromobenzene. Regioselective alkynylation at positions 1 and 4 afforded dibromophenyldiacetylene (**13**), which was then monoalkynylated at

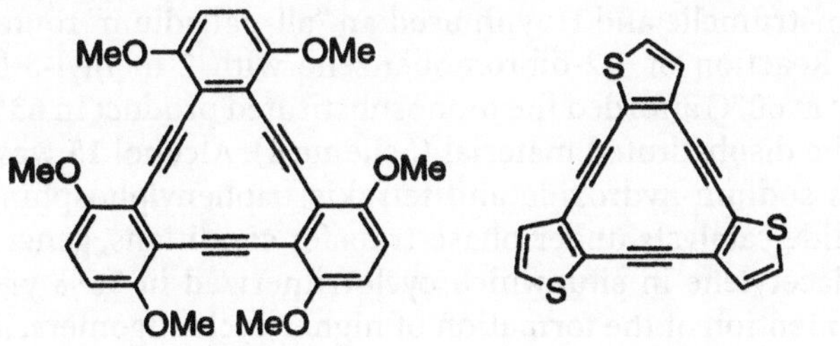

9   **10**

position 2 to give **14** (Scheme 5). The bromide was converted to the iodide by treatment with BuLi and subsequent quenching with molecular iodine. Selective desilylation with potassium carbonate in methanol furnished **12**. Conversion to the copper acetylide followed by reflux in pyridine gave cyclotrimer **11** in ca. 10% overall yield. X-ray structural analysis of the cyclohexylmethyl-substituted dehydroannulene (**11**: R=CH$_2$C$_6$H$_{11}$) showed a highly distorted PAM core due to severe congestion of the exocyclic substituents. Although the triple bonds were significantly distorted from linearity, on average by 5.7° (cyclic) and 4.5° (exocyclic), the overall molecular structure is similar to that found in the parent macrocycle.

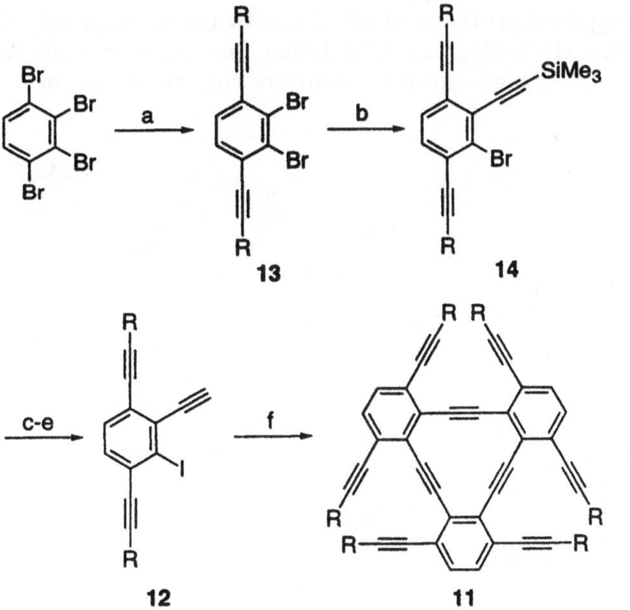

R=SiMe$_2$(Thexyl), Pr, CH$_2$C$_6$H$_{11}$

Scheme 5. a) RC≡CH (2 equiv), PdCl$_2$(PPh$_3$)$_2$, CuI, Et$_3$N; b) TMSA (1 equiv), PdCl$_2$(PPh$_3$)$_2$, CuI, Et$_3$N; c) BuLi, Et$_2$O; d) I$_2$, Et$_2$O; e) K$_2$CO$_3$, MeOH; f) CuCl, NH$_4$OH, EtOH, py

In 1988, Linstrumelle and Huynh used an "all-palladium" route to construct PAM **4** [21]. Reaction of 1,2-dibromobenzene with 2-methyl-3-butyn-2-ol in triethylamine at 60 °C afforded the monosubstituted product in 63 % yield along with 3 % of the disubstituted material (Scheme 6). Alcohol **15** was then treated with aqueous sodium hydroxide and tetrakis(triphenylphosphine)palladium-copper(I) iodide catalysts under phase-transfer conditions, generating the terminal phenylacetylene in situ, which cyclotrimerized in 36 % yield. Although there was no mention of the formation of higher cyclooligomers, it is likely that this reaction did produce these larger species, as is typically seen in Stephens-Castro coupling reactions [22].

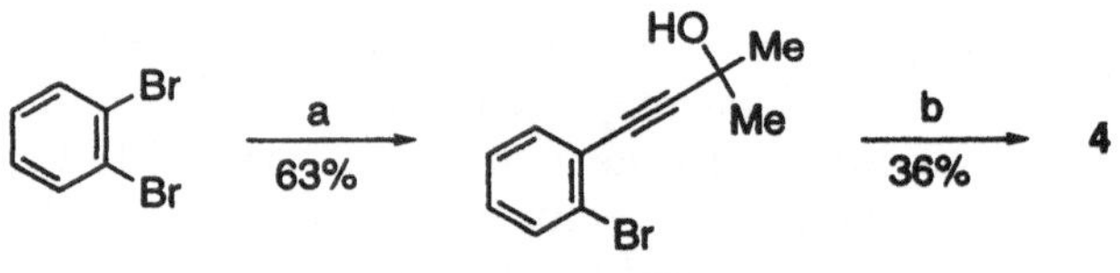

Scheme 6. a) 2-methyl-3-butyn-2-ol, $Pd(PPh_3)_4$, CuI, Et_3N; b) $Pd(PPh_3)_4$, CuI, aq NaOH, PhH, $BnEt_3N^+Cl^-$

In order to minimize the formation of side products, PAM **4** can be assembled via an intramolecular approach [23]. The Sonogashira protocol [15] and conversion of masked iodides [24] comprises most of the chemistry involved in Scheme 7. Using these proven methods, diyne **16** and subsequently triyne **17** can be prepared quickly. Iodination, desilylation, and intramolecular alkynylation with $Pd(dba)_2$ under high dilution conditions furnished **4** as the sole product.

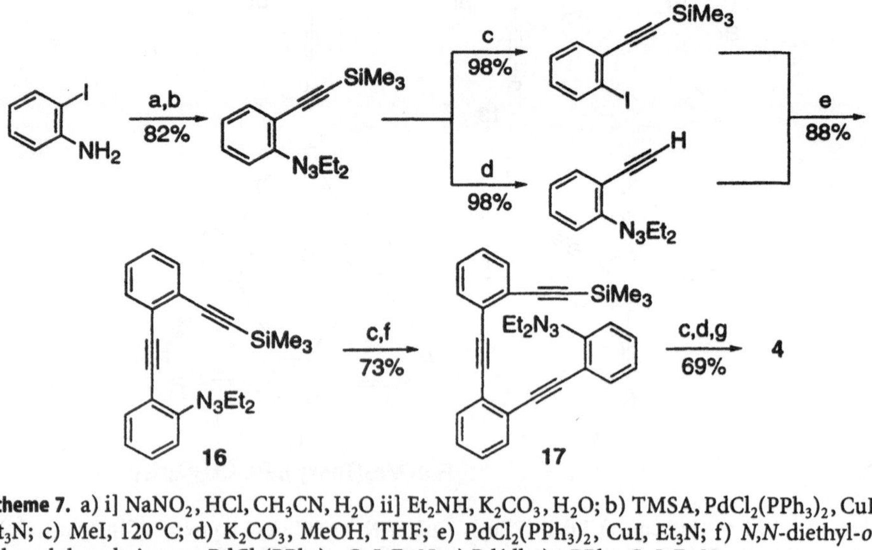

Scheme 7. a) i] $NaNO_2$, HCl, CH_3CN, H_2O ii] Et_2NH, K_2CO_3, H_2O; b) TMSA, $PdCl_2(PPh_3)_2$, CuI, Et_3N; c) MeI, 120 °C; d) K_2CO_3, MeOH, THF; e) $PdCl_2(PPh_3)_2$, CuI, Et_3N; f) *N,N*-diethyl-*o*-ethynylphenyltriazene, $PdCl_2(PPh_3)_2$, CuI, Et_3N; g) $Pd(dba)_2$, PPh_3, CuI, Et_3N

Although this route had several more steps than the two-step pathway, the overall yield was higher – 35% versus the reported 23% yield for Scheme 6. More importantly, larger macrocycles were not detected, thus making product isolation and purification relatively facile.

The above success suggested that construction of more complex *ortho*-PAMs might be achieved through intramolecular routes. For example, molecules 18 and 19 are impossible to prepare via the cyclooligomerization path, and other intermolecular strategies failed to give the desired macrocycles [23]. Scheme 8 illustrates the synthesis of 18 [23]. Using the same general sequence of reactions, the carbon backbone (16 → 20) was efficiently assembled. Closure of 20, after vacuum sublimation, gave a yellow solid that proved to be only sparingly soluble in organic solvents. This problem prevented the acquisition of NMR data; nevertheless, mass spectrometry, IR, and UV-Vis data confirmed the formation of 18. Even though a solubility problem was anticipated in the case of larger planar PAMs, its severity was surprising at this stage. It is quite probable that the low yield of the cyclization step (< 15%) is due mainly to this complication. Unfortunately, repetition of the synthetic sequence with solubilizing substituents on the arenes has failed so far to furnish derivatives of 18 [25].

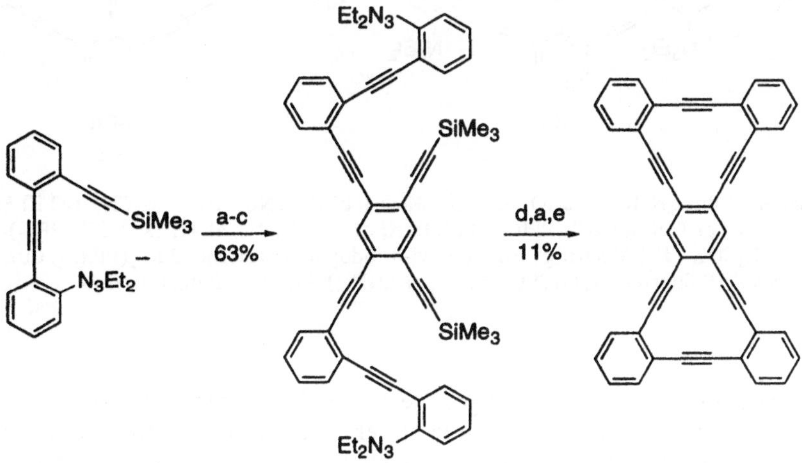

Scheme 8. a) K_2CO_3, MeOH, THF; b) 1,5-dibromo-2,4-diiodobenzene, $PdCl_2(PPh_3)_2$, CuI, Et_3N; c) TMSA, $PdCl_2(PPh_3)_2$, CuI, Et_3N; d) MeI, 120 °C; e) $Pd(dba)_2$, PPh_3, CuI, Et_3N

Given the solubility problems described above, construction of 19 was accomplished using strategically placed *tert*-butyl groups [26]. Starting with *p*-*tert*-butylaniline, iododiyne 21 was prepared in multigram quantities (Scheme 9). Part of the material was carried forward to 22, which was then desilylated and coupled to the remaining 21, thus affording 23. Iodination, desilylation, and double intramolecular coupling gratifyingly furnished 19 as a bright yellow solid. To date, attempts to obtain X-ray diffracting crystals have been unsuccessful; nevertheless, the spectral properties were fully in accordance with the proposed structure. Although the overall yield was quite low (0.6% for 13 steps), the Pd-catalyzed intramolecular coupling appears to be the sole approach to PAMs containing two or more complete macrocycles.

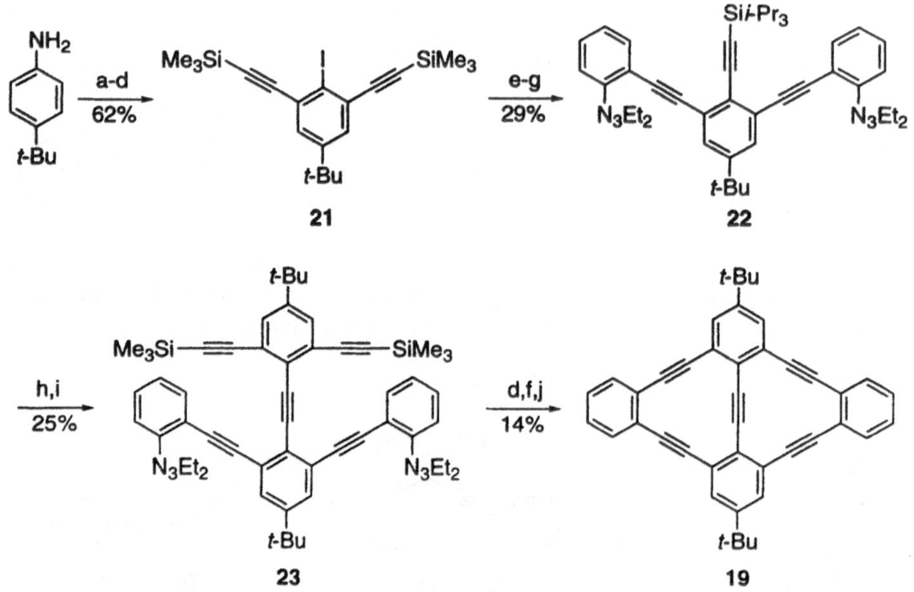

Scheme 9. a) BnEt$_3$N$^+$ICl$_2^-$, CaCO$_3$, CH$_2$Cl$_2$, MeOH; b) i] NaNO$_2$, HCl, CH$_3$CN, H$_2$O ii] Et$_2$NH, K$_2$CO$_3$, H$_2$O; c) TMSA, PdCl$_2$(PPh$_3$)$_2$, CuI, Et$_3$N; d) MeI, 120°C; e) TIPSA, PdCl$_2$(PPh$_3$)$_2$, CuI, Et$_3$N; f) K$_2$CO$_3$, THF, MeOH; g) N,N-diethyl-o-iodophenyltriazene, PdCl$_2$(PPh$_3$)$_2$, CuI, Et$_3$N; h) Bu$_4$NF, THF, EtOH; i) 21, PdCl$_2$(PPh$_3$)$_2$, CuI, Et$_3$N; j) Pd(dba)$_2$, PPh$_3$, CuI, Et$_3$N

Considerably larger *ortho*-oligo(phenylacetylene)s can be prepared by reacting α,ω-diiodide **24** with polyyne **25** under Pd-catalyzed coupling conditions to give the 40-membered PAM **26** in 25% yield [Eq.(1)] [27]. The higher derivatives of C$_{160}$H$_{80}$, C$_{240}$H$_{120}$, C$_{320}$H$_{160}$ and C$_{400}$H$_{200}$ were also isolated in small amounts and characterized by TOF-MS [28].

The ready availability of PAM **4** via several syntheses has allowed elucidation of the chemistry displayed by the molecule. Youngs and coworkers, the principle protagonists in this subarea, have utilized the π-electron-rich nature of **4** to prepare a variety of organometallic species in which **4** participates as a ligand. For example, treatment with dicobaltoctacarbonyl yielded the sixty-six-electron cluster **27** [29]. The structure of complex **27** was analyzed by X-ray crystallography, from which the authors inferred a resemblance to the postulated transition state during metal-mediated [2 + 2 + 2] cyclotrimerization of alkynes. An alternative mode of metal bonding was observed with the Ni(0) complex (**28**), gen-

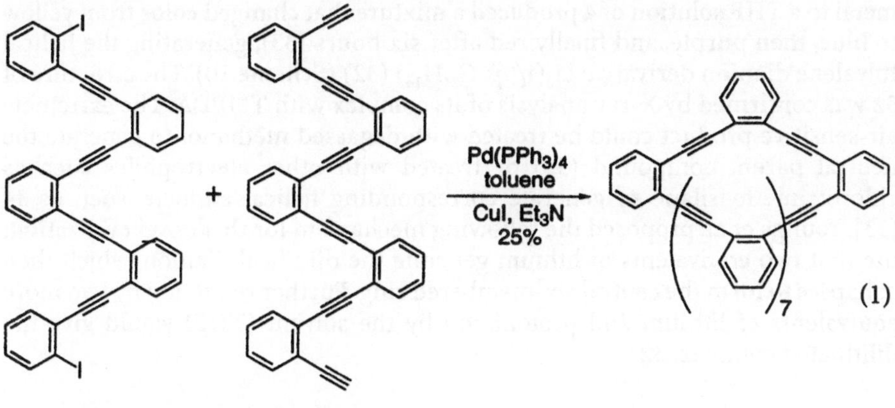

$$\mathbf{24} + \mathbf{25} \xrightarrow[\substack{CuI, Et_3N \\ 25\%}]{\substack{Pd(PPh_3)_4 \\ toluene}} \mathbf{26} \tag{1}$$

erated from a benzene solution of $Ni(COD)_2$ and **4** [30]. Youngs recognized that the distance from the center of the 12-membered dehydroannulene to the center of the carbon-carbon triple bonds approximated 1.2 Å, thus providing a good environment for π-alkylene/transition-metal bonding interactions. This resulted in the nickel atom of **28** residing in the plane of the macrocycle. In treating **4** with copper triflate, either **29** or **30** could be isolated, depending on the stoichiometry [31]. Use of the slightly larger silver cation gave rise to the first PAM sandwich complex, **31** [32].

Beside complexation with transition-metals, PAM **4** exhibited unusual chemistry when treated with an alkali metal. Addition of four equivalents of lithium

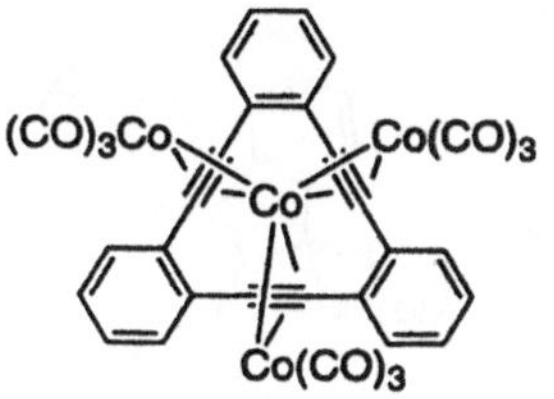

27

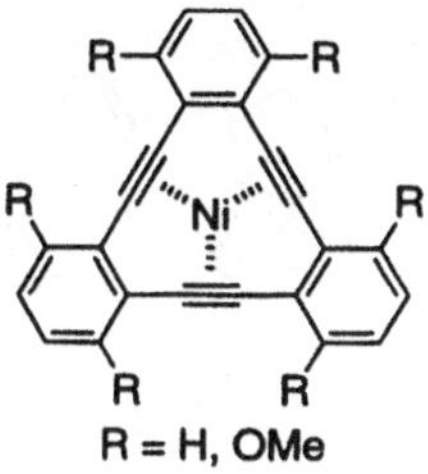

R = H, OMe

28

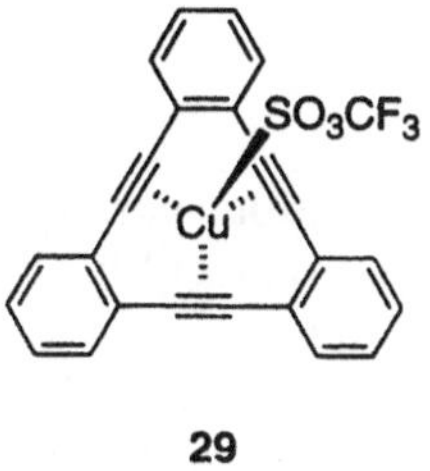

29

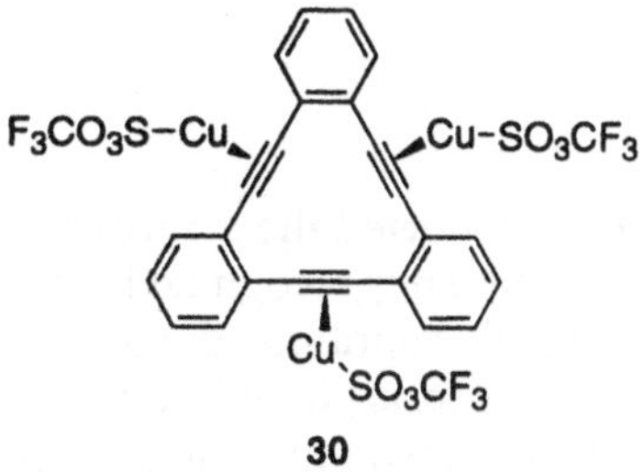

30

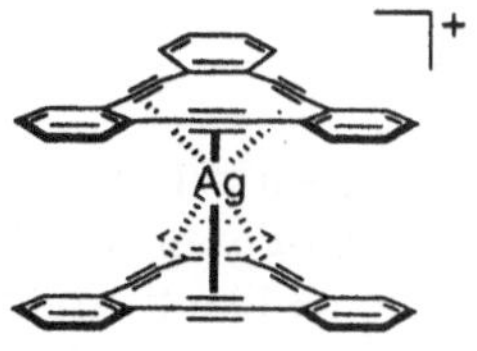

31

metal to a THF solution of **4** produced a mixture that changed color from yellow to blue, then purple, and finally red after six hours [33], generating the helical fulvalene dianion derivative $Li_2(\eta^5\eta^5\text{-}C_{24}H_{14})$ (**32**) (Scheme 10). The structure of **32** was confirmed by X-ray analysis of its complex with TMEDA. The extremely air-sensitive product could be treated with degassed methanol to generate the neutral parent compound (**33**) or treated with other electrophiles such as chlorotrimethylsilane to generate corresponding helical adducts, such as **34** [33]. Youngs et al. proposed the following mechanism for this novel cyclization: the first two equivalents of lithium generate the diradical dianion, which then collapses to form the central six-membered ring. Further reduction by two more equivalents of lithium and protonation by the solvent (THF) would give the dilithiated complex, **32**.

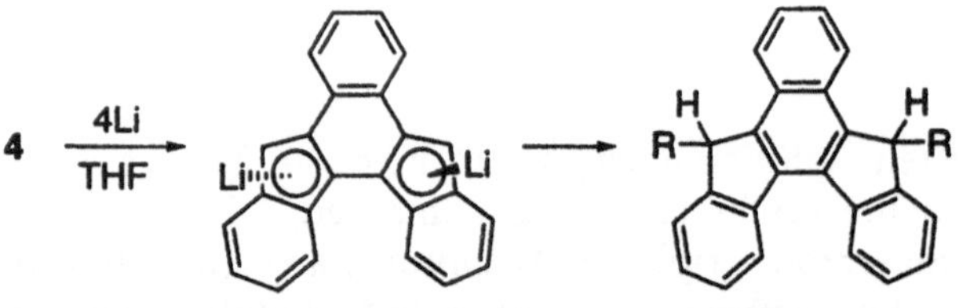

Scheme 10

32   **33 R=H**
                         34 R=SiMe₃

An analogous type of lithium-induced "zipper" cyclization was observed with PAM **5**, affording the *meso*/DL pair of helicenes (Scheme 11) [34]. PAM **5** also reacted with dicobaltoctacarbonyl to give a tetracobalt cluster in which only two of the triple bonds have been complexed [35].

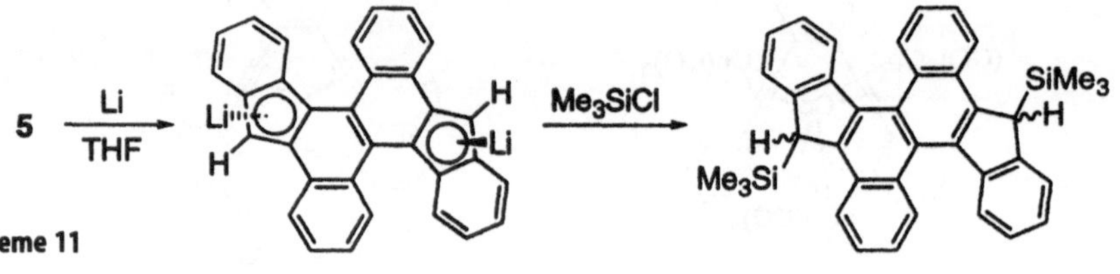

Scheme 11

3.2
Meta

In the early 1990s, Moore et al. reported the syntheses of a tremendous variety of *meta*-connected PAMs [5b, 36]. They recognized that the structural rigidity of systems like **8** could be useful in supramolecular chemistry. The convergent, stepwise approach of linear oligomeric phenylacetylene sequences developed in the Moore laboratory permitted absolute control over chain length, order of monomer addition, and substituent placement [37]. Appropriately derivatized phenylacetylene monomers (e.g. 35–37) were sequentially added and cyclized to generate geometrically well defined phenylacetylene macrocycles in high overall yields [38]. This strategy was particularly advantageous, compared to the more commonly used one-pot cyclooligomerization route, in terms of efficiency, product purity, and control over ring size. Furthermore, additional transformation of functional groups could be performed efficiently.

35   **36**   **37**

R=H, OH, CH₂OH, *t*-Bu, OBu, CO₂Bu

Monomers contained terminal phenylacetylenes protected with trimethylsilyl groups and aryl iodides masked as *N,N*-dialkyltriazenes. A typical PAM synthesis involved three basic transformations (Scheme 12):

(1) reaction of the dialkyltriazene moiety in **35** with iodomethane at 110 °C gave aryl iodide **38**, often in very high yields;
(2) protiodesilylation of trimethylsilyl-protected acetylene in **35** with methanol and catalytic potassium carbonate afforded terminal phenylacetylene **39**;
(3) Pd-catalyzed cross-coupling of aryl iodide **38** and terminal phenylacetylene **39** gave a dimer (e.g. **40**) containing both trimethylsilyl-protected acetylene and aryltriazene moieties.

This sequence of three reactions could be repeated an appropriate number of times to generate the desired linear phenylacetylene oligomer (e.g. **41**). The strength of this stepwise assembly of linear oligo(phenylacetylene)s is due to the fact that trimethylsilylacetylene and aryldialkyltriazene are very effective protecting and masking groups for a terminal acetylene and for an aryl iodide, respectively. Each could be removed selectively in the presence of other functionalities under relatively simple reaction conditions in nearly quantitative yield.

PAMs were prepared by utilizing conditions for intramolecular cross-coupling of linear oligomers. Unmasking the triazene unit and deprotection of the trimethylsilyl group in **41** afforded a difunctional terminal phenylacetylene/aryl iodide intermediate. The cyclization reaction was most effective under pseudo-high-dilution conditions: slow addition via syringe pump of a 1:1 benzene:triethylamine solution of deprotected **41** to a solution of Pd-catalyst under N₂ atmosphere. Under these conditions, only the desired product was isolated; no evidence of catenane or higher oligomer formation was observed.

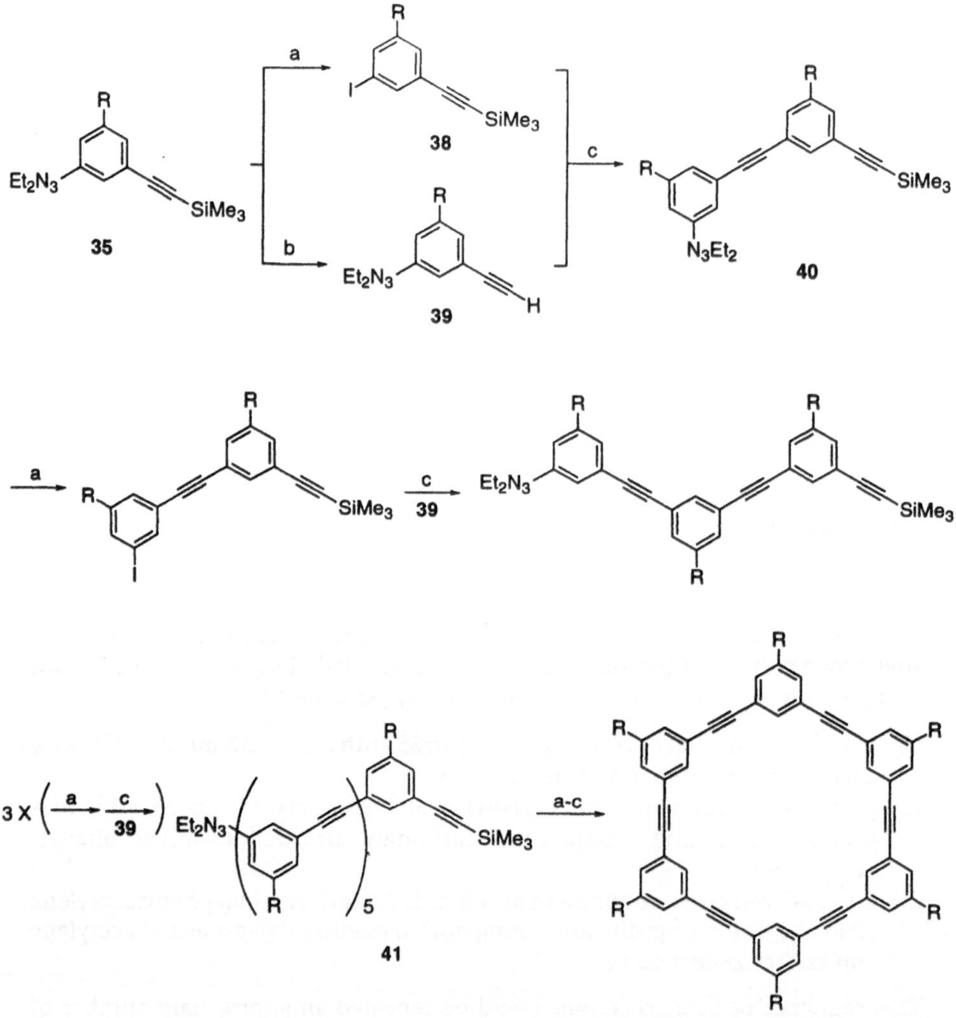

Scheme 12. a) MeI, 110 °C; b) K$_2$CO$_3$, CH$_2$Cl$_2$, MeOH; c) Pd(dba)$_2$, PPh$_3$, CuI, Et$_3$N

Cyclization of substituted phenylacetylene sequences afforded functional-ized macrocycles that were amenable to subsequent manipulation. For ex-ample, transesterification of **42** with octanol in the presence of 18-crown-6 ether and potassium carbonate gave the corresponding ester in 85 % yield (Scheme 13). The ester functionalities could be reduced by DIBALH to give the hydroxymethyl-substituted macrocycle (**43**) in 61 % yield. The low yield of this particular transformation is attributed to mechanical losses during purifica-tion, due to the highly polar nature of the product. Macrocycle **43** could then be treated with alkyl bromides to give a group of benzyl ether derivatized PAMs.

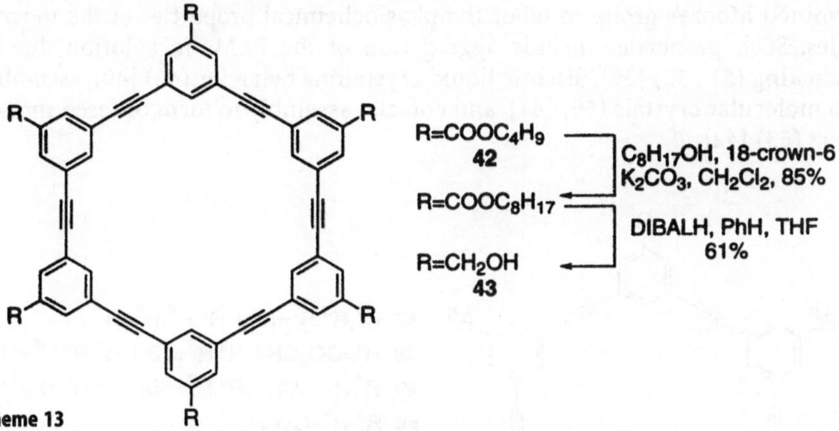

$R=COOC_4H_9$
42

$R=COOC_8H_{17}$ — $C_8H_{17}OH$, 18-crown-6
K_2CO_3, CH_2Cl_2, 85%

DIBALH, PhH, THF
61%

$R=CH_2OH$
43

Scheme 13

The phenyl ether linkage of butoxy substituted **44** was cleaved with boron tribromide to give a hexaphenol derivative (**45**) in 60% yield (Scheme 14). The hexaphenol derivative was found to be remarkably soluble in most polar organic solvents, such as ethanol, DMF, DMSO, and dioxane, as well as in aqueous base solution. Condensation with octanoic acid in the presence of DCC and catalytic H^+ afforded the inverse ester (**46**) in high yield.

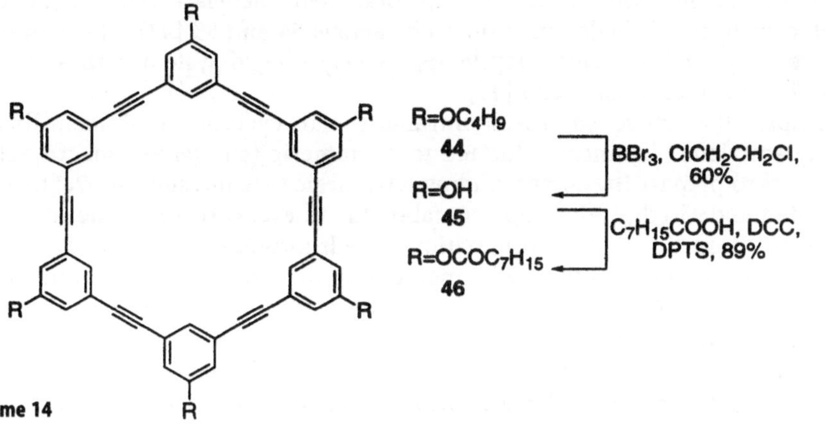

$R=OC_4H_9$
44

$R=OH$
45

BBr$_3$, ClCH$_2$CH$_2$Cl,
60%

$C_7H_{15}COOH$, DCC,
DPTS, 89%

$R=OCOC_7H_{15}$
46

Scheme 14

By using monomers with different substituents, multiple functionalities could be introduced into the phenylacetylene oligomer at any desired position along the sequence backbone, resulting in macrocycles with a wide variety of symmetries (e.g. **47–53**). In principle, this versatile synthetic method should allow construction of PAMs in which any particular group could be placed at any particular site. Judicious choice of the type and placement of functionalities has permitted Moore's group to tailor the physiochemical properties of the macro-cycles. Such properties include aggregation of the PAMs in solution due to π-stacking (**51**–**52**) [39], discotic liquid crystalline behavior (**46**) [40], assembly into molecular crystals (**50**) [41], and cofacial assembly to form ordered mono-layers (**53**) [42].

47 R^1,R^3,R^5=t-Bu R^2,R^4,R^6=H

48 R^1=CO_2CH_3 R^3,R^5=t-Bu R^2,R^4,R^6=H

49 R^1=CO_2CH_3 R^4,R^6=t-Bu R^2,R^3,R^5=H

50 R^1-R^6=CO_2H

51 R^1-R^3=OBu R^4-R^6=CO_2Bu

52 R^1,R^3,R^5=OBu R^2,R^4,R^6=CO_2Bu

53 R^1-R^4=t-Bu, R^5-R^6=COO^- $^+NBu_4$

An impressive achievement of this strategy has been the construction of three-dimensional structures. Utilizing branched phenylacetylene sequences, double cyclization yielded macrobicyclic arrays **54** and **55** [43]. The zenith of Moore's approach is macrotricycle **56**, a freely hinged system with a sizable $36 \times 12 \times 12$ Å molecular cavity [44].

Despite the above successes, intramolecular cyclization of *meta*-linked phenylacetylene oligomers is limited to pentameric (and larger) macrocycles. Attempts to prepare the corresponding tetrameric systems, such as **57**, failed to provide the desired products, presumably due to excess strain in the resultant molecule. Resorting to the bromination/dehydrobromination route, Oda et al. successfully synthesized the target macrocycle (Scheme 15) [45]. Treatment of dialdehyde **58** with low-valent titanium, generated from $TiCl_4$ and Zn in DME, furnished a mixture of the (Z,Z,Z,E)- and (E,Z,E,Z) isomers of **59** in 35–40% yield; neither the (Z,Z,Z,Z) isomer nor the intramolecular coupling product were detected. Reaction with excess bromine presumably gave the octabromin-ated macrocycle. Subsequent treatment of the crude mixture with t-BuOK afforded **57** in 55% yield.

X-ray structure analysis showed that macrocycle **57** was essentially planar, with the twist angle of the benzene rings from the plane of the macrocycle being less than 2°. Most of the strain was seemingly contained in the triple bonds, as these were bent from linearity by 10.1° to 12.3°. Despite its strained nature, the macrocycle showed remarkable stability. Decomposition occurred above 300°C on attempted melting. No reaction was observed between **57** and cyclopenta-diene at room temperature.

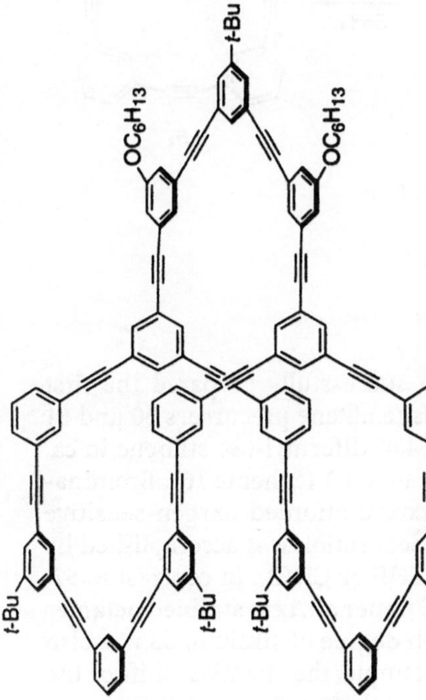

55

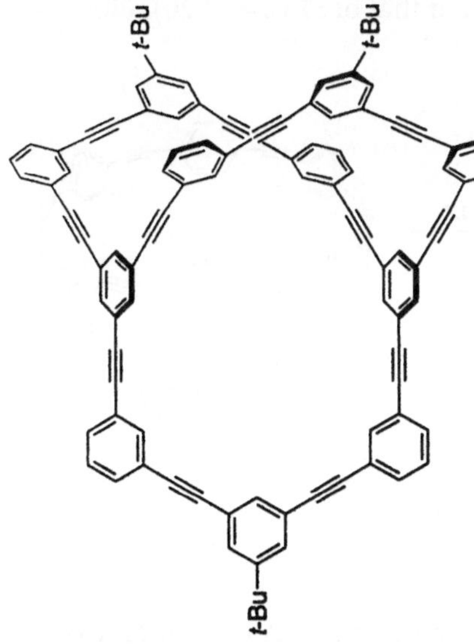

54

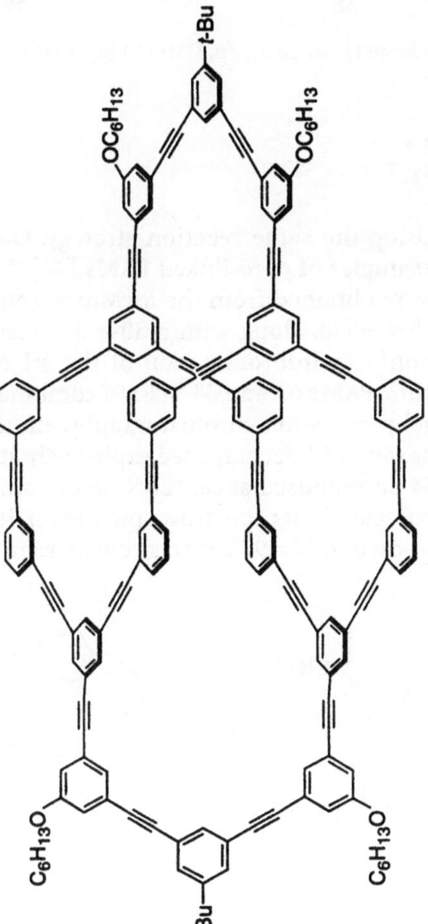

56

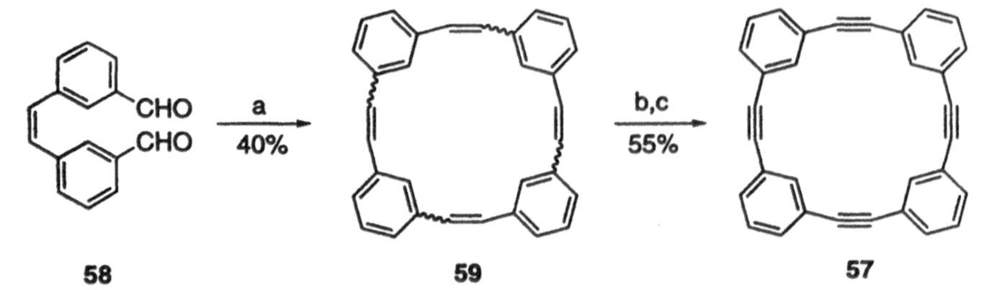

Scheme 15. a) TiCl$_4$, Zn, DME; b) Br$_2$, CHCl$_3$; c) t-BuOK, THF

3.3
Para

Using the same reaction strategy, Oda's team successfully prepared the first examples of *para*-linked PAMs [46]. The requisite alkene precursors **60** and **61** were obtained from the McMurry coupling of 4,4'-diformyl-(Z)-stilbene in ca. 20% yield, along with a 30–40% yield of tetraene **62** (Scheme 16). Bromination/dehydrobromination of the 4:1 **60:61** mixture afforded oxygen-sensitive *para*-PAMs **63** and **64** in 85% combined yield. Separation was accomplished by gel permeation chromatography, eluting with THF or CHCl$_3$. In contrast to **57**, hexamer **63** decomposed explosively at ca. 80 °C under air. Less strained octamer **64** decomposed at ca. 120 °C under air. The high degree of strain in **63** was also reflected in its spectroscopic properties. For example, the chemical shift of the *sp*-carbon ($\delta = 97.65$) was considerably lower than that of **57** ($\delta = 92.20$).

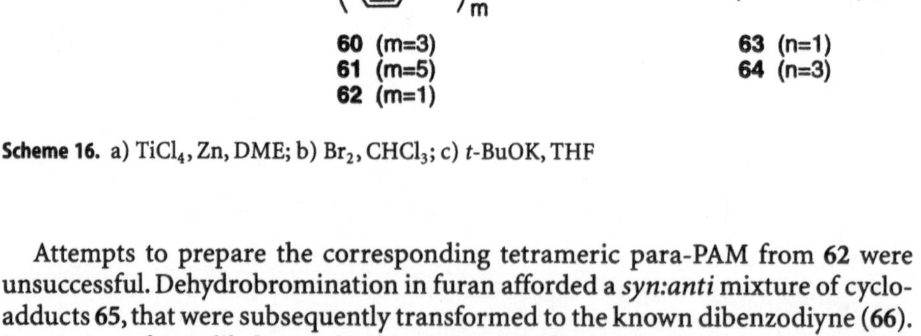

Scheme 16. a) TiCl$_4$, Zn, DME; b) Br$_2$, CHCl$_3$; c) t-BuOK, THF

Attempts to prepare the corresponding tetrameric para-PAM from **62** were unsuccessful. Dehydrobromination in furan afforded a *syn:anti* mixture of cyclo-adducts **65**, that were subsequently transformed to the known dibenzodiyne (**66**). Formation of **65** is likely to arise from stepwise elimination/cycloaddition rather than to involve the intermediacy of the highly strained tetrameric PAM.

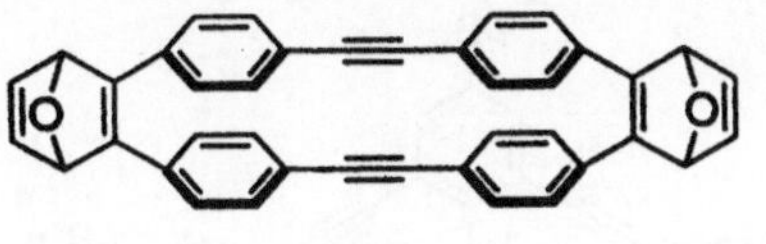

65 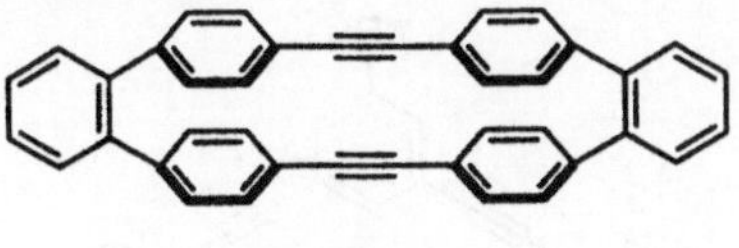   **66**

3.4
Mixed

Using combinations of *ortho-*, *meta-*, and *para-*connected phenylacetylene monomers, Moore's group has been able to assemble a number of unstrained monocyclic geometries based on a trigonal lattice. PAMs **67–69** are three examples that clearly illustrate the enormous potential of this route to the preparation – not only of a large variety of hydrocarbon frameworks – but also of a tremendous number of *endo-* and *exo-*functionalized structures for each framework [47]. A unique application of this method was the synthesis of a "molecular turnstile" [48]. Modeling of turnstile **70** showed that the inner ring freely rotates; however, substituted derivatives **71** and **72** might be sufficiently hindered for rotation to be observed. Indeed this was the case; whereas **72** was conforma-

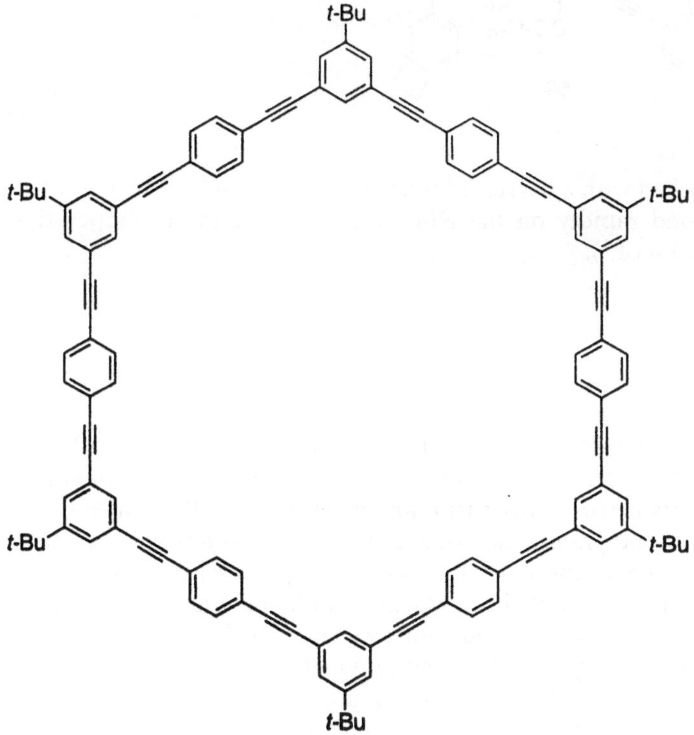

67

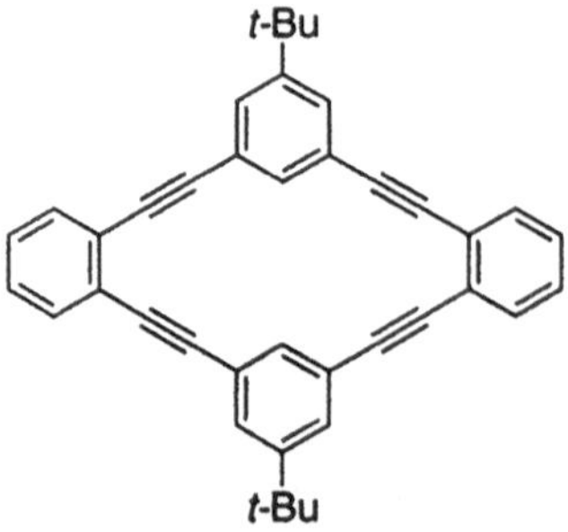

68

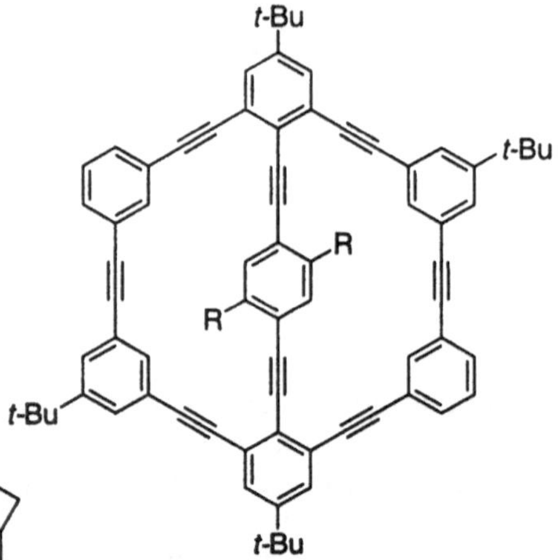

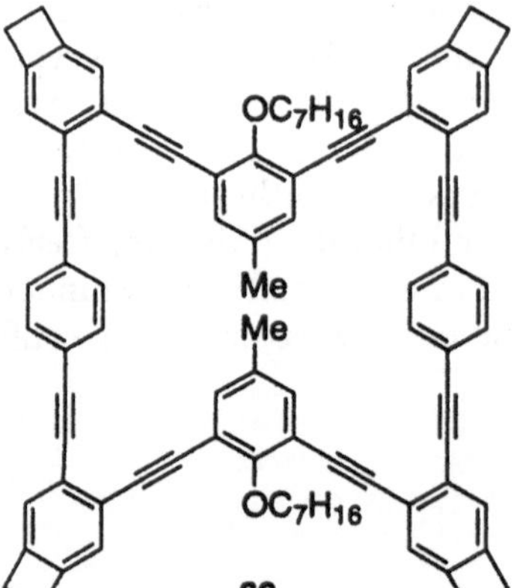

69

70 R=H
71 R=CH$_2$OCH$_3$
72 R=CH$_2$O—

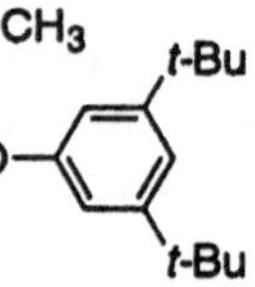

tionally locked with a barrier of rotation greater that 20.6 kcal mol^{-1}, the spindle in **71** rotated rapidly on the NMR time scale at room temperature (barrier ca. 13.4 kcal mol^{-1}).

4
Phenyldiacetylenes

In the thirty years following Eglinton's groundbreaking work on **3** [8], no reports were published on the preparation of related macrocyclic phenyldiacetylene systems. This is somewhat surprising given the fact that, since the mid-1970s, topochemical polymerization of diacetylene monomers has been recognized as a means of generating crystalline polymers possessing conjugated backbones [49]. If this reactivity could be extended to include molecules such as **1** and **3**, PDMs might serve as precursors for a variety of technologically important compounds, such as highly conjugated polymers, nonlinear optical materials, and allotropes of carbon [5c, 50]. Given Eglinton's difficulties (vide supra), better synthetic methods would have to be employed. Fortunately, organotransition metal chemistry developed over the intervening thirty year period provided the wherewithal for PDM construction. This allowed for an "explosive" revival of PDM chemistry starting in the early 1990s.

4.1
Ortho

Armed with new synthetic techniques and modern chemical instrumentation, Swager et al. revisited the *ortho*-diethynylbenzene cyclooligomerization route to PDMs in 1994 [51]. This group was intrigued by the possibility of employing cyclic diacetylene molecules as monomers, the polymerization of which would produce conjugated networks of *sp* and sp^2 carbon atoms. Presaging their interest, several publications had appeared in the mid to late 1980s predicting that such networks could display useful electronic, optical, and physical properties [52].

In addition to diyne 2, the Swager team investigated the copper-promoted cyclooligomerizations of several 4,5-disubstituted-*o*-diethynylbenzenes (73). Preparation of the requisite *o*-dialkylbenzenes was accomplished by cross-coupling the Grignard salt of the desired alkane with *o*-dichlorobenzene in the presence of catalytic (1,3-bis(diphenylphosphino)propane)nickel(II) chloride (Scheme 17). Diiodination was achieved with a mixture of iodine and sodium iodate in glacial acetic acid. Double alkynylation with trimethylsilylacetylene, using Sonogashira conditions followed by protiodesilylation with catalytic KOH in 5:1 methanol:THF (v:v), furnished the dialkylated analogs of 73. The corresponding didecoxy compound was assembled in an analogous manner; however, standard Sonogashira coupling conditions produced only trace amounts of the monoethynylated product. The diethynylated product could be produced in excellent yield (88%) through utilization of two equivalents of CuI at a higher reaction temperature (70°C). Although the exact origin of this effect remains unclear, it is possible that the lone pairs on the heteroatom function as Lewis bases in the formation of weakly bound aryl-copper complexes, thereby preventing formation of the requisite cuprous acetylide coupling partner.

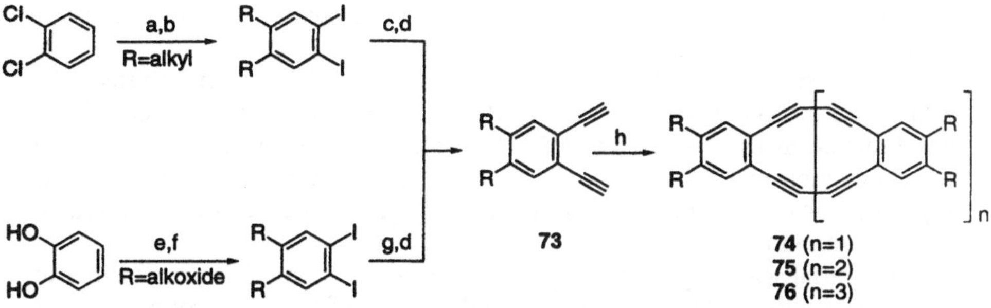

Scheme 17. a) H(CH$_2$)$_n$MgBr, (dppp)NiCl$_2$, Et$_2$O; b) I$_2$, NaIO$_3$, H$_2$SO$_4$, AcOH; c) TMSA, PdCl$_2$(PPh$_3$)$_2$, CuI, *i*-Pr$_2$NH; d) aq KOH, THF, MeOH; e) C$_{10}$H$_{21}$Br, K$_2$CO$_3$, acetone; f) Hg(OAc)$_2$, I$_2$, CH$_2$Cl$_2$; g) TMSA, PdCl$_2$(PPh$_3$)$_2$, xs CuI, *i*-Pr$_2$NH, 70°C; h) CuCl, TMEDA, O$_2$, ODCB

With a variety of 4,5-disubstituted-*o*-diethynylbenzenes **73** in hand, the somewhat unstable arenes were then oxidatively cyclooligomerized under Hay conditions (CuCl, TMEDA, O_2) [53], to give a mixture of dimeric (**74**), trimeric (**75**), and tetrameric (**76**) PDMs as well as some polymeric material. The enhanced solubility conferred by the alkyl and alkoxy substituents proved crucial in facilitating isolation of **75** and **76**. Purification of the different products from the resultant reaction mixture was feasible – but nontrivial, especially for the larger carbocycles. Only through repetitive chromatography and fractional crystallization could the trimers and tetramers be obtained as pure materials. Yields of a given PDM varied widely, depending on the length of the hydrocarbon tails, the scale of the reaction, and the exact experimental conditions. Several trends became clear over time. For example, derivatization with shorter alkyl groups (i.e. butyl, hexyl) and alkoxy groups tended to produce higher yields of the larger macrocycles in addition to more diacetylenic polymer. Conversely, functionalization with longer hydrocarbon moieties improved the yields of the dimeric product. Additionally, the cyclooligomerization reaction was more efficient at converting starting material to macrocyclic products when the hydrocarbon tails were longer than six carbons. For example, when $R=C_{12}H_{25}$, the combined yield of macrocycles totaled 85 % of theory. Otherwise, formation of macrocyclic products was optimized when cyclooligomerization was conducted on a small scale with short reaction times (e. g. 30 min). Moreover, when setting up Hay conditions, the order of reagent addition was found to influence both overall yield of each product and the ratio of one macrocycle to the next. All in all, this type of cyclooligomerization is an extremely sensitive reaction, giving complex mixtures of products that then require great skill and patience to separate into pure macrocycles.

Due to an interest in studying their unusual reactivity (vide infra), several attempts were made to maximize yields of the strained dimers **74**. Lengthening reaction times and decreasing substrate concentrations in the cyclooligomerization experiments proved fruitless. In response to this situation, a stepwise synthesis of the tetrahexyl-substituted dimer was developed as shown in Scheme 18. Surprisingly, Hay coupling of **77** resulted in an improved yield of the tetramer (45 vs 8 %) and a substantial decrease in the yield of dimer (13 vs 30%). This product distribution was unexpected, since intramolecular reactions are typically much faster than intermolecular reactions.

An X-ray structure analysis of **74** ($R=C_4H_9$) revealed that the unsaturated portion of the molecule was planar, with the angles between adjacent acetylenic bonds deviating by 13 – 15° from 180°, the value for a strain-free molecule. Since the connection of the alkyne moieties to the aromatic rings was only shifted slightly (2 – 3°), distortion of the acetylene linkages appears as the major source of instability in these macrocycles.

Differential scanning calorimetry (DSC) experiments on the various dimeric carbocycles indicated that, depending on the length of the alkyl groups, thermal polymerization had occurred between 100 and 125 °C as an abrupt, exothermic process. The narrow temperature range for each exotherm was suggestive of a chain reaction; however, IR spectroscopy revealed the absence of acetylene functionalities in the polymerized material. Consequently, none of the substi-

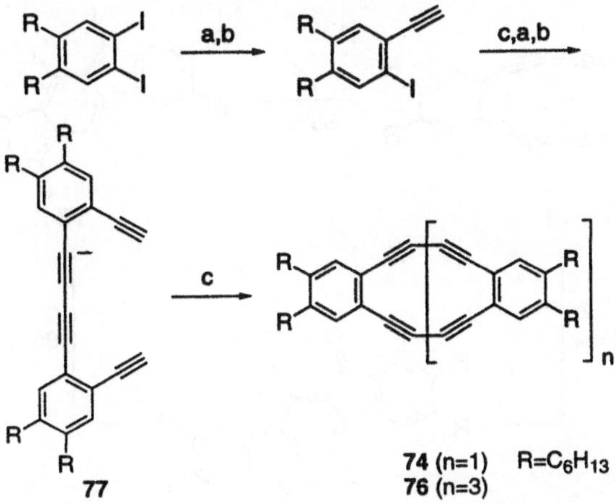

Scheme 18. a) TMSA, PdCl$_2$(PPh$_3$)$_2$, CuI, i-Pr$_2$NH; b) aq KOH, THF, MeOH; c) CuCl, TMEDA, O$_2$, ODCB

tuted dimers seemed to have undergone a well defined topochemical diacetylene polymerization. X-ray powder diffraction gave rudimentary structural information on the polymers derived from **74**. Beyond indications of a layered morphology, the material was too disordered for more detailed information to be obtained. Solid state ^{13}C NMR failed to provide useful structural data due to the predominance of carbon-centered radicals.

Analysis of DSC experiments on various alkyl-substituted trimers gave even more disappointing results. Although more thermally resilient, these macrocycles polymerized with very broad exotherms. For the hexyl-substituted trimer, melting occurred around 150 °C, while polymerization extended from ca. 170 to 230 °C. This pattern was thought to be indicative of a random polymerization process. Overall, polymerization of trimeric macrocycles occurred at sufficiently high temperatures that the resultant materials were intractable brown tars.

Fortunately, the reactivity of the dimeric PDMs was much easier to follow in solution. As shown in Scheme 19, addition of elemental iodine to a benzene solution of **74** (any derivative) under argon afforded the tetraiodinated 6–5–6–5–6 fused ring system **78** in 50–67 % isolated yield. Eglinton had reported formation of the same type of fused ring system upon treating **3** with Na/NH$_3$ in a reductive hydrogenation experiment [8]. The reaction is thought to proceed through radical intermediates, though efforts to trap such species failed. Exposure of **78** to oxygen or iodination of **74** in the presence of oxygen afforded dione **79**. Labeling experiments with ^{18}OH$_2$ showed that incorporation of the carbonyl oxygen did not occur via hydrolysis; rather, direct oxidation of **79** by molecular oxygen is presumably how the carbonyl oxygens were introduced.

In Swager's study, exposure of **2** to Hay conditions also failed to yield **1** or the corresponding tetrameric PDM (**76**, R=H). Preparation of the latter molecule, however, was reported almost simultaneously. Starting with known o-iodo-

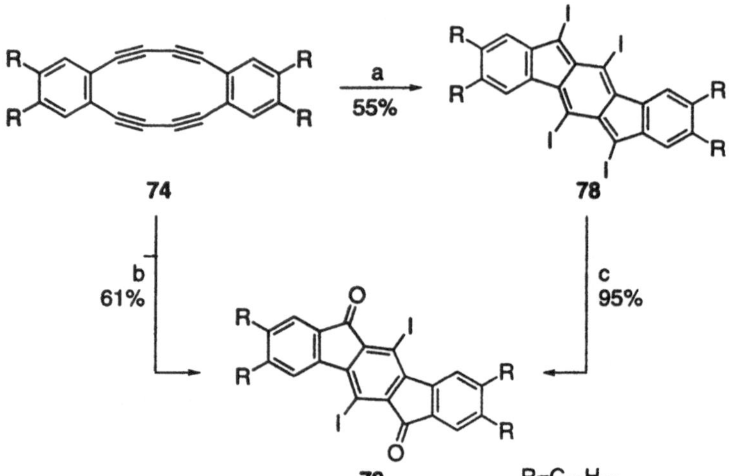

Scheme 19. a) I_2, PhH; b) I_2, O_2, PhH; c) O_2, PhH

ethynylbenzene, Youngs et al. assembled the macrocycle by an intermolecular reaction, utilizing chemistry similar to that shown in Scheme 18 [54]. As above, the exact copper coupling conditions used proved crucial for product formation. Eglinton conditions gave large amounts of insoluble polymer and only a small quantity of 76 (R=H). Since this reaction was conducted under fairly dilute conditions (0.01 mol L^{-1}), the predominance of polymer was somewhat surprising. However, by bubbling air through a very dilute (0.001 mol L^{-1}) pyridine solution of 77 (R=H) in the presence of 120 equivalents of CuCl, the desired carbocycle was formed in about 20% isolated yield. Interestingly, the modified Glaser procedure, which gave the highest yield of tetrameric PDM also furnished Eglinton's dimeric macrocycle 3 in about 50% isolated yield.

Recrystallization of 76 (R=H) from CH_2Cl_2 provided crystals adequate for X-ray structural determination. The molecule was found to be saddle shaped with a phenyl ring at each vertex and nadir. The alkynyl bonds were found to be essentially linear and to possess a mean length of 1.194 Å, typical for the length of triple bonds in free butadiyne. Although 76 is a dehydrobenzoannulene possessing a 4n π-electron circuit, the nonplanarity of the macrocycle alleviated much of the strain associated with a flat structure and thus precluded the possibility of anti-aromatic ring currents.

In late 1995, a team led by Vollhardt and Youngs reported their work on the strained PAM/PDM hybrid 80 [55]. Whereas the synthesis of 80 was not remarkable [Eq. (2)], the solid-state behavior of the molecule was. X-ray crystallography revealed that the macrocycle was moderately strained, with the monoynes bent inward toward the center of the macrocycle by 3.9–11.5° and the diyne unit bent outward by 8.6–11.2°. More importantly, crystal packing revealed that the diyne moieties were aligned in the prerequisite fashion for topochemical diacetylene polymerization to occur. Indeed, irradiation of crystals of 80 produced a violet surface layer exhibiting a metallic luster. Laser desorption time-of-flight mass spectrometry of **80** demonstrated that polymerization was occurring, since oligomers as large as nine units were detected. Complete polymerization could be accomplished either by thermal annealing (ca. 150 °C by DSC) or by high pressure (20,000 psi, 1 h). Solid state ^{13}C NMR of the violet polymer displayed two new peaks at 145.1 and 150.0 ppm, corresponding to the alkene carbons in the polydiacetylene chain. Taken together, this is convincing evidence for occurrence of a topochemical diacetylene polymerization, the product of which is possibly a tube-like structure with the fused annulenes adopting an all-*syn* configuration.

$$(2)$$

80

Confirmation of this hypothetical structure by synthesis and polymerization of more soluble derivatives of **80**, while seemingly the next logical step, will likely be problematic. Previous studies of topochemical polymerizations of relatively strain-free diacetylenic macrocycles concluded that solid-state packing was more important than ring-strain effects [56]. Swager's work with **74** seems to corroborate this statement. Even though the molecules were more distorted and polymerized at lower temperatures, the resultant products were sufficiently disordered that further characterization was not possible. The packing diagram of **74** (R=C$_4$H$_9$) reflected the poor overlap of the diyne moieties. In addition to ring strain, subtle changes of structure can alter molecular packing dramatically and thus affect the solid state behavior. For example, replacement of the phenyl ring between the monoyne units in **80** with a *cis*-ethene moiety produced a crystal containing staggered diacetylenic units, thus negating the potential for a topochemical reaction pathway [57]. Based on the above results and work with other PDMs and PAM/PDM hybrids, it is reasonable to conclude that each annulenic macrocycle behaves independently and that no generalization based on annulene topology can be proposed at this time for accurately predicting thermoproduct structure. Short of growing crystals (an insurmountable task for some macrocycles) and obtaining crystallographic data (a costly and time-consuming process) for each and every diacetylenic system, the ability to overcome these vexing problems appears doubtful at the present time.

An alternative mode for dehydrobenzoannulene decomposition was recently reported by Vollhardt et al. [58]. Non-planar hybrid **81**, prepared in low yield via cyclodimerization of known triyne **82** [Eq.(3)], reacted explosively at ca. 250 °C to give a nearly pure carbon residue. Solvent extraction of the black powder failed to yield soluble materials such as fullerenes; however, analysis of the residue by TEM showed formation of "bucky onions" and "bucky tubes" [59], in addition to copious amounts of amorphous carbon and graphite. Although the amount of "bucky" materials was quite small (1 – 2 %), this result was clear validation of the argument that PAMs and PDMs might function as precursors to carbon allotropes.

introducing an interesting permutation of a familiar theme, Rubin and co-workers cyclooligomerized 1,2,3,4/5,6-differentially terminated hexaethynyl-benzenes (HEB, e.g. 83), to create PDM derivatives 84 – 86 [60]. These systems could be viewed as molecular fragments of the all-carbon network *graphdiyne*, a novel allotrope in which the graphitic motif is expanded by butadiyne linkages [52b, 61]. Such networks have attracted considerable interest of late, but continue to remain elusive from a synthetic perspective [5c]. Unlike other PDMs, Rubin's macrocyclic products feature a fringe of *tert*-butyl capped ethynyl groups around the periphery. Not only does this feature increase the potential for nonlinear optical activity by extending the conjugation network [50], it also serves to substantially stabilize the characteristically reactive dimer 84 (vide supra). However, perethynylation necessitated more intricate synthetic maneuvering prior to cyclooligomerization. The requisite substitution pattern of differentially terminated alkynes on 83 was unobtainable through Pd-medi-ated cross-coupling procedures. Instead, a large number of HEB precursors could be prepared through Diels-Alder reactions between suitably derivatized tetraalkynylcyclopentadienones [62] and disubstituted acetylenes.

As fate would have it, the direct approach to the TMS-protected version of the desired HEB failed. Rather than providing a symmetrical adduct, the Diels-Alder reaction between *tert*-butyl-capped tetraethynylated cyclopentadienone 87 and bis(trimethylsilyl)hexatriyne gave the adduct arising from cycloaddition to one of the outer triple bonds. Although this is a surprising result, computer modeling (PM3) suggested that the Diels-Alder reaction across the central triple bond was inhibited in the transition state by steric crowding of the bulky dieno-phile with the large ethynyl capping groups on the dienone. Accordingly, Rubin's team turned to an oblique approach for installation of the two remaining ethynyl moieties.

The successful construction of **83** is depicted in Scheme 20. Reaction of **87** with diethoxybutynal afforded cycloadduct **88** in 61% yield. Whereas the dialdehyde arising from **88** proved to be unstable, stepwise homologation via standard procedures furnished **83** in modest overall yield. Cyclooligomerization of **83** under Hay conditions produced a mixture of highly fluorescent PDMs which could only be partially resolved. Column chromatography on silica gel resulted in a 25% yield of pure [12]annulene **84** and an inseparable mixture of trimer **85** and tetramer **86** (13% combined yield). X-ray crystallography of **84** showed the diyne units in the molecule to be distorted to the same extent as **74**; however, in strong contrast to other dimeric PDMs, the steric bulk of the eight *tert*-butyl groups imparted unusually high stability for **84**.

Just prior to Rubin's publication, another article appeared focusing on substructures of graphdiyne [63]. Like the other researchers in the PDM area, the Haley team was intrigued by the predictions of useful materials properties and technological applications for this and similar carbon-rich systems [5c, 50, 52]. In particular, topochemical polymerization of a crystalline substructure of this network could produce an environmentally robust material with a large third-

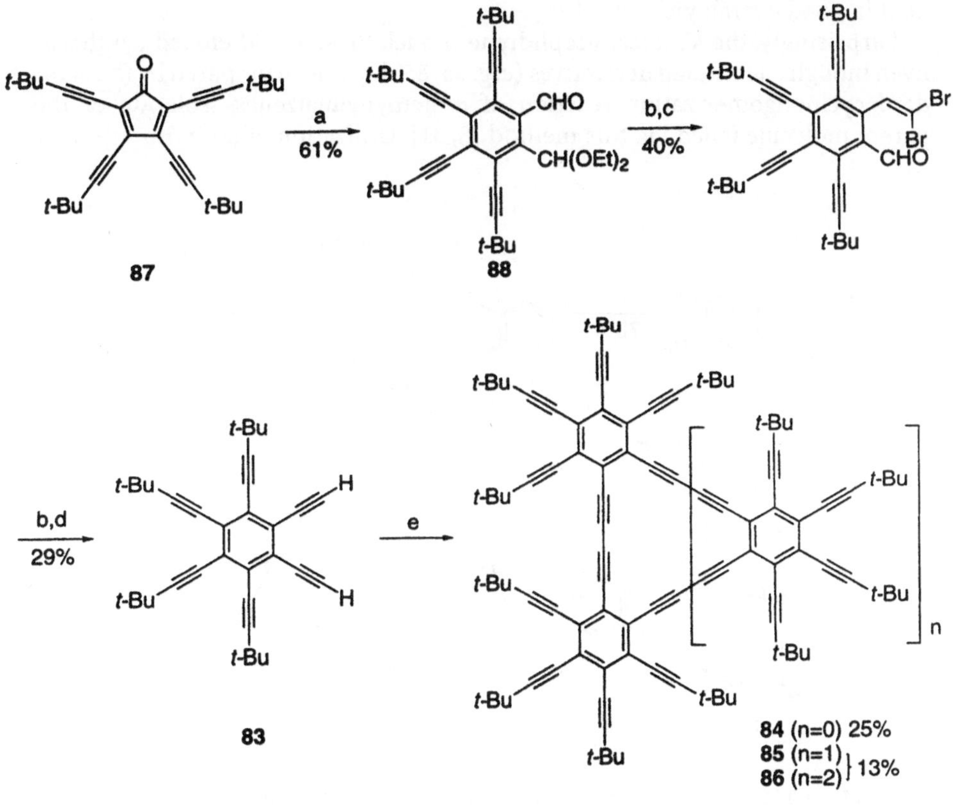

Scheme 20. a) OHCC≡CCH(OEt)$_2$, PhH; b) CBr$_4$, PPh$_3$, CH$_2$Cl$_2$; c) SiO$_2$; d) i] LDA, THF ii] aq NH$_4$Cl; e) CuCl, TMEDA, O$_2$, acetone

order nonlinear optical susceptibility. Such materials are of considerable economic interest, since the light mediated flow of information required for the development of photonic technology is contingent on high speed optical switching [50]. Additionally, the diyne units in graphdiyne result in a pore size of about 2.5 Å, a cavity that can easily incorporate atoms as large as cesium. This should lead to interesting redox applications, as the large holes in the planar sheets could accommodate through-sheet transport of metal ions and possibly provide a unique method of dopant storage by intrasheet intercalation.

The fundamental synthetic difference between the route followed by the Haley group and much of what had been previously reported for PDM construction was assembly of the final structures by an intramolecular reaction. All previous methods of preparing substructures of diacetylenic carbon networks had relied upon cyclooligomerization of perethynylated monomers [5c, 60a]. As already demonstrated, this route invariably produced complex PDM mixtures which made separation of the dimers, trimers, and tetramers very difficult and resulted in low isolated yields of a given macrocycle. Additionally, the variation in product structures was severely limited by the ease of construction, or lack thereof, of the starting o-diethynylbenzene. Intramolecular coupling of a suitable α,ω-polyyne, on the other hand, was likely to ensure formation of a single product in good overall yield.

Surprisingly, the simplest graphdiyne model, PDM 1, had eluded synthesis. Even though substituted derivatives (e.g. 75, 85) have been prepared by the standard cyclooligomerization reaction of o-diethynylbenzenes, isolation of the parent molecule failed via this method [8, 51]. Utilization of an intramolecular

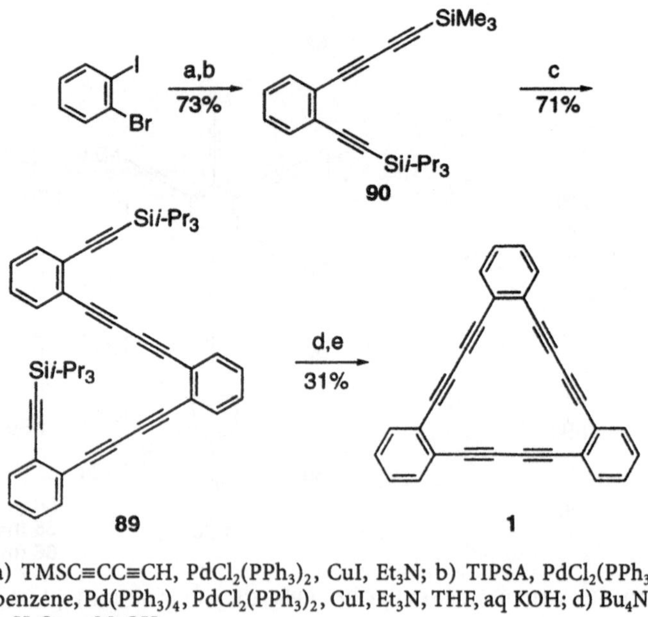

Scheme 21. a) TMSC≡CC≡CH, PdCl$_2$(PPh$_3$)$_2$, CuI, Et$_3$N; b) TIPSA, PdCl$_2$(PPh$_3$)$_2$, CuI, Et$_3$N; c) o-diiodobenzene, Pd(PPh$_3$)$_4$, PdCl$_2$(PPh$_3$)$_2$, CuI, Et$_3$N, THF, aq KOH; d) Bu$_4$NF, THF, EtOH; e) Cu(OAc)$_2$·H$_2$O, py, MeOH

approach would require hexayne **89**, which in turn would necessitate use of a suitably functionalized phenylbutadiyne such as **90** (Scheme 21). The parent molecule, 1-phenyl-1,3-butadiyne, is a highly reactive compound which polymerizes rapidly when neat or in concentrated solution; even a dilute solution at −20 °C polymerizes within a few hours [64]. Not surprisingly, then, all attempts to use monodeprotected **90** in Pd-catalyzed alkynylation reactions provided intractable polymeric gums [65].

The answer to this problem was in situ generation of the free phenylbutadiyne under standard Pd-coupling conditions [66]. Addition of a few milliliters of a concentrated KOH solution provided the bis-coupled product in 71 % yield. Desilylation and use of high dilution conditions in the oxidative coupling reaction gave **1** as the sole product in moderate yield. Compound **1** was poorly soluble in common organic solvents; nevertheless, all of the spectral data (NMR, IR, UV, MS) supported the assigned structure. The minimal solubility of the product was no-doubt responsible for the low isolated yield.

Whereas PDM **1** in theory could be assembled via the cyclooligomerization reaction, larger, more complex graphdiyne substructures like **91-94** could be constructed only via the intramolecular cyclization route. Scheme 22 illustrates the preparation of **91** [63]. From the outset, the need for solubilizing substituents was recognized; thus, the required building blocks (**95**) were readily prepared by standard transformations. Fourfold in situ desilylation/alkynylation gave the

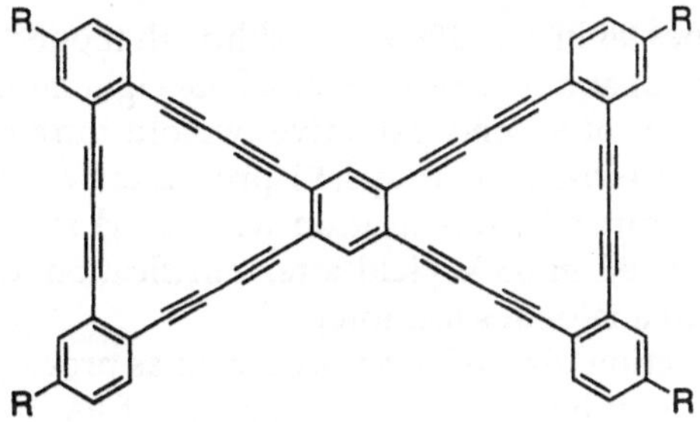

91

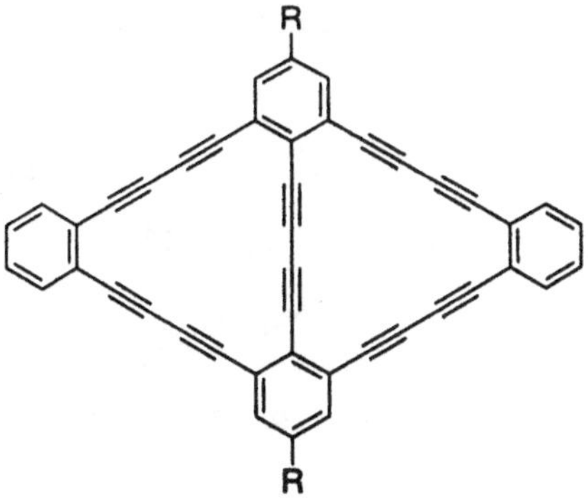

92

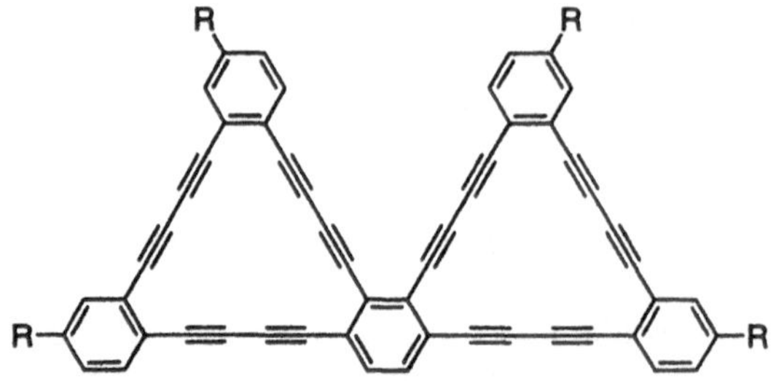

93

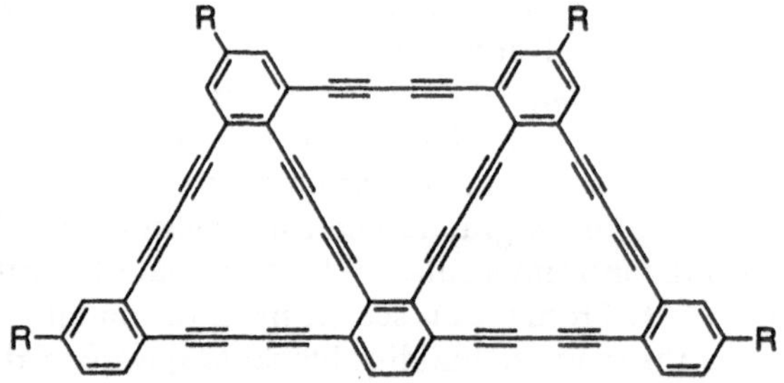

94

fluorescent dodecaynes **96** in ca. 35% yield. Although modest, this yield implies an average conversion of about 90% for each of the eight necessary transformations. Protiodesilylation of **96** and oxidative twofold intramolecular coupling gave **91**. Even with four *tert*-butyl groups, **91** proved to be virtually insoluble in common solvents. Fortunately, substitution with decyl moieties worked quite well, with **91** now isolated in 83% yield after purification. Compounds **92–94** were synthesized in an analogous fashion [67].

The in situ deprotection/alkynylation protocol has proven to be exceedingly useful. In addition to graphdiyne models, an array of PAM/PDM hybrids have been prepared (**76** (R=H), **81**, **97–99**) [66]. Extension of the simple, one-pot procedure to various iodoarenes has allowed preparation of a series of bis(triisopropylsilyl)-protected α,ω-polyynes in very good yields. Subsequent depro-

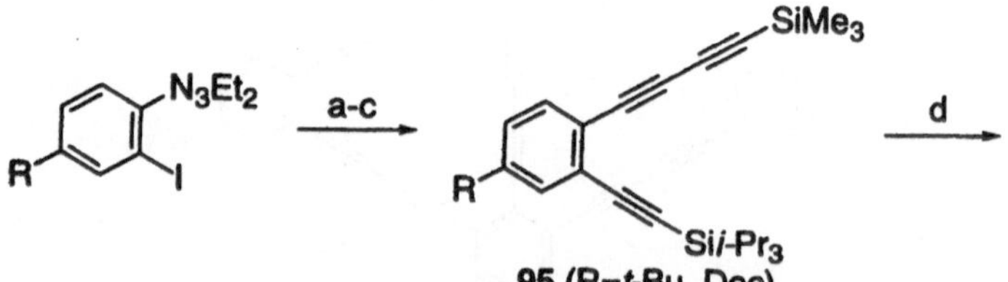

95 (R=t-Bu, Dec)

Scheme 22. a) TIPSA, $PdCl_2(PPh_3)_2$, CuI, Et_3N; b) MeI, 125 °C; c) TMSC≡CC≡CH, $PdCl_2(PPh_3)_2$, CuI, Et_3N; d) 1,2,4,5-tetraiodobenzene, KOH, $PdCl_2(PPh_3)_2$, $Pd(PPh_3)_4$, CuI, Et_3N, THF

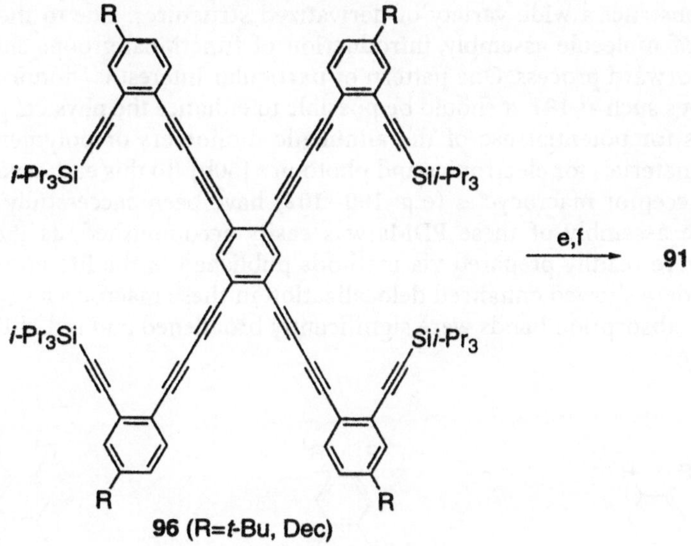

96 (R=t-Bu, Dec)

Scheme 22. e) Bu$_4$NF, THF, EtOH; f) Cu(OAc)$_2$, py, MeOH, Et$_2$O

tection and cyclization under pseudo-high dilution conditions provided dehydrobenzoannulene topologies that were either inaccessible by traditional routes (97–99) or previously available only in low yield (76 (R=H) [54], 81 [58]). Fortunately, product solubility was not an issue for these macrocycles, due to their lower symmetry (e.g. 97 vs. 1) and/or non-planarity (76, 81, 98–99).

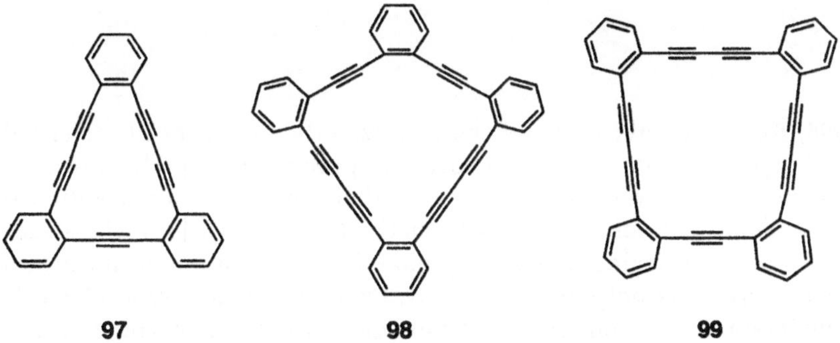

97   **98**   **99**

Another limitation of the traditional Cu-mediated cyclooligomerization reaction is generation of differentially substituted PDMs. In the above case, the substitution pattern in the starting o-diethynylbenzene must be maintained on each and every benzene moiety in the oligomeric mixture of PDMs that is produced. Thus, it is impossible to prepare less symmetric systems like 100 via this route. With the intramolecular synthetic approach, however, it should be possible to construct a wide variety of derivatized structures. Due to the stepwise pattern of molecule assembly, introduction of functional groups should be a straightforward process. One pattern of particular interest is "donor-acceptor" derivatives such as **101**. It should be possible to enhance the physical properties of PDMs for potential use of the annulenic monomers or polymers as conjugated materials for electronics and photonics [50b]. To this end, several donor and/or acceptor macrocycles (e.g. **100–102**) have been successfully prepared [68]. The assembly of these PDMs was easily accomplished, as the starting arenes were readily prepared via methods published in the literature. UV-Vis spectral data showed enhanced delocalization in these macrocycles (compared to **1**): the absorption bands were significantly broadened and red-shifted up to 200 nm [68].

100
R=Dec, OMe, OOct
RR=-O(CH$_2$CH$_2$O)$_4$-

101
R^1=Dec, OMe, OOct
R^2=NO$_2$

102
R^1=R^3=NBu$_2$ R^2=R^4=NO$_2$
R^1=R^4=NBu$_2$ R^2=R^3=NO$_2$

4.2
Meta

In 1996, Tobe et al. prepared and characterized the first *meta*-PDMs and then studied their solution-phase self-association properties [69]. This work builds upon concepts advanced by Moore et al. in their investigation of solution-state aggregation behavior of various functionalized *meta*-PAMs [39]. The attractive force at work, π-stacking of aromatic moieties, plays an important organizational role in supramolecular chemistry. Examples of this type of localized binding interaction abound in nature; biological systems are particularly reliant on this organizing principle. For example, the existence of stable DNA double helices is contingent upon the weak associative forces between vertically stacked adjacent base pairs. π-Stacking also influences the packing orientation of aromatic molecules in the crystalline state and the intercalation of small, planar molecules between nucleotides (e.g. camptothecins, enediyne anti-tumor antibiotics). Consequently, understanding the structural characteristics that contribute to a molecule's ability to engage in this type of behavior has important real-world implications.

With an eye toward this goal, Tobe and coworkers assembled *exo*-functionalized macrocycles of type **103**, as portrayed in Scheme 23 [69]. Sequential application of standard Pd-cross-coupling techniques furnished differentially protected diethynylbenzene monomers **104** and **105** in good yields. Selective protiodesilylation of the *tert*-butyl derivative, oxidative dimerization, and removal of the TIPS group gave dimeric tetrayne **106**. Cyclooligomerization furnished rigid tetrameric PDM **103** (25%) along with conformationally flexible octamer **107** (13%). In the case of **105**, extended exposure to aqueous LiOH in THF removed the TMS group and hydrolyzed the ester. Treatment of the crude carboxylate with $SOCl_2$ followed by hexadecyl alcohol gave the long chain ester. Analogous transformations afforded first tetrayne **108**, then PDM **109** in 17% yield. Curiously, the cyclooligomerization reaction failed to produce isolable amounts of hexameric PDM, the product of cyclotrimerization.

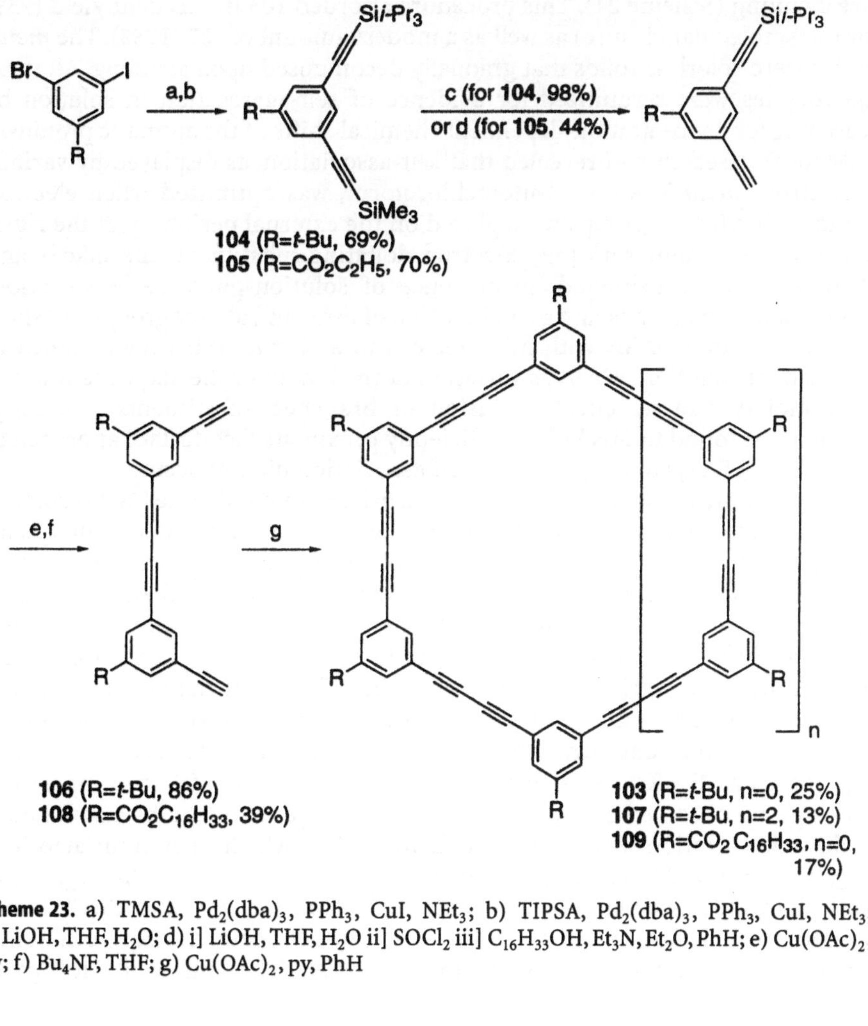

Br
I
R
a,b
Si*i*-Pr3
SiMe3
c (for 104, 98%)
or d (for 105, 44%)
104 (R=*t*-Bu, 69%)
105 (R=CO2C2H5, 70%)
e,f
g
n
106 (R=*t*-Bu, 86%)
108 (R=CO2C16H33, 39%)
103 (R=*t*-Bu, n=0, 25%)
107 (R=*t*-Bu, n=2, 13%)
109 (R=CO2C16H33, n=0, 17%)

Scheme 23. a) TMSA, $Pd_2(dba)_3$, PPh_3, CuI, NEt_3; b) TIPSA, $Pd_2(dba)_3$, PPh_3, CuI, NEt_3; c) LiOH, THF, H_2O; d) i] LiOH, THF, H_2O ii] $SOCl_2$ iii] $C_{16}H_{33}OH$, Et_3N, Et_2O, PhH; e) $Cu(OAc)_2$, py; f) Bu_4NF, THF; g) $Cu(OAc)_2$, py, PhH

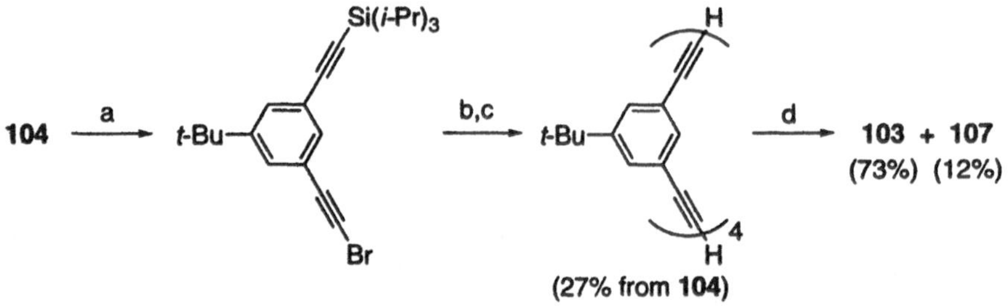

Scheme 24. a) NBS, AgNO$_3$, acetone; b) **106**, Pd$_2$(dba)$_3$, CuI, 1,2,2,6,6-pentamethylpiperidine, LiI, HMPA, PhH; c) Bu$_4$NF, THF; d) Cu(OAc)$_2$, py, PhH

PDM **103** was also prepared by building up the appropriate α,ω-polyyne and then employing very dilute Eglinton conditions to effect intramolecular oxidative coupling (Scheme 24). This procedure afforded **103** in excellent yield (73% for intramolecular closure) as well as a modest amount of **107** (12%). The *meta*-PDMs were colorless solids that gradually decomposed upon standing. All three macrocycles were scrutinized for evidence of self-aggregation in solution by searching for concentration-dependent chemical shifts of the aromatic protons.

Moore's research had revealed that self-association, as displayed by various derivatized *meta*-PAMs in deuterochloroform, was optimized when electron withdrawing (ester) groups were placed on the external periphery of the rigid, planar carbon framework [39]. Electron-donating substituents like alkoxy and alkanoate groups exhibited no evidence of solution-phase self-association, though macrocycles bearing a combination of ester and alkoxy groups did show weak association. PAMs with mixtures of *exo* and *endo* substituents failed to aggregate in solution. Whereas variation of the length of the alkyl chain in the ester moiety had no effect, inclusion of branched substituents (*tert*-butyl groups) was found to preclude a sufficiently proximate face-to-face approach of two solvated PAMs to the point that self association did not occur.

Based on the above criteria, it is not surprising that only ester **109** displayed behavior indicative of self-association through π–π stacking interactions. Relative to the variation in the chemical shift of the aromatic protons observed by Moore et al. in similarly functionalized PAM **42** [39], the shift changes noted by Tobe and coworkers were comparatively small. For example, Moore saw upfield chemical shift differences of 0.9–1.0 ppm in the aromatic protons of **42** as the concentration was varied by a factor of 128 in CDCl$_3$. If it can be assumed that the monomer-dimer equilibrium is the predominant self-association process in this solvent in the concentration range under study, the dimerization constant, K_{assoc}, of **42**, at 293 K was calculated to be ca. 60 M^{-1}. For **109**, in comparison, Tobe's group observed upfield chemical shift changes between 0.1 and 0.2 ppm as the concentration was varied by a factor of 394, which they translated into $K_{assoc} = 38$ M^{-1} at 293 K. Thermodynamic parameters, obtained through van't Hoff plots, indicated that self-association of PAMs and PDMs in CDCl$_3$ is not favored entropically, but is driven by a small, favorable enthalpy difference (-5.0 kcal mol^{-1} for **42** vs -5.8 kcal mol^{-1} for **109**). The same van't Hoff plots

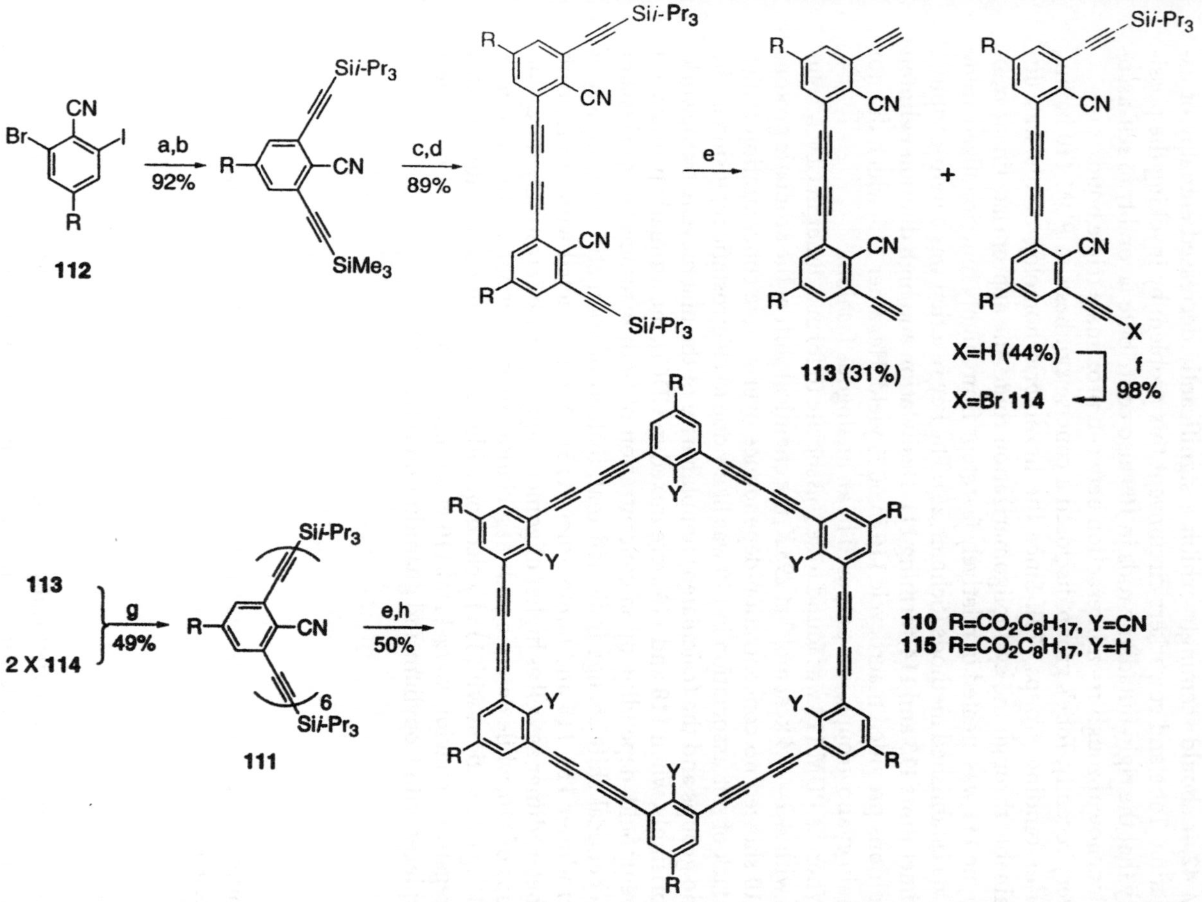

Scheme 25. a) TIPSA, Pd₂(dba)₃, PPh₃, CuI, Et₃N; b) TMSA, Pd₂(dba)₃, PPh₃, CuI, Et₃N; c) LiOH, THF, H₂O; d) CuCl, TMEDA, O₂, acetone; e) Bu₄NF, THF, H₂O; f) NBS, AgNO₃, acetone; g) Pd₂(dba)₃, PPh₃, CuI, i-Pr₂NH, PhH; h) Cu(OAc)₂, py, PhH

yielded corresponding ΔG values for self-aggregation: -2.4 kcal mol^{-1} for **42** vs -2.1 kcal mol^{-1} for **109**. This close agreement is somewhat surprising, since the π-stacking interaction is thought to occur primarily through the face-to-face proximity of phenyl groups. Since PDM **109** has two fewer phenyl rings than PAM **42**, it should seemingly exhibit a significantly decreased enthalpy of association. Tobe and coworkers circumvent this problem by invoking the possibility that the rigid butadiyne rods in **109** also contribute favorably to self-association, now through π–π interaction between proximate triple bonds.

Very recently, Tobe's group disclosed a concise synthesis of PDM **110** having interior binding groups [70]. Since the hexameric homolog of **109** was unobtainable through the cyclooligomerization route, the appropriate linear α,ω-polyyne **111** was created in relatively few steps from highly functionalized arene **112** via established methods (Scheme 25). The key reaction was hetero-coupling of dimer units **113** and **114**, forming **111**. Desilylation and high dilution Eglinton conditions provided macrocycle **110** in 50% yield. Hexamer **115**, which did not contain cyano groups, was prepared in an analogous fashion.

Whereas PDM **115** was found to self-associate to form a dimer in CDCl$_3$ solution with $\Delta G=-3.4$ kcal mol^{-1} at 293 K, the chemical shift of the aromatic protons of **110** showed no concentration dependence over a wide concentration range. The lack of self-association in **110** was likely due to electrostatic repulsion of the cyano groups and the (calculated) nonplanarity of the macrocyclic framework. Surprisingly, when **110** and **115** were mixed in CDCl$_3$, the aromatic protons did move upfield, depending upon concentration of both macrocycles. Analysis of the chemical shift change indicated that, instead of competitive formation of heterodimer **110·115** and homodimer **(115)$_2$**, **110** interacted with **115** to form the heterodimer as well as higher oligomeric aggregates. Interestingly, the cyano groups of **110**, while deterring self-association, serve to enhance attractive π–π stacking interactions with **115** by electron withdrawal. In addition to this unusual aggregation behavior, *meta*-PDM **110** was found to form 1:1 and 2:1 host:guest complexes with tropylium and guanidinium cations [70].

4.3
Para

A survey of the literature revealed that no synthesis of *para*-PDMs has been reported.

4.4
Mixed

The shape-persistent, structurally well-defined nature of PAMs and PDMs make them attractive models for binding guest molecules within their cavities. In 1995, Höger and Enkelmann reported the construction of the first *meta/para*-PAM/PDM hybrid designed to possess hydrophobic and hydrophilic substituents for subsequent use in host:guest chemistry [71]. Macrocyclic amphiphile **116** was assembled via the straightforward manner depicted in Scheme 26.

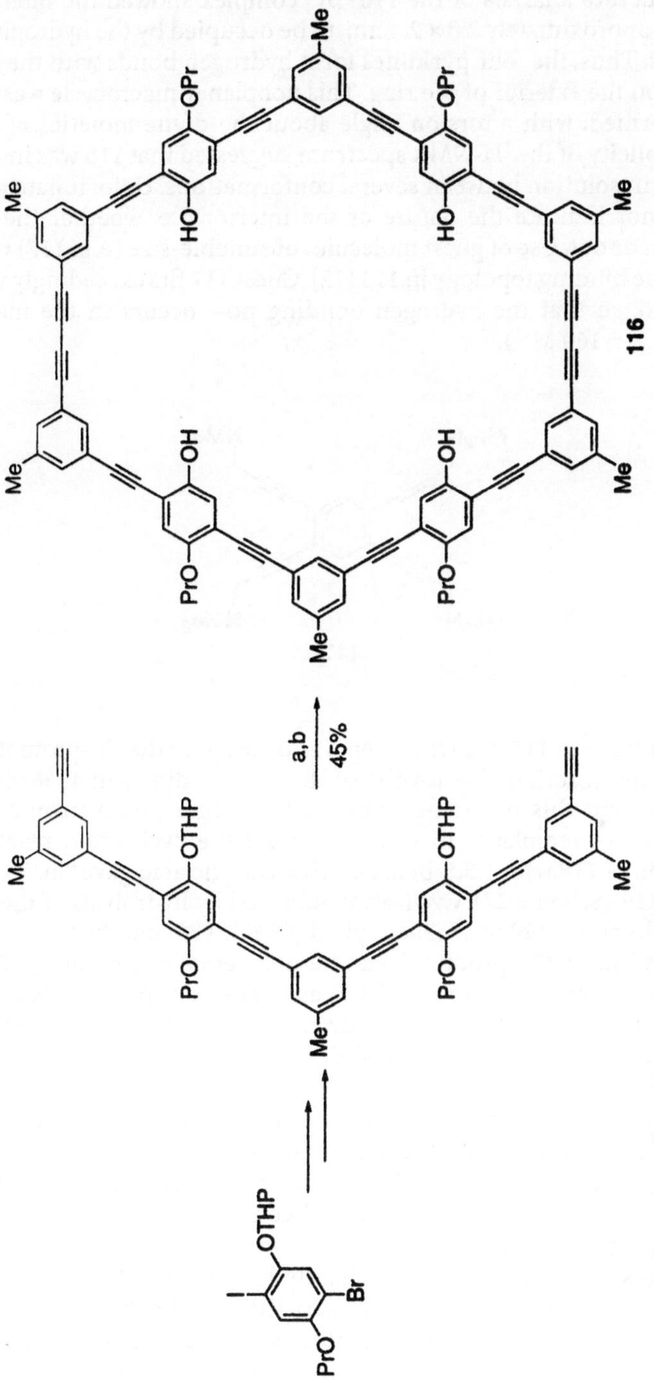

Scheme 26. a) CuCl, CuCl$_2$, py; b) CH$_2$Cl$_2$, MeOH, H$^+$

X-ray structure analysis of the $116 \cdot py_4$ complex showed the interior of the macrocycle, approximately 2.0×2.4 nm, to be occupied by the hydrophobic propoxy groups. Thus, the four pyridines form hydrogen bonds with the phenolic-OH groups on the exterior of the ring. This nonplanar macrocycle was asymmetrically deformed, with a torsion angle about the diyne moieties of $6.7°$. The relative simplicity of the ^{1}H-NMR spectrum suggested that 116 was interconverting rapidly in solution between several conformations. Unfortunately, solution data could not indicate the nature of the interior, i.e. whether the hydroxyl groups are in or out. Use of guest molecules of suitable-size (e.g. 117) resulted in reversal of the binding topology in 116 [72]. Guest 117 fits exceedingly well in the cavity of 116, so that the hydrogen bonding now occurs in the macrocyclic interior ($K_{assoc} = 160$ M^{-1}).

117

In the synthesis of 116, the weak step was once again the Cu-promoted cyclooligomerization reaction, that furnished the desired dimer in 45% yield. Seeking to circumvent this recurring problem, the Höger group turned to use of covalently bound templates, in order to direct the cyclization reaction [73]. Condensation of 118 with 1,3,5-benzenetricarboxylic acid gave "inside" templated triester 119 (Scheme 27). Cyclization followed by hydrolysis of the template afforded cyclotrimer 120 in excellent yield (89%). This number is in stark contrast to the yield of 120 produced via the non-templated pathway (20–25%). Moreover, the latter route produced 120 as part of an inseparable cyclooligomeric mess. Replacement of the 3-hydroxypropyl tether with a substantially longer 11-hydroxyundecyl unit made little difference in the isolated yield (84%) of the macrocycle. This result suggested that the impressive yield of the cyclization reaction is due mainly to the low concentration of 119 (high dilution conditions) but high local concentration of terminal acetylenes (intramolecular reaction as a result of templation). Thus, as long as the geometry is preorganized (offered by the template), proximal spatial constraint is not necessary for the high-yield cyclization to occur.

Use of "outside" templates gave a slightly lower yield (86%) of macrocycle [Eq. (4)]. Subsequent experiments showed that, in the "outside" case, tether length played a role in product yield [73b]. Assuming that cyclization is a stepwise process, use of too short a tether, while facilitating the first dimerization, would geometrically restrict the second. This indeed proved to be the case: shortening the tether of the acyclic precursor, in Eq. 4 by ten carbons, resulted in a

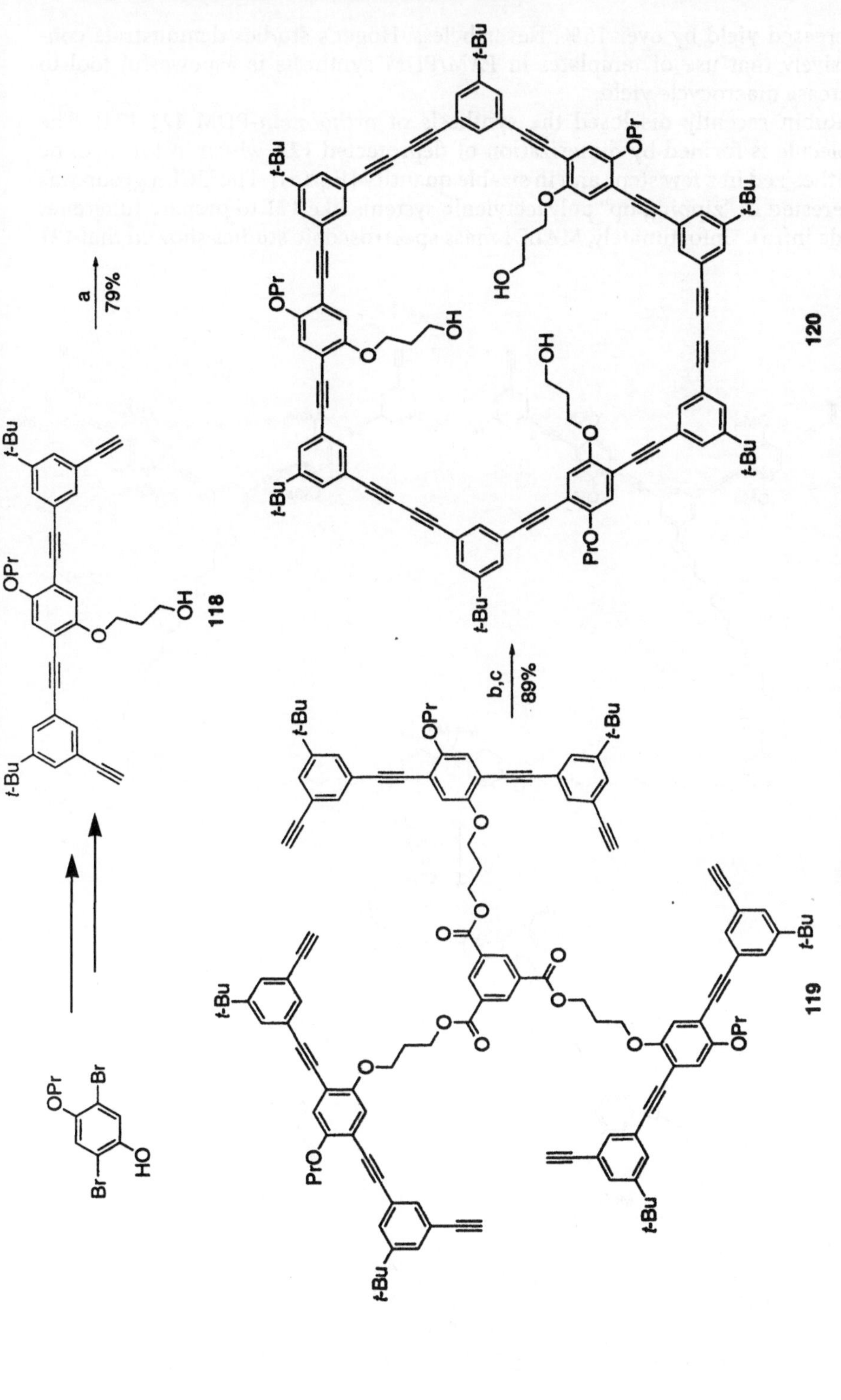

Scheme 27. a) 1,3,5-benzenetricarboxylic acid, DEAD, PPh$_3$, THF; b) CuCl, CuCl$_2$, py; c) NaOH, LiOH, MeOH, H$_2$O, THF

decreased yield by over 15%. Nevertheless, Höger's studies demonstrate conclusively that use of templates in PAM/PDM synthesis is a powerful tool to increase macrocycle yield.

Rubin recently disclosed the synthesis of *ortho/meta*-PDM **121** [74]. The molecule is formed by dimerization of deprotected **122**, which in turn can be synthesized in a few steps and in sizable quantity [Eq. (5)]. The UCLA group was interested in "zipping up" polyacetylenic systems like **121** to prepare fullerenes (vide infra). Unfortunately, MALDI mass spectroscopic studies showed that **121**

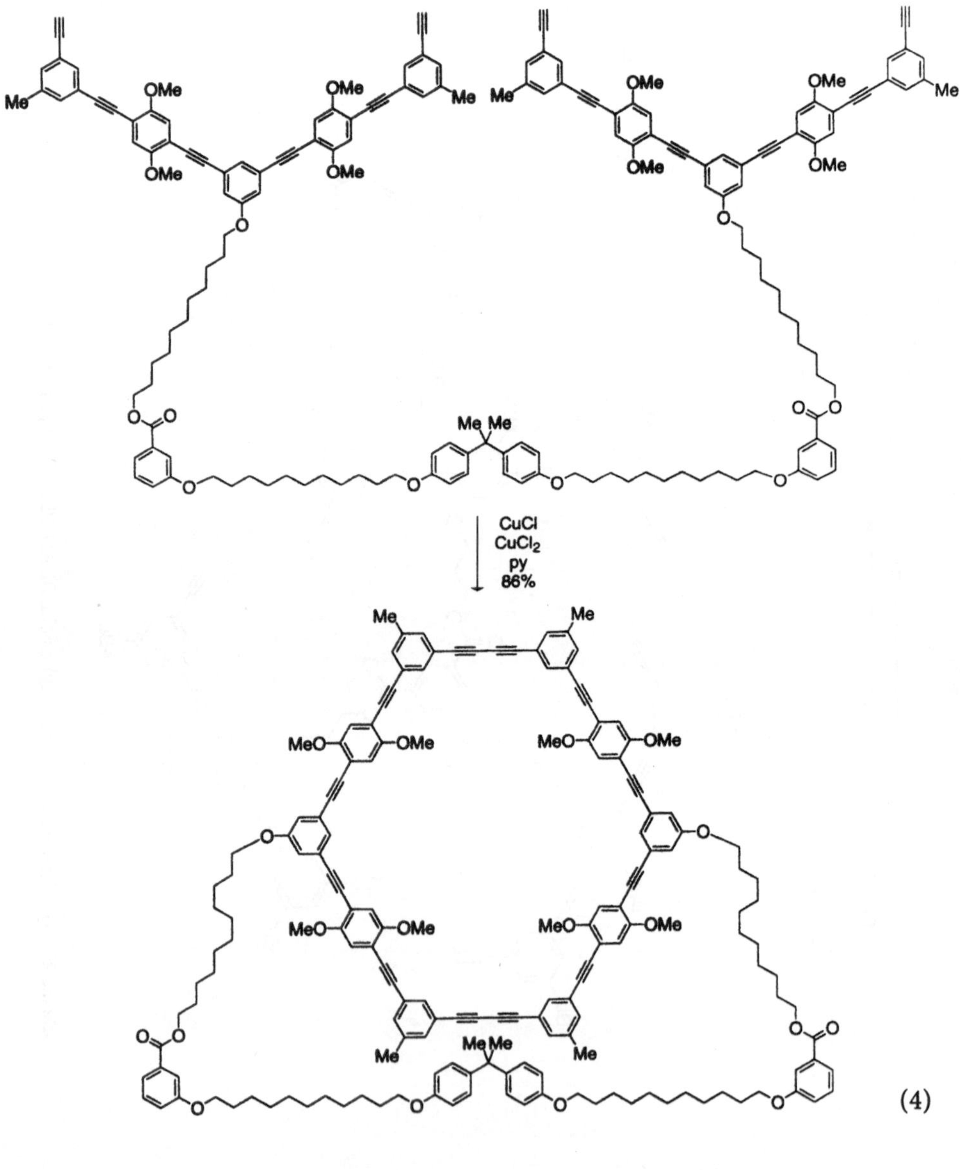

(4)

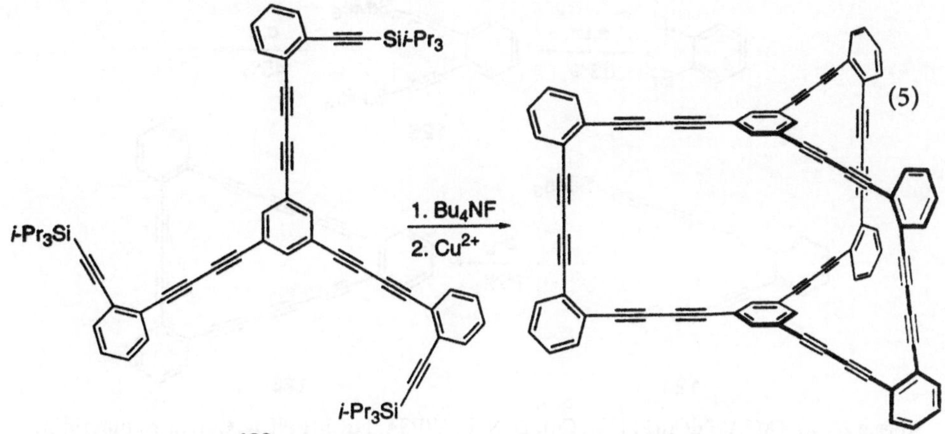

122   **121**   (5)

and its ethylenic analog (benzene replaced by ethene) possessed little propensity to lose hydrogen [75].

5
Phenyltriacetylenes

Extension of the Haley in situ deprotection/alkynylation process to substituted phenyltriacetylenes led to the formation of the first triyne-connected annulenes (e.g. **123**) [76]. The assembly of such macrocycles was not feasible via cyclooligomerization chemistry. Additionally, standard metal-mediated cross-coupling reactions were out of the question: although the synthesis of phenylhexatriynes has been reported previously, the molecules were so unstable that they were only characterized by UV-Vis spectroscopy [77]. The requisite building block **124** was prepared with some difficulty from **125** (Scheme 28), an intermediate in the syntheses of **97** and **98** that is readily available in ten gram quantities from 1-bromo-2-iodobenzene. Modified Cadiot-Chodkiewicz coupling [64,75] using monodesilylated **125** and TMS-C≡CC≡C-Br gave **124** in moderate yield, which in turn was converted into **123** by standard procedures. As with annulene **97**, **123** proved to be readily soluble. Additionally, since this type of synthesis produced only single macrocycles, extensive separation and purification procedures that result in material loss were avoided.

Using similar methodology, macrocycle **126** was prepared, as well as the unusual monoene **127** [76]. Considerable debate in the literature over the last thirty years has focused on whether dehydrobenzoannulenes are able to sustain induced ring currents [5a]. Although fusion of arenes to the annulenic core provides rigidity and stability, this also weakens the diatropicity/paratropicity of the macrocycle significantly. Until quite recently, the number of planar systems available for study was limited; however, with the the addition of **123** and **126**, the series of alkyne-linked, tribenzo-fused dehydroannulenes is complete from

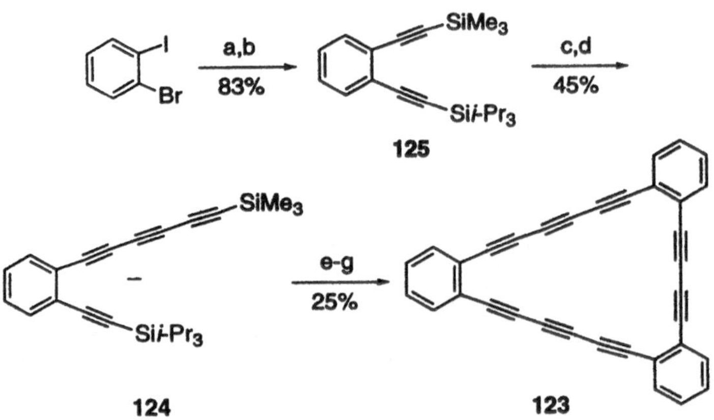

Scheme 28. a) TMSA, PdCl$_2$(PPh$_3$)$_2$, CuI, Et$_3$N; b) TIPSA, PdCl$_2$(PPh$_3$)$_2$, CuI, Et$_3$N, piperidine; c) K$_2$CO$_3$, MeOH; d) i] BuLi ii] CuBr iii] TMSC≡CC≡CBr, py; e) o-diiodobenzene, aq KOH, Pd(PPh$_3$)$_4$, PdCl$_2$(PPh$_3$)$_2$, CuI, Et$_3$N, THF; f) Bu$_4$NF, EtOH; g) Cu(OAc)$_2$, CuCl, py

[12]- to [22]annulene. NMR spectroscopic studies showed that the arene protons of the $4n + 2$ Hückel-type systems (**1, 80, 123**) possess distinct downfield shifts compared to model compounds, whereas the $4n$ systems (**4, 97**) exhibited opposite but attenuated behavior [76]. The chemical shifts of arene protons in **126**, on the other hand, were virtually unchanged compared to the acyclic models, suggesting that this macrocycle is atropic. Although it is well known that ring currents lessen with increasing macrocycle size and that paratropicity diminishes faster than diatropicity [5a], it is somewhat surprising that the 20-membered macrocycle seems to lack a ring current.

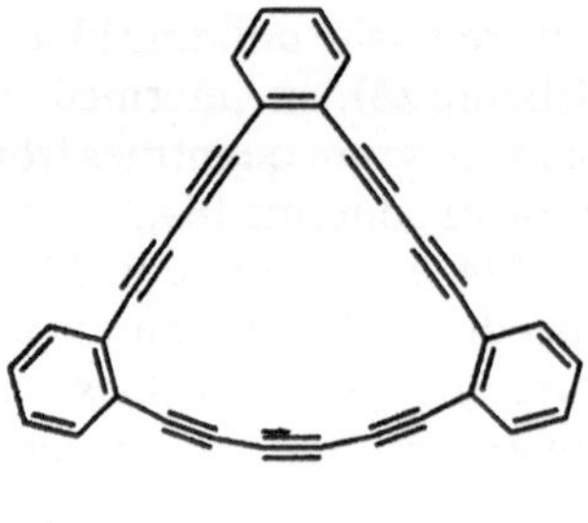

6
Phenyltetraacetylenes

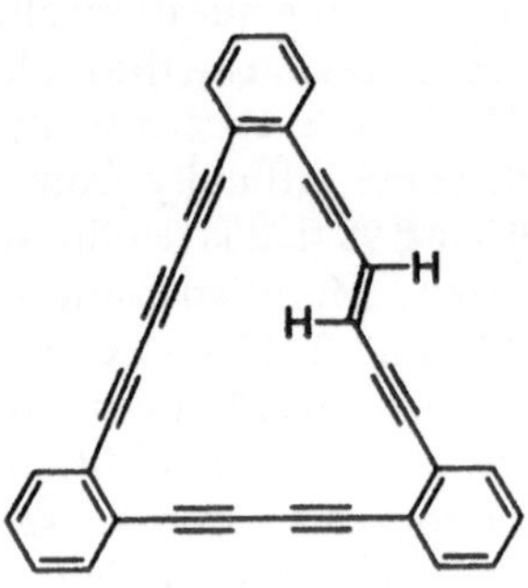

The success of generating phenylbutadiynes and phenylhexatriynes in situ under Pd-coupling conditions suggested that preparation of even larger macrocycles should be possible. Given the problems with constructing **124**, other methods were investigated for the synthesis of tetrayne-linked systems. One subtle modification that worked particularly well was addition of excess K_2CO_3 to standard Eglinton oxidative acetylene-coupling conditions. This worked for a series of monoynes and diynes, accomplishing desilylation and alkyne dimerization in a single pot with yields as high as 98% [78]. In the case of **90**, exposure to the same conditions afforded tetrayne **128** as a bright yellow solid (Scheme 29). Subsequent removal of the TIPS groups and cyclization with $CuCl/Cu(OAc)_2$ under pseudo-high dilution conditions provided orange cyclodimer **129** as the sole product (51%); neither higher cyclooligomers nor the highly strained carbocycle arising from intramolecular ring closure were detected. Despite being non-planar, compound **129** proved to be only marginally soluble in common solvents. Repetition of the synthetic sequence utilizing **95** (R=*t*-Bu) furnished derivative **130**. Inclusion of four *tert*-butyl moieties noticeably improved product solubility and thus the isolated yield (64%). To date, attempts to grow suitable crystals for X-ray structure studies have been unsuccessful.

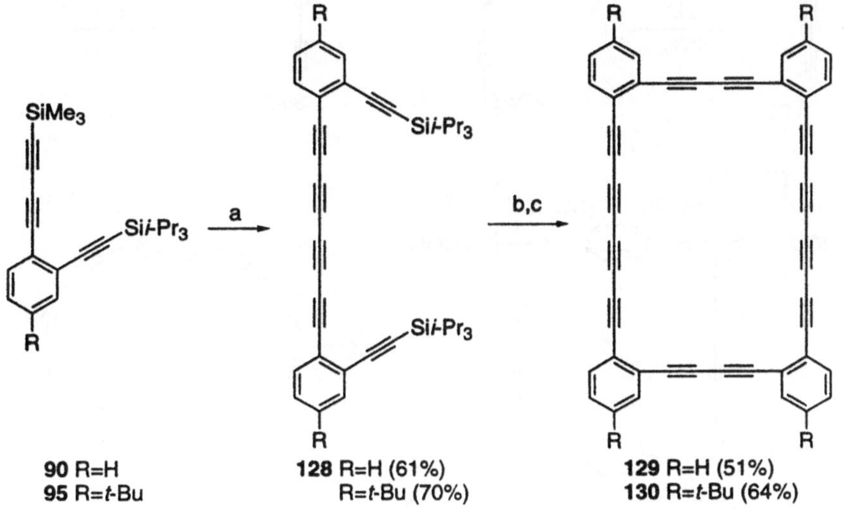

Scheme 29. a) $Cu(OAc)_2 \cdot H_2O$, K_2CO_3, py, MeOH; b) Bu_4NF, EtOH, THF; c) $Cu(OAc)_2$, CuCl, py

An alternate approach to tetrayne-linked systems focused on the use of organometallic fragments to stabilize a highly strained annulene, followed by liberation of the hydrocarbon. Despite significant efforts on the part of several research groups [1, 5c], systems containing contiguous, bent triynes and higher polyynes have proven elusive; only strained monoyne and diyne derivatives (e.g. **3, 7, 63, 80**) are known. Although triyne- and tetrayne-linked systems have been prepared (vide supra), calculations showed these to be relatively strain-free. The most "highly strained" is compound **126**. Even there, the largest deviation of the triple bonds from linearity was calculated to be only 5.1°.

With *sp* bond angles calculated to be around 162°, macrocycle **131** would be highly strained and was therefore expected to be quite reactive [79]. The octa-cobalt complex **132**, on the other hand, should be readily isolable. Indeed, **132** was prepared easily from **133** in five steps, and was isolated as stable, deep maroon crystals (Scheme 30). All spectroscopic data supported formation of the strain-free dimeric structure. Unfortunately, all attempts to liberate **132** from the cobalt units led only to insoluble materials. Diederich et al. observed similar problems when trying to prepare the cyclocarbons [5c]. Whether the failure to prepare these two classes of macrocycles is due to the extreme reactivity of the distorted polyyne moiety or to the lack of a viable synthetic route is not certain. Thus, isolation and characterization of smaller bent hexatriyne– and octatetrayne-containing systems is an important goal that should help answer these questions.

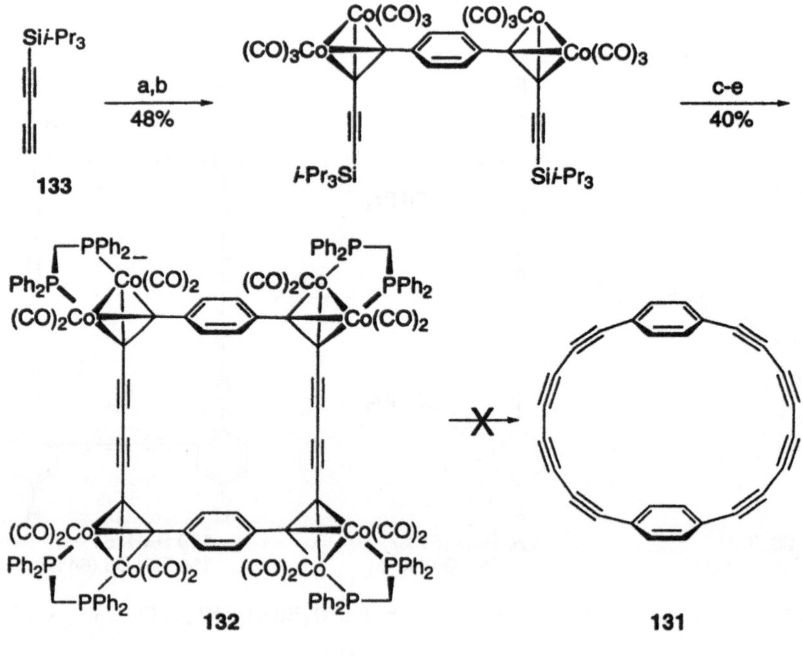

Scheme 30. a) *p*-diiodobenzene, PdCl$_2$(PPh$_3$)$_2$, CuI, Et$_3$N; b) Co$_2$(CO)$_8$, Et$_2$O; c) dppm, toluene; d) Bu$_4$NF, THF, EtOH; e) Cu(OAc)$_2$ · H$_2$O, py

7
Phenyloligoacetylenes

The penultimate example of macrocycles based on phenyl and acetylenic units has been the very recent report by Tobe [80] and Rubin [81] of cyclophane **134**. Both groups generated **134** in the mass spectrometer by laser desorption of hexa-protected polyynes **135** (robust) and **136** (unstable), respectively (Scheme 31).

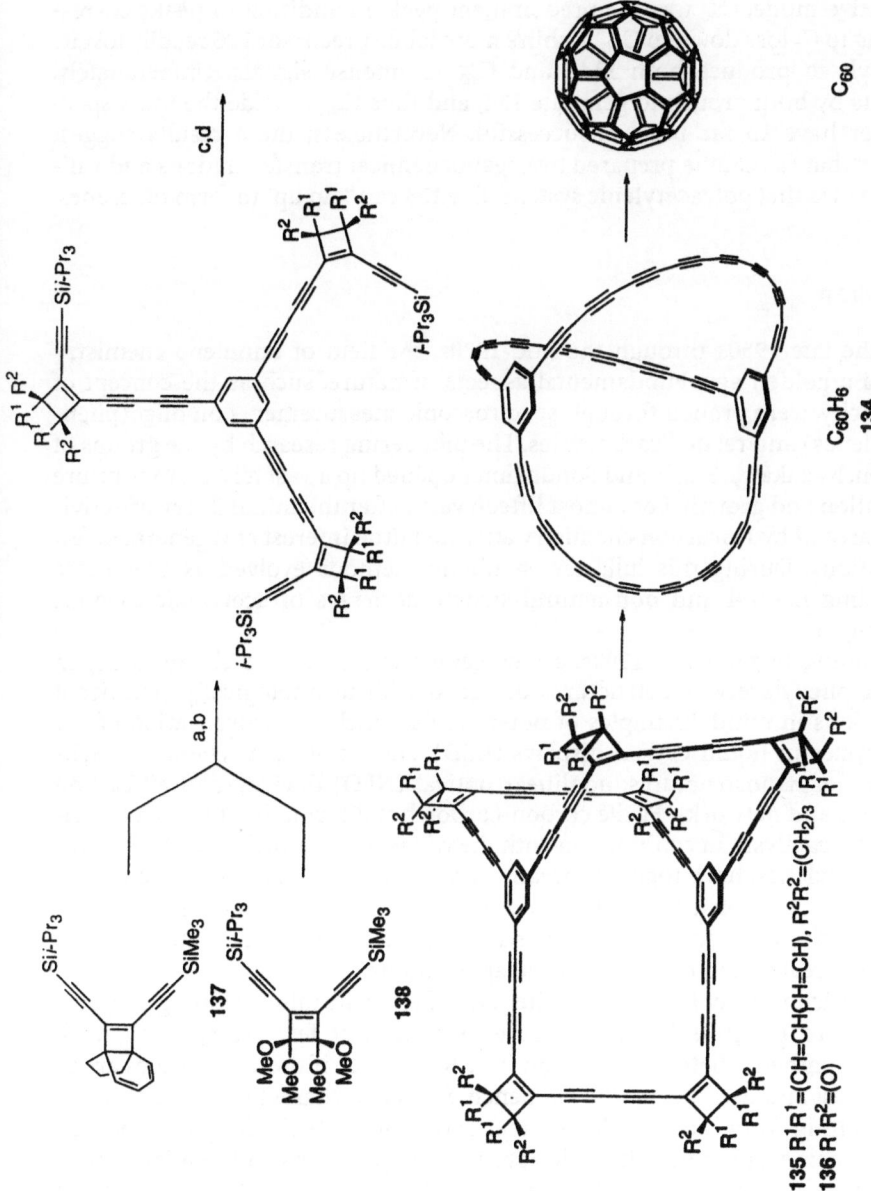

Scheme 31. a) K$_2$CO$_3$, MeOH; b) for **137**: 1,3,5-tris(bromoethynyl)benzene, Pd$_2$(dba)$_3$, PPh$_3$, CuI, 1,2,2,6,6-pentamethylpiperidine, PhH; for **138**: i] LHMDS, THF ii] CuBr iii] 1,3,5-tris(bromoethynyl)benzene, py; c) Bu$_4$NF, THF; d) for **135**: Cu(OAc)$_2$, py; for **136**: i] CuCl, TMEDA, ODCB ii] TFA

The precursors were prepared starting from either **137** or **138**, themselves the products of somewhat involved syntheses.

In Tobe's TOF-MS experiments, negative mode data clearly showed generation of **134**$^-$ from **135**; in addition, a small peak attributed to C_{60}^- was observed. In positive mode, C_{60}^+ was the predominant peak, in addition to peaks corresponding to C_2 loss down to C_{50}^+. Rubin's more labile precursor **136** readily lost its carbonyls to produce both **134**$^{\pm}$ and $C_{60}^{\cdot-}$ as intense signals. Unfortunately, attempts by both groups to generate **134**, and thus C_{60}, outside the mass spectrometer have so far been unsuccessful. Nevertheless, these results suggest strongly that C_{60} can be prepared by organochemical transformations and validate the idea that polyacetylenic systems like **134** can "zip up" to form fullerenes.

8
Conclusion

From the late 1950s through the mid-1970s, the field of annulene chemistry enjoyed a golden age. Fundamental aspects of nature, such as the concept of aromaticity, were probed through spectroscopic measurements on oligo(phenylacetylenes) and related carbocycles. The pioneering research by the groups of Eglinton, Nagakawa, Staab, and Sondheimer opened up a vast territory for future exploration and growth. For almost fifteen years after the initial flurry of activity, this area of hydrocarbon chemistry attracted little interest and generated few publications. During this lull, new synthetic methods evolved as ever more demanding natural and non-natural structural arrays of acetylenic carbons were sought.

Beginning in the early 1990s, a new generation of alkyne chemists began crafting phenylacetylene structures designed with technologically significant applications in mind. Examples of potential uses include incorporation of the macrocycles in liquid crystal displays (LCD), supramolecular chemistry, self-assembly of nanostructures, nonlinear optical (NLO) devices, and all carbon molecules and networks. Facile carbon-carbon bond formation through transition metal catalysis, in conjunction with the widespread availability of silyl-protected acetylenes, have together greatly accelerated the rate at which complex structures can be constructed. These developments now make it possible to tune and tailor the physiochemical properties of macrocycles, a definite first step toward the realization of technological applications.

It is difficult to say just exactly where this field is heading. Although significant forward progress has been made over the last few years, much work remains. Synthetic methods must continue to advance if these molecules are to find interest and use beyond the lab bench. Similarly, characterization of materials resulting from macrocycle polymerization, while a challenging prospect, should become a high priority. Solutions to these problems will require a multidisciplinary approach by chemists and materials scientists alike, offering fresh perspectives to old ideas. The future direction of phenyacetylene macrocycles, while uncertain, is rich with possibilities. One thing is for certain – the current spate of publications and discoveries can truly be said to mark a renaissance of this exciting field.

9
References

1. Stang PJ, Diederich F (eds) (1995) Modern Acetylene Chemistry, VCH, Weinheim
2. a) Collman JP, Hegedus LS, Norton JR, Finke RG (1987) Principles and applications of organotransition metal chemistry, University Science, Mill Valley, CA; b) Diederich F, Stang PJ (eds) (1998) Metal-catalyzed cross-coupling reactions, Wiley-VCH, Weinheim; c) Davies SG, Murahashi S (eds) (1998) Transition metal catalysed reactions – IUPAC Monographs Chemistry for the 21st Century, Blackwell Science, Oxford
3. a) Heck RF (1985) Palladium reagents in organic syntheses, Academic, London; b) Bräse S, de Meijere A (1998) In: Diederich F, Stang PJ (eds) Metal-catalyzed cross-coupling reactions, Wiley-VCH, Weinheim, p 99
4. Haley MM (1998) Synlett 557
5. For related reviews see a) Balaban AT, Banciu M, Ciorba V (1987) Annulenes, benzo-, hetero-, homo- derivatives and their valence isomers, CRC, Boca Raton, FL, Vol 2, p 146; b) Moore JS (1997) Acc Chem Res 30:402; c) Diederich F (1994) Nature 369:199
6. Eglinton G, Galbraith AR (1957) Proc Chem Soc 350
7. a) Eglinton G, Galbraith AR (1956) Chem & Ind 737; b) Eglinton G, Galbraith AR (1959) J Chem Soc 889
8. a) Behr OM, Eglinton G, Raphael RA (1959) Chem & Ind 699; b) Eglinton G, Galbraith AR (1960) J Chem Soc 3614
9. Behr OM, Eglinton G, Lardy IA, Raphael RA (1964) J Chem Soc 1151
10. Stephens RD, Castro CE (1963) J Org Chem 28:3313
11. Campbell ID, Eglinton G, Henderson W, Raphael RA (1966) J Chem Soc Chem Commun 87
12. a) Staab HA, Graf F (1966) Tetrahedron Lett 751; b) Staab HA, Graf F (1970) Chem Ber 103:1107; c) Irngartinger H, Leiserowitz L, Schmidt GMJ (1970) Chem Ber 103:1119
13. Wong HNC, Garratt PJ, Sondheimer F (1974) J Am Chem Soc 96:17
14. Staab HA, Neunhoeffer K (1974) Synthesis 424
15. a) Takahashi S, Kuroyama Y, Sonogashira K, Hagihara N (1980) Synthesis 627; b) Austin WB, Bilow N, Kelleghan WJ, Lau KSY (1981) J Org Chem 46:2280; c) Review: Sonogashira K In: Diederich F, Stang PJ (eds) Metal-catalysed cross-coupling reactions, Wiley-VCH, Weinheim, p 203
16. a) Barton JW, Shephard MK (1984) Tetrahedron Lett 25:4967; b) Diercks R, Vollhardt KPC (1986) J Am Chem Soc 108:3150
17. Solooki D, Ferrara JD, Malaba D, Bradshaw JD, Tessier CA, Youngs WJ (1997) Inorg Synth 31:122
18. Kinder JD, Tessier CA, Youngs WJ (1993) Synlett 149
19. a) Solooki D, Kennedy VO, Tessier CA, Youngs WJ (1990) Synlett 427; b) Solooki D, Bradshaw JD, Tessier CA, Youngs WJ (1994) Organometallics 13:451
20. Eickmeier C, Junga H, Matzger AJ, Scherhag F, Shim M, Vollhardt KPC (1997) Angew Chem Int Ed Engl 36:2103
21. Huynh C, Linstrumelle G (1988) Tetrahedron 44:6337 See also: Iyoda M, Vorasingha A, Kuwatani Y, Yoshida M (1998) Tetrahedron Lett 39:4701
22. In our hands this route produced a mixture of cyclooligomers; Haley MM, Pham S unpublished observations
23. Haley MM, Kehoe JM unpublished results
24. Moore JS, Weinstein EJ, Wu Z (1991) Tetrahedron Lett 32:2465
25. Haley MM, English JJ, Johnson CA, Petersen RC unpublished results
26. Haley MM, Kiley JH unpublished results
27. Baldwin KP, Bradshaw JD, Tessier CA, Youngs WJ (1993) Synlett 853
28. Baldwin KP, Simons RS, Rose J, Zimmerman P, Hercules DM, Tessier CA, Youngs WJ (1994) J Chem Soc Chem Commun 1257
29. Djebli A, Ferrara JD, Tessier-Youngs C, Youngs WJ (1988) J Chem Soc Chem Commun 548

30. a) Ferrara JD, Tessier-Youngs C, Youngs WJ (1985) J Am Chem Soc 107:6719; b) Ferrara JD, Tanaka AA, Fierro C, Tessier-Youngs CA, Youngs WJ (1989) Organometallics 8:2089; c) Youngs WJ, Kinder JD, Bradshaw JD, Tessier CA (1993) Organometallics 12:2406
31. a) Ferrara JD, Tessier-Youngs C, Youngs WJ (1987) Organometallics 6:676; b) Ferrara JD, Tessier-Youngs C, Youngs WJ (1988) Inorg Chem 27:2201
32. Ferrara JD, Djebli A, Tessier-Youngs C, Youngs WJ (1988) J Am Chem Soc 110:647
33. a) Youngs WJ, Djebli A, Tessier CA (1991) Organometallics 10:2089; b) Malaba D, Djebli A, Chen L, Zarate EA, Tessier CA, Youngs WJ (1993) Organometallics 12:1266
34. Bradshaw JD, Solooki D, Tessier CA, Youngs WJ (1994) J Am Chem Soc 116:3177
35. Solooki D, Bradshaw JD, Tessier CA, Youngs WJ, See RF, Churchill M, Ferrara JD (1994) J Organomet Chem 470:231
36. Zhang J, Pesak DJ, Ludwick JL, Moore JS (1994) J Am Chem Soc 116:4227
37. Zhang J, Moore JS, Xu Z, Aguirre R (1992) J Am Chem Soc 114:2273
38. Moore JS, Zhang J (1992) Angew Chem Int Ed Engl 31:922
39. Shetty AS, Zhang J, Moore JS (1996) J Am Chem Soc 118:1019
40. Zhang J, Moore JS (1994) J Am Chem Soc 116:2655
41. Venkataraman D, Lee S, Zhang J, Moore JS (1994) Nature 371:591
42. Shetty AS, Fischer PR, Stork KF, Bohn PW, Moore JS (1996) J Am Chem Soc 118:9409
43. Wu Z, Lee S, Moore JS (1992) J Am Chem Soc 114:8730
44. Wu Z, Moore JS (1996) Angew Chem Int Ed Engl 35:297
45. Kawase T, Ueda N, Darabi HR, Oda M (1996) Angew Chem Int Ed Engl 35:1556
46. Kawase T, Darabi HR, Oda M (1996) Angew Chem Int Ed Engl 35:2664
47. Young JK, Moore JS (1995) In: Stang PJ, Diederich F (eds) Modern Acetylene Chemistry, VCH, Weinheim, p 415
48. Bedard TC, Moore JS (1995) J Am Chem Soc 117:10662
49. Bloor D, Chance RR (eds) (1985) Polydiacetylenes: synthesis, structure, and electronic properties. Martinus Jijhoff, Boston
50. a) Prasad RN, Reinhardt BA (1990) Chem Mater 2:660; b) Nalwa HS (1993) Adv Mater 5:341
51. Zhou Q, Carroll PJ, Swager TM (1994) J Org Chem 59:1294
52. a) Hoffmann R, Hughbanks T, Kertesz M, Bird PH (1983) J Am Chem Soc 105:4831; b) Baughman RH, Eckhardt H, Kertesz M (1987) J Chem Phys 87:6687
53. a) Hay AS (1960) J Org Chem 25:1275; b) Hay AS (1962) J Org Chem 27:3320
54. Guo L, Bradshaw JD, Tessier CA, Youngs WJ (1994) J Chem Soc Chem Commun 243
55. Baldwin KP, Matzger AJ, Scheiman DA, Tessier CA, Vollhardt KPC, Youngs WJ (1995) Synlett 1215
56. a) Baughman RH, Yee KC (1974) J Polym Sci: Polym Chem 12:2467; b) Enkelmann V, Graf HJ (1978) Acta Crystallogr B34:3715
57. Haley MM, Naruse H, Weakley TJR unpublished results
58. Boese R, Matzger AJ, Vollhardt KPC (1997) J Am Chem Soc 119:2052
59. Dresselhaus MS, Dresselhaus G, Ecklund PC (1996) Science of Fullerenes and Carbon Nanotubes, Academic, San Diego
60. a) Tovar JD, Jux N, Jarrosson T, Khan SI, Rubin Y (1997) J Org Chem 62:3432; b) correction (1997) J Org Chem 62:5656; c) correction (1998) J Org Chem 63:4856
61. In a similar manner, molecules 18 and 19 could be envisaged as substructures of the acetylenic all-carbon analog graphyne.
62. a) Jux N, Holczer K, Rubin Y (1996) Angew Chem Int Ed Engl 35:1986; b) Tobe Y, Kubota K, Naemura K (1997) J Org Chem 62:3430
63. Haley MM, Brand SC, Pak JJ (1997) Angew Chem Int Ed Engl 36:835
64. Brandsma L (1971) Preparative acetylenic chemistry, Elsevier, Amsterdam. The first edition contained the preparation of 1-phenyl-1,3-butadiyne, but stated "...the compound proved to be very unstable, (and) can be stored at –20°C for a very limited period." This procedure has subsequently been deleted from the second edition (1988).
65. Other groups have reported similar problems using free phenylbutadiynes in synthesis: Godt A (1997) J Org Chem 62:7471

66. Haley MM, Bell ML, English JJ, Johnson CA, Weakley TJR (1997) J Am Chem Soc 119:2956
67. Haley MM, Wan WB, Brand SC manuscript in preparation
68. Haley MM, Pak JJ, Weakley TJR submitted
69. Tobe Y, Utsumi N, Kawabata K, Naemura K (1996) Tetrahedron Lett 37:9325
70. Tobe Y, Utsumi N, Nagano A, Naemura K (1998) Angew Chem Int Ed 37:1285
71. Höger S, Enkelmann V (1995) Angew Chem Int Ed Engl 34:2713
72. Morrison DL, Höger S (1996) J Chem Soc Chem Commun 2313
73. a) Höger S, Meckenstock A-D, Pellen H (1997) J Org Chem 62:4556; b) Höger S, Mecken-stock A-D (1998) Tetrahedron Lett 39:1735
74. a) Parker TC, Khan SI, Holliman CL, McElvany SW, Rubin Y manuscript in preparation; b) Rubin Y (1997) Chem Eur J 3:1009
75. Rubin Y, Parker TC, Khan SI, Holliman CL, McElvany SW (1996) J Am Chem Soc 118:5308
76. Haley MM, Wan WB, Kimball DB (1998) Tetrahedron Lett 39:6795
77. Eastmond R, Walton DMR (1972) Tetrahedron 28:4591
78. Haley MM, Bell ML, Brand SC, Kimball DB, Pak JJ, Wan WB (1997) Tetrahedron Lett 38:7483
79. Haley MM, Langsdorf BL (1997) J Chem Soc Chem Commun 1121
80. Tobe Y, Nakagawa N, Naemura K, Wakabayashi T, Shida T, Achiba Y (1998) J Am Chem Soc 120:4544
81. Rubin Y, Parker TC, Pastor SJ, Jalisatgi S, Boulle C, Wilkins CL (1998) Angew Chem Int Ed 37:1226
82. Kawase T, Ueda N, Oda M, (1997) Tetrahedron Lett 38:6681

Note Added in Proof

Oda et al. recently reported the synthesis of the highly strained trimeric *meta*-PAM **139** (Scheme 32) [82]. The required triene **140** was prepared from α,ω-dial-dehyde **141** by an intramolecular McMurry coupling reaction. Bromination/dehy-drobromination of **140** furnished **139** as moderately stable, colorless crystals which decomposed above 180 °C. The greater degree of strain in **139** compared to

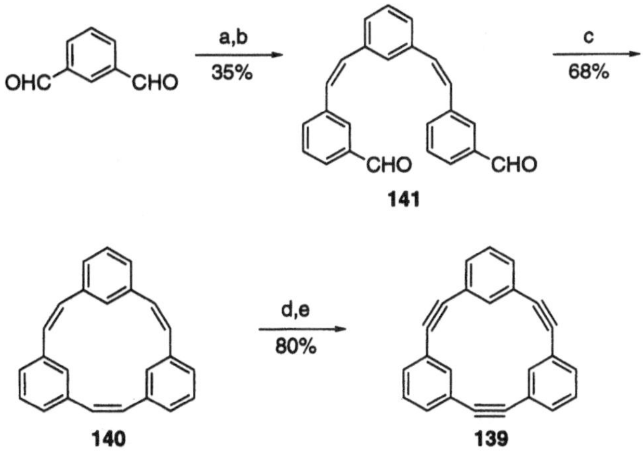

Scheme 32. a) m-BrC$_6$H$_4$CH$_2$PPh$_3^+$ Br$^-$, t-BuOK, DMSO; b) i] BuLi, THF ii] DMF; c) TiCl$_4$, Zn, DME; d) Br$_2$, CHCl$_3$; e) t-BuOK, Et$_2$O

tetramer **57** was clearly reflected in the spectroscopic properties of the molecule: 1) the inner protons of **139** appeared at $\delta = 8.38$, 0.31 ppm lower field than **57** ($\delta = 8.07$) and significantly shifted compared to 1,3-bis(phenylethynyl)benzene ($\delta = 7.72$); 2) the sp carbon atoms of **139** ($\delta = 99.86$) resonated at 7.7 ppm lower field than those of **57** ($\delta = 92.20$); and 3) the Raman frequency of the triple bonds in **139** ($\nu = 2155$ cm^{-1}) was appreciably shifted to that found for **57** ($\nu = 2202$ cm^{-1}).

Like **57**, X-ray structure analysis of **139** showed that the macrocycle was essentially planar with the benzene rings deformed from a regular hexagon by ca. 3.3°. Torsion angles between the benzene rings and triple bonds were less than 3.5°. The alkynes were the primary location of the strain in the macrocycle and thus were highly distorted from linearity. The average sp bond angle of **139** (158.6°) was about 10° smaller than those of **57** and was comparable to those of **7** (155.8°). [13] Unlike **57, 139** added two equivalents of cyclopentadiene at room temperature to give a 1:1 syn:anti mixture of diadducts; neither mono– nor triadducts were obtained.

Carbon-Rich Molecular Objects
from Multiply Ethynylated π-Complexes

Uwe H. F. Bunz

Department of Chemistry and Biochemistry, University of South Carolina, Columbia, SC 29208, USA. *E-mail: Bunz@psc.sc.edu*

Dedicated to Professor Klaus Hafner

The synthesis and properties of a series of multiply ethynylated π-complexes of iron, manganese, and cobalt are described. These complexes were shown to be useful as modules for the construction of larger organometallic carbon-rich molecular objects which can be regarded as segments out of either a planar organometallic all-carbon net or an expanded fullerene, a fullerenyne. Screening of the material properties of these objects showed that some of them form organometallic liquid crystalline phases or stable Langmuir-Blodgett films at the air-water interface. The methodology for the construction of these organometallic modules should lead to the future synthesis of larger segments of organometallic all-carbon nets of different topologies.

Keywords: Organometallic alkynes, Carbon-rich compounds.

Topics in Current Chemistry, Vol. 201
© Springer-Verlag Berlin Heidelberg 1999

1
Introduction

The Nobel prize in Chemistry for the year 1996 was awarded for the discovery of the fullerenes, the third allotropic form of carbon, with C_{60} and C_{70} as the two most prominent representatives. While the fullerenes of course are the epitome of carbon-rich molecular compounds, it is an irony that their synthesis is more of a physical phase transition, taking place under drastic conditions [1].

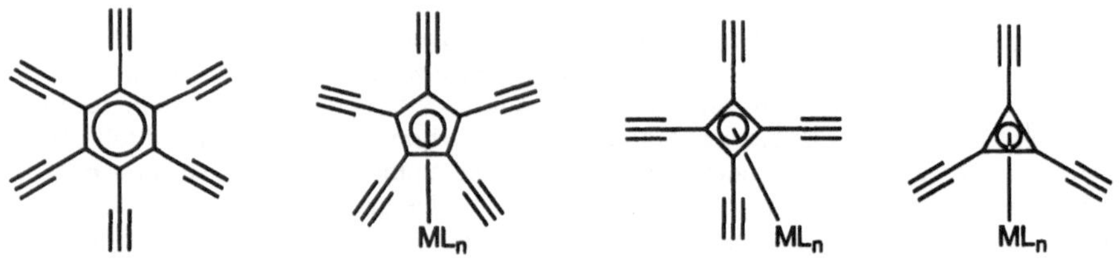

Along with the discovery of the fullerenes, starting in the mid-1980s, the chemistry of alkynes and their use as building blocks for novel molecular structures, such as hexaethynylbenzene 1 [2], tetraethynylethylene [3], and cyclic dehydroannulenes [4], as well as pericyclynes [5] and oxocarbons [6], experienced an "explosive" revival, which was greatly stimulated by the (probably correct) belief that a total synthesis of fullerene-like cages would proceed through alkyne intermediates [7]. Yet not only cage-type or bowl-shaped targets such as corannulene and derivatives [8], but also graphite segments [9], alkyne-bridged graphite (graphyne [10, 11]) and other more exotic carbon-rich tetrahedral and porphyrine-carrying species [11 c] and cycles [12, 13] were considered attractive synthetic goals and have been made since then.

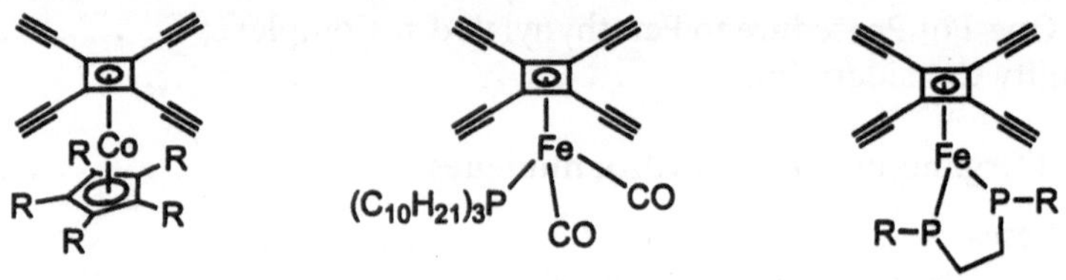

While the topic of *purely* organic carbon-rich compounds has attracted quite a number of different groups, the "organometallic arm" of the enterprise has, with some exceptions [11 b, d], been much less developed until recently. This is due to the anticipated synthetic difficulties and supposed lack of stability in synthesis and manipulation of ethynylated organometallic compounds. Both of these concerns can be obviated, if the target molecules, such as 2 and 3, are chosen carefully.

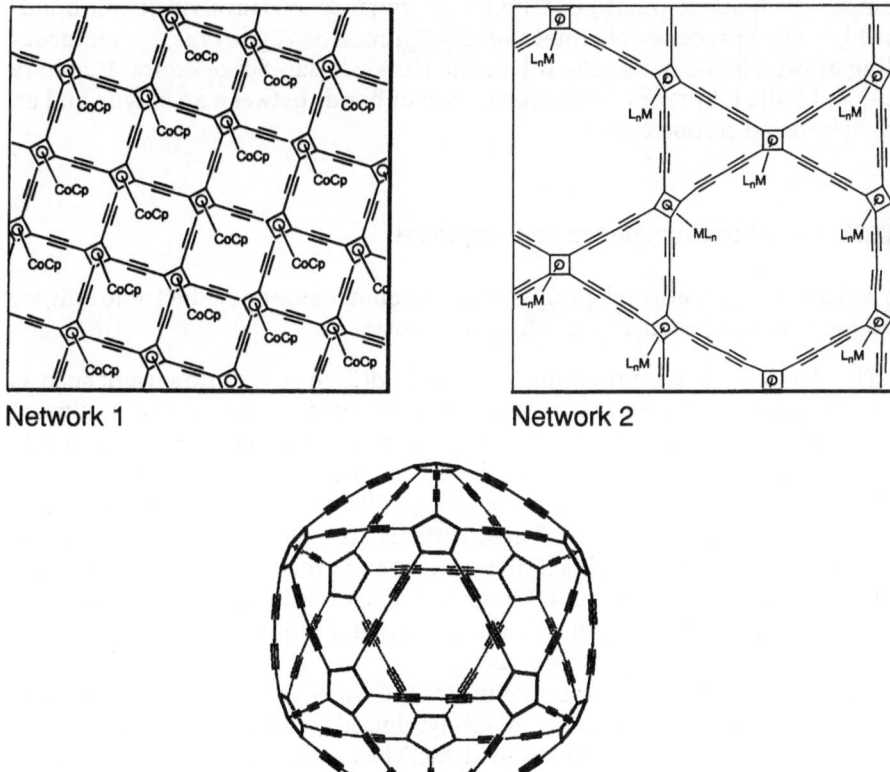

8

If one thinks of organometal-fragment-supported all-carbon networks in comparison to their purely organic congeners, differences are immediately evident. The first difference is that a wider variation of geometries may be accessible, such as shown in networks 1 and 2: cyclobutadiene rings ligated by organometallic 14-electron fragments are the central building blocks in networks with either trigonal or tetragonal symmetry. Other geometries, unfavorable in organic networks, such as obtained by incorporation of cyclopropenylium moieties 4, should be accessible in their organometallic variants as well. A second advantageous difference of organometal-ligated carbon networks would be the in-

Network 1   Network 2

Fullerenyne

Fig. 1. Different alkyne bridged networks and a fullerenyne

creased solubility compared to their organic counterparts, originating from the organometallic ligate or its spectator ligands bearing solubilizing alkyl chains such as in 5–7. The number of solubilizing groups would be larger or equal to the number of organometallic fragments, while in organic systems the attachment of solubilizing groups is restricted to the "rim" of the molecule.

Recent calculations by Burdett [14] suggest that even one-dimensional extended segments out of the networks 1, such as depicted in 8 should be metallic, granted that an unsaturated 12-electron fragment ligates the cyclobutadiene. If one chooses a bulky enough bisphosphine ligand in 8, the resulting environment should render these unsaturated species stable. Such stable species would be very attractive in the wake of molecular electronics [15], where the observance of single-molecule conductivity and true single molecule wires represent "holy grails" in the field.

While it is at the moment not possible to synthesize extended 2D-segments out of the organometallic all-carbon nets 1 and 2, the obvious way to attack that problem is to construct different small modules at first and then assemble them into larger structures. In our case these modules are multiply ethynylated π-complexes, connected to yield oligomers of differing topology and size. An advantage of alkynyl-bridged modules is that (a) the alkyne unit is rigid and comparatively stable and (b) C–C bond forming reactions involving triple bonds are high-yield processes of either the Hay, Eglinton or Vögtle type for homocoupling of two alkyne units [16–18], or the Heck–Cassar–Sonogashira–Hagihara [19] and Stille [20] types for the formation of bonds between an alkyne and an sp^2-hybridized carbon center.

2
Syntheses of Diethynylated π-Complexes

In order to connect ethynyl groups to the cyclobutadiene or cymantrene unit, we had four catalyst systems from which to choose:

- The classic Stille systems, utilizing $(PPh_3)_2PdCl_2$ in an aprotic solvent such as DMF, toluene or benzene to couple tin-substituted alkynes to aryl or alkenyl bromides or iodides. The disadvantage of this catalyst system is the relatively high temperature (80–120°C) needed to conduct the coupling [20a].
- The Farina variant of the Stille coupling, utilizing $AsPh_3$ or $P(furfuryl)_3$ and Pd_2dba_3 and additional CuI. These systems are much more reactive (up to 1000 times) than the original Stille systems and proceed at ambient temperatures with sufficient rates in DMF or THF. The disadvantages of this coupling system is that $AsPh_3$ is relatively toxic and that it is often difficult to separate catalyst residues from products [20b].
- The "ligandless" Beletskaya catalysts, $PdCl_2(CH_3CN)_2$ used in DMF or acetone with tin-substituted alkynes and sp^2-hybridized iodides and bromides. This system is very reactive and relatively cheap. The disadvantage is that it decomposes quickly under development of catalytically inactive Pd-black particles [20c].
- The Heck–Cassar–Sonogashira–Hagihara system, which couples terminal alkynes and vinylic or aromatic halides in an amine solvent, typically piperi-

dine or triethylamine, with added $Pd(PPh_3)_2Cl_2$ and CuI. This catalytic system is very powerful but not every organometallic substrate will endure the basic reaction conditions. Aromatic iodides react much faster than bromides, and aromatic chlorides are unreactive [19] as in all of the coupling systems described here.

2.1
Cyclopentadienyl Complexes

An attempt by Manriquez et al. [21] to prepare alkynylated cyclopentadienes as ligands was not met by success, due to the fact that these hydrocarbons were too sensitive. Stille and Sterzo published two landmark papers [22] describing the synthesis of ethynylated cyclopentadienyl complexes by a metallation-iodination coupling strategy, starting from cymantrene 9 or other half-sandwich complexes. The ease of metallation allows the functionalization of the cymantrene Cp ring without any problem, while coupling of 10 in the subsequent alkynylation step proceeds in high yields to furnish 11 – 13.

Yet interestingly enough, no di- or multi-alkynylated Cp-complexes had been made by extension of this methodology. Metallation of 13 under strict temperature control (– 78 °C) with added (tetramethyl)ethylenediamine (TMEDA), iodination by 1,2-diiodoethane and subsequent coupling with (trimethylsilyl) (trimethylstannyl)ethyne utilizing the "Beletskaya" type catalyst $PdCl_2(CH_3CN)_2$ in DMF gave rise only to the formation of the *ortho*-diethynylated cymantrene 16 [20]. A metallation experiment conducted in ethyl ether at – 40 °C took a different course, furnishing a mixture of the corresponding *ortho*- and *meta*-dialkynylated cymantrenes 16 and 17 after workup with diiodoethane and subsequent Beletskaya coupling. It was not possible to characterize the corresponding intermediate iodoalkynyl-substituted cymantrenes 14 and 15, isolated as viscous oils. A trace of a paramagnetic impurity, inseparable from 14 and 15 by chromatography, rendered NMR spectroscopy as an analytical tool in this case virtually useless.

Both dialkynylated cymantrenes 16 and 17 have the same symmetry and similar NMR spectra, so that the ultimate structure elucidation had to rest upon X-ray crystallography. It was reasoned that the dialkynylcymantrene with the smaller $J(H,H)$ coupling of the cyclopentadienyl protons should be 17, an interpretation reinforced [23a] by the result of the single-crystal structure.

Attempts to directly dimetallate the cymantrene nucleus, a reaction feasible for CpReCl(CO)(NO) [23b], failed completely, despite the strong electron-withdrawing effect of the $Mn(CO)_3$ group. The use of an acetal of formylcymantrene 18, should give 2,5-dialkynylated cymantrenes after metallation, iodination, coupling and manipulation of the acetal function, but attempts to metalate 18 (using a procedure described by a Russian group [23 c]) with *sec*-BuLi instead of BuLi (as described in the preparation) initially gave rise to only minor amounts of the corresponding dimetallated cymantrene 19 in addition to the anticipated monometallation product. By adding 2.5 equivalents of *sec*-BuLi to 18, 19 was formed and trapped by 1,2-diiodoethane, chlorotrimethylsilane and methyl disulfide, respectively. The presence of the two acetal oxygens in 18

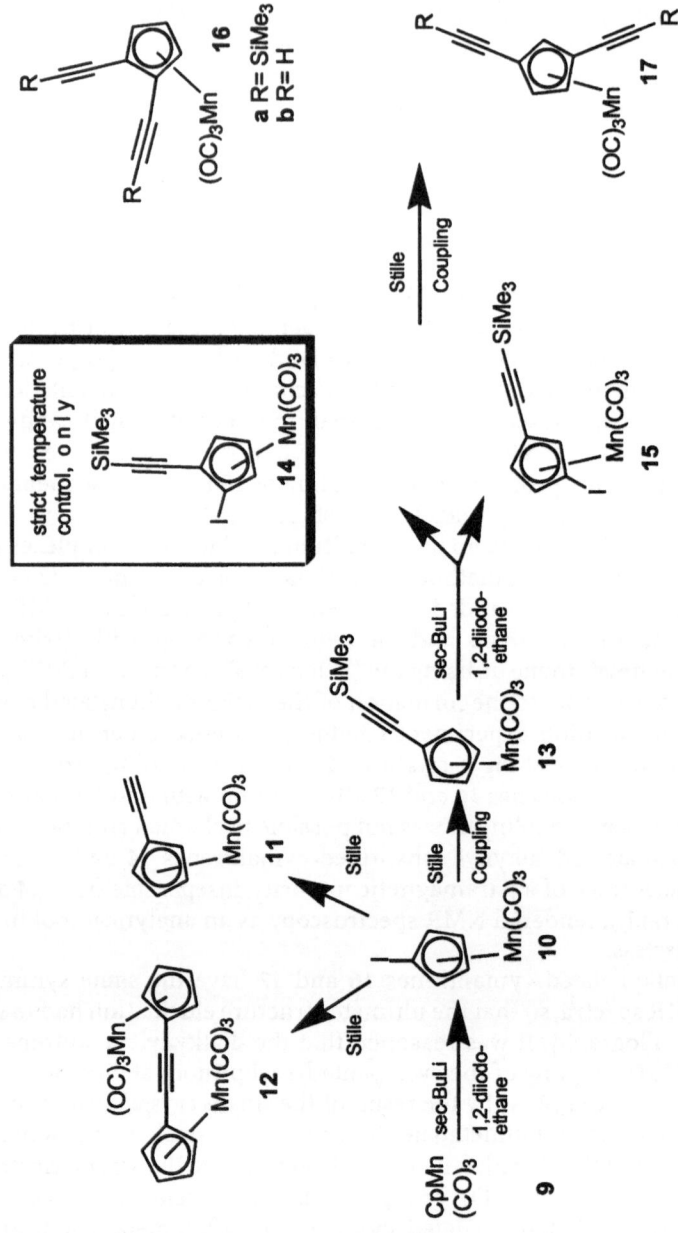

apparently has a strong enough activating effect to make **19** accessible by double deprotonation.

Beletskaya-coupling of **20** gave rise to the isolation of **21** in good yields [23 d] which should be an attractive module for the synthesis of cymantrene-containing oligomers and polymers.

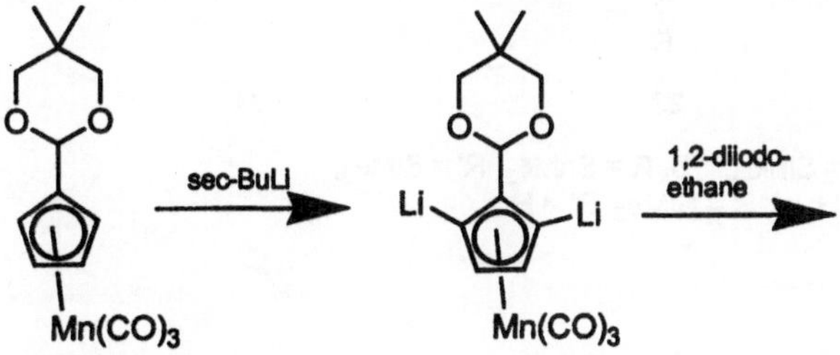

18   **19**

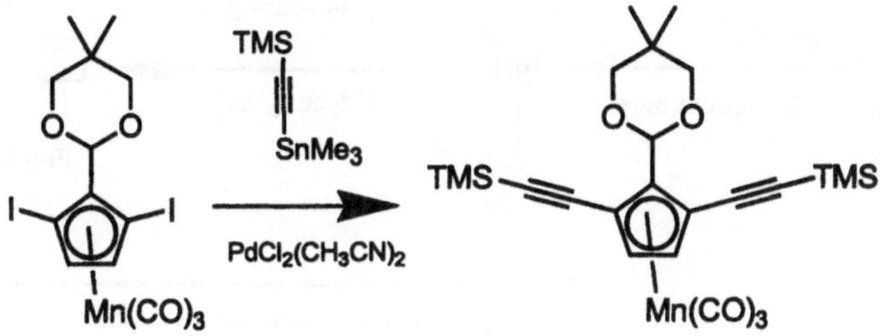

20   **21**

2.2
Cyclobutadiene Complexes

Until 1992, the only ethynylated cyclobutadiene complexes pertinent in the literature were **22 – 24**, prepared by Fritch and Vollhardt using [2 + 2]-cycloaddition of suitable polyynes over CpCo(CO)₂ [24]. No alkynylated derivatives of **25**, however, had been prepared.

The analogous dimerization of alkynes over Fe(CO)₅ is not applicable, so clearly a different route towards alkynylated derivatives of **25** was needed. Comparison of **25** to cymantrene suggests that metallation of the hydrocarbon ligand should be the route of choice for the synthesis of novel substituted cyclobutadienes. In the literature, addition of organolithium bases (MeLi, BuLi) to the CO ligands with concomitant rearrangement had been observed [25]. But the utilization of LiTMP (lithium tetramethylpiperidide, Hafner [26]) or *sec*-BuLi as effectively non-nucleophilic bases led to clean deprotonation of the cyclobuta-

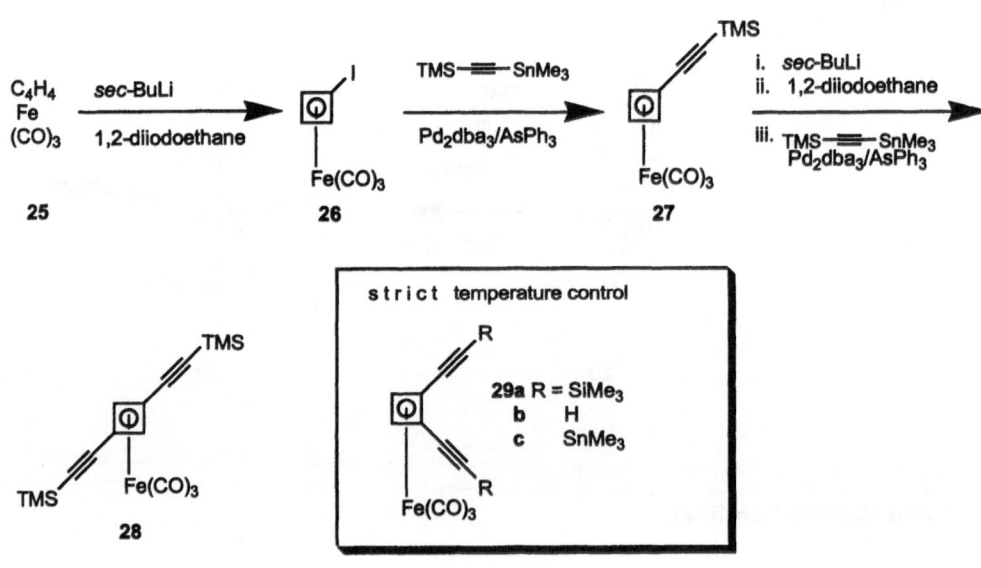

22, 23 a, R = H, R' = SiMe$_3$;   **b**, R = SnMe$_3$, R' = SiMe$_3$;
c, R = SnMe$_3$, R' = H; **d**, R = SiMe$_3$, R' = H

diene ligand and to a lithio compound which could be reacted with diiodoethane to furnish **26** in 69%.

Reaction of **26** under typical Stille-coupling conditions developed by Farina and Krishnan [20c] made the ethynylated cyclobutadiene complex **27** accessible [27]. The reaction is broadly applicable and not restricted to tin-substituted alkynes carrying trimethylsilyl groups. Even substituted stannylated butadiynes and hexatriynes can be coupled to **26**. Functionalization of the second position of the cyclobutadiene ring is performed without any problem, similar to the cymantrene system, giving rise to 1,2-functionalized cyclobutadienes when rigorous temperature control is exerted, i.e. the *sec*-BuLi is precooled. Addition of *sec*-BuLi to **27** without prior cooling promotes the formation of the corresponding *para*-substituted product in addition to the statistically preferred *ortho*-iodide. We assume that the observed *ortho*-selectivity in the deprotonation step is caused by complexation of the lithium cation to the adjacent triple bond, which would function as an internal ligand. Farina-type coupling to **43** and **44** respectively gives the dialkynylated cyclobutadienes **28** and **29** in good to excellent yields [28]. In this case the intermediate iodides **43** and **44** could be isolated, fully characterized, and did not show paramagnetic impurities [28].

Attempts to use in the alkynylation reaction not tin-substituted alkynes, but to couple **26** directly under the conditions developed by Heck, Cassar, Sonogashira, and Hagihara, surprisingly enough gave rise to the formation of the corresponding amino-substituted cyclobutadiene complex **30** in good yields.

The palladium catalyst is essential in this reaction, as was shown in control experiments to make sure that this was not a direct nucleophilic addition of the amine to the electron-poor (regarding the low lying LUMO!) cyclobutadiene ligand. A series of amino-substituted cyclobutadiene complexes have been synthesized by this methodology [29].

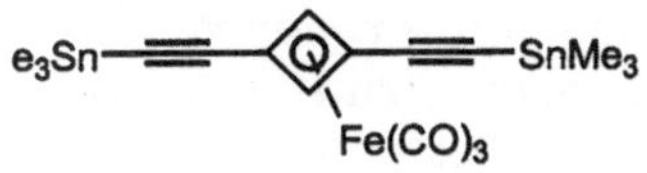

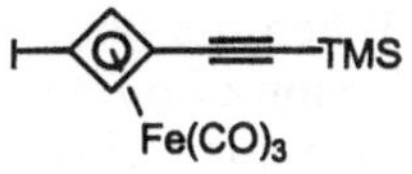

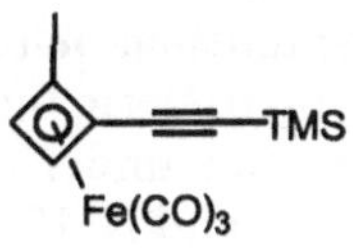

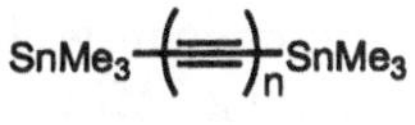

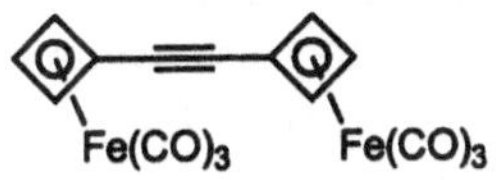

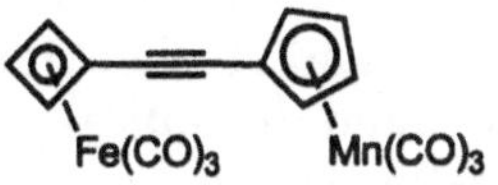

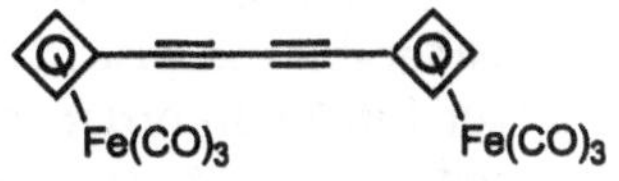

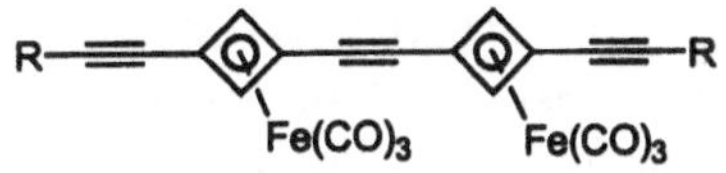

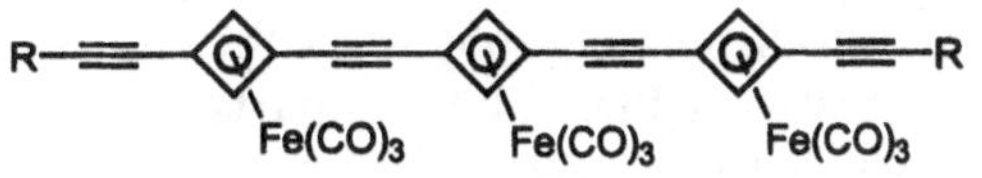

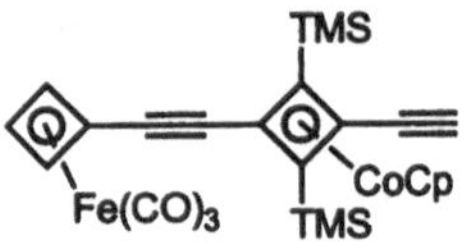

51

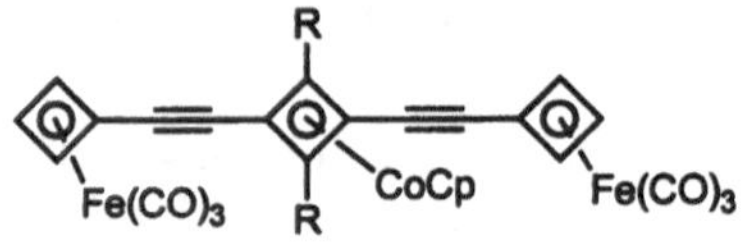

52a R = H: **b** R = SiMe₃

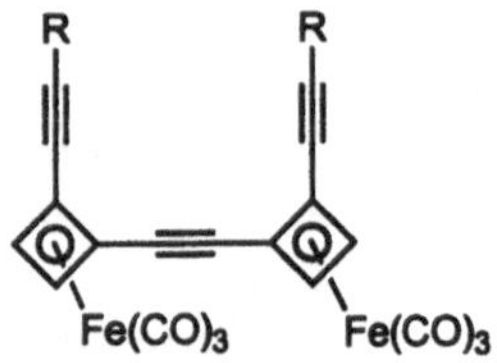

53 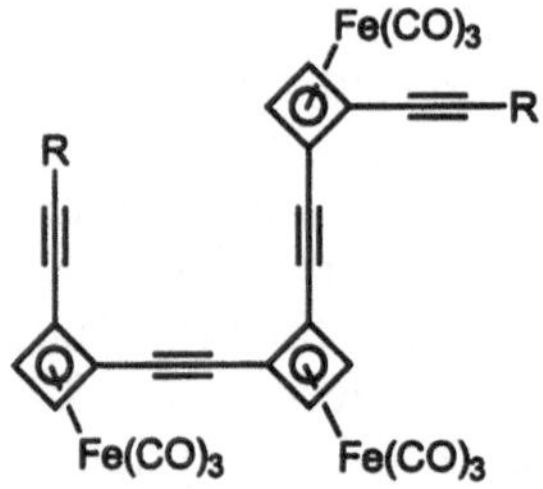   **54** 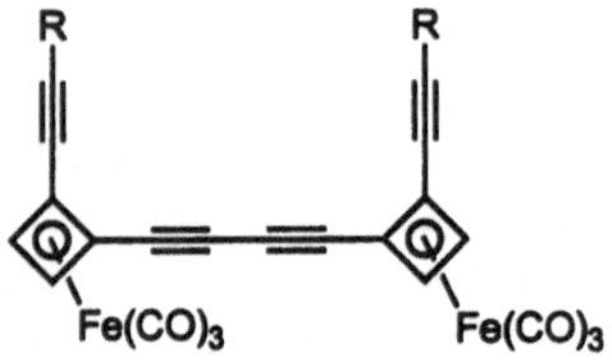   **55**

2.3
Linear Oligomers and Homopolymers

The complex **23** was regarded to be a suitable module for the synthesis of linear polymers **31** consisting of cyclobutadiene and butadiyne units. The TMS groups, useful relics of the synthetic path to **23**, should provide the necessary solubility to the formed polymer chain. Our initial attempt for the synthesis of **31** involved a Hay-coupling reaction conducted at ambient temperature. Surprisingly, only traces of **32** and over 95 % starting material were obtained. A tentative explanation for this rather unusual behavior is that the TMS groups shield the alkyne functionalities by their steric bulk. When the same reaction was conducted in boiling TMEDA for 18 h under admission of pure oxygen, the formation of **31** was observed by GPC and NMR spectroscopy. This result indicates that certain organometallic monomers and polymers can be very stable and survive even harsh oxidizing conditions [30].

The degree of polymerization in **31** is $P_n \approx 15$ according to the analytical data. The UV/vis spectrum of **31** in comparison to that of **23** is remarkable, revealing a considerable increase in the ε-value and a bathochromic shift of the transitions recorded in **23**. In order to understand the optical properties of this system, it seemed desirable to synthesize defined oligomers: lowering of the temperature to 78 °C and shortening the reaction time from 18 h to 4 h in the Hay-coupling of **23** led to the formation of a product mixture comprised of oligomers instead of polymer **31** as evidenced by GPC [31]. Separation of **32 – 34** was achieved by column chromatography, while the higher oligomers **35 – 39** had to be isolated by preparative HPLC. The comparison of the UV/vis spectra of the oligomers shows that the spectra of **37** and **39** are very similar to each other and almost superimposable on the spectrum of the polymer **31**. From these data it can be

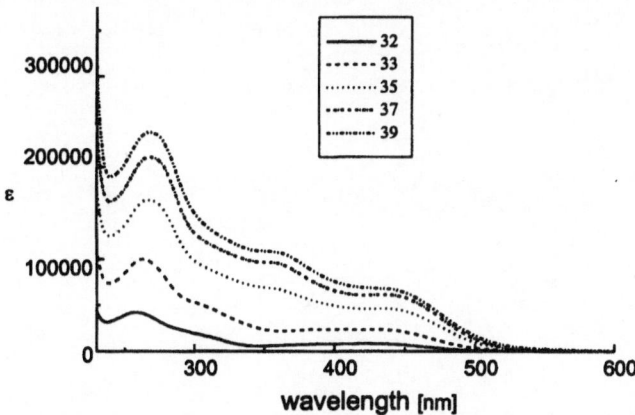

Fig. 2. UV/vis spectra of 32–39

concluded that electronic saturation of the oligomeric sequence is reached with the hexamer **36** or the heptamer **37**, a situation not uncommon for ethynyl-spaced, conjugated polymers [31 c].

It is conspicuous that the feature located at 300 nm in **32** shifts to 360 nm in **37**, indicating that it is probably attributable to a transition involving the large hydrocarbon ligand. The band at 450 nm, first observed in **32**, does not change position in the whole series, indicating that it may be a localized charge-transfer transition. This implies that the variable band at 300–360 nm in the sequence of oligomers **32–39** is analogous to λ_{max} in organic systems.

In an attempt to further exploit the Hay coupling, the corresponding elongated systems were synthesized by a double chlorovinylation of **23** to give **40**, which was subjected to dehydrohalogenation. Trapping of the formed bisanion with chlorotrimethylsilane furnished **41** as a stable and isolable compound. In order to prepare **42**, the TMS groups residing at the butadiyne termini of **41** were

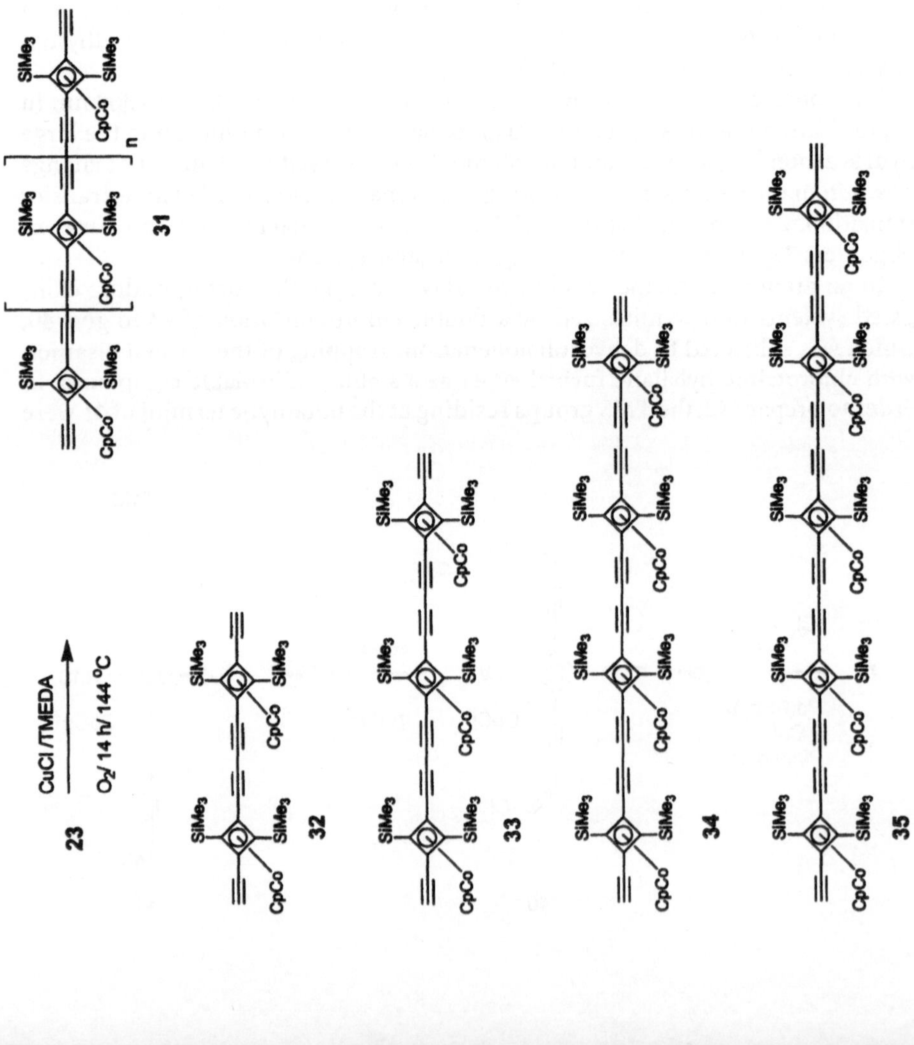
23
CuCl /TMEDA
O2/ 14 h/ 144 °C
31
32
33
34
35
SiMe3
CpCo
n

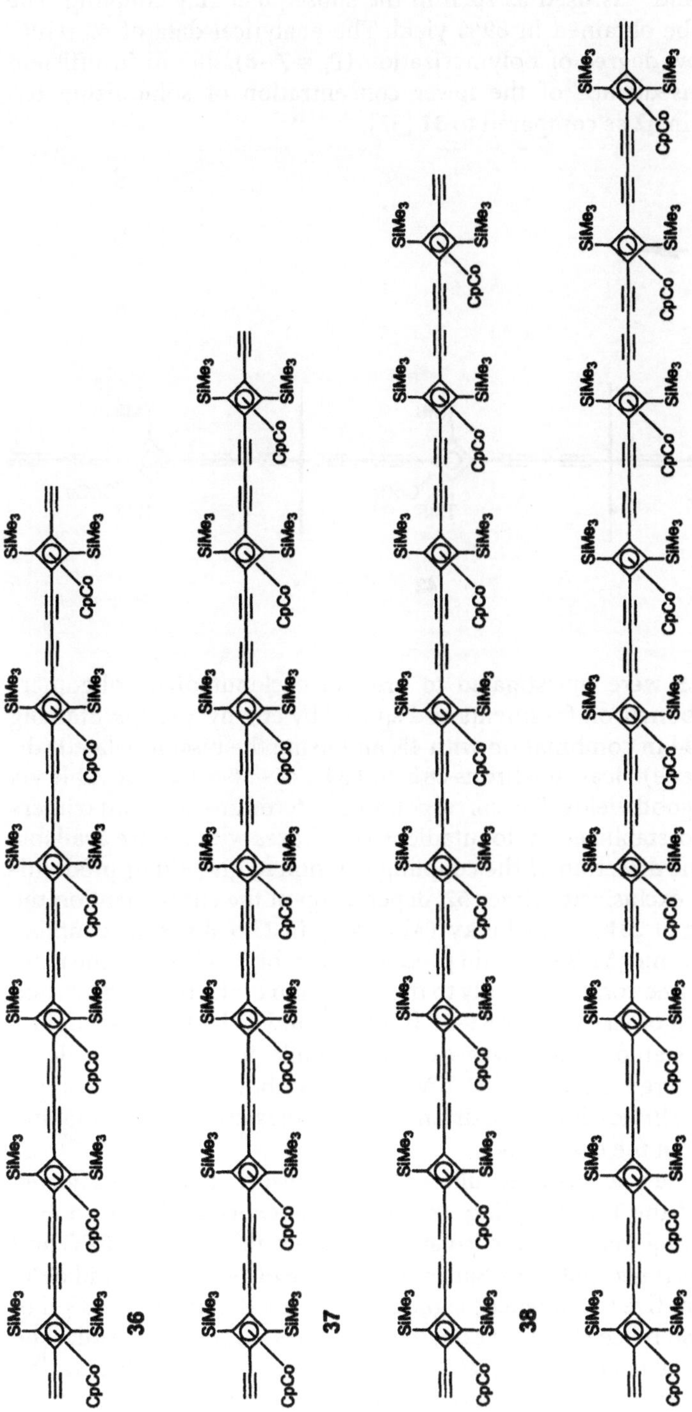

36 37 38 39

cleaved off by potassium carbonate in ethanol. The free bisdiyne was only stable in dilute solution and was used as such in the subsequent Hay coupling. The polymer **42** could be obtained in 89 % yield. The analytical data of **42** (GPC, NMR) showed a low degree of polymerization ($P_n \cong 7-8$), due to insufficient solubility as a consequence of the lower concentration of solubilizing trimethylsilyl groups in **42** as compared to **31** [32].

i. KOH/MeOH

41 $\longrightarrow$

ii. CuCl/TMEDA
   O$_2$/acetone

42

Derivatives of **25** were investigated to prepare cyclobutadiene oligomers stabilized by tricarbonyliron fragments and spaced by ethynyl groups. Starting from either **26, 43, 44** in combination with **45**, and using the bisstannylated (diethynylcyclobutadiene)tricarbonylirons **28b** and **29c, 46 – 50** were accessible via Farina-coupling in good yields. The corresponding heterodimers **51** and trimers **52**, combining CpCo-stabilized cyclobutadiene complexes with **26**, are available as well. Note, that the direction of the coupling reaction to give either predominantly dimer **51**, or exclusively trimer **52**, depends upon the substituent on the cyclobutadiene ring of **23 b, c**. The bulky TMS group in **23 b** slows the coupling reaction so that the dimer **51** is the main product, while for R = H (**23 c**) the reaction led under the same conditions only to the formation of the trimer **52**. When preparing kinked oligomers such as **53–55** from **29 c** and **44** by Farina-coupling the occurrence of stereoisomers was expected. While it was impossible to separate these diastereomers, the ^{13}C NMR spectra show that some of the resonances are split, indicating the existence of two diastereomers for **53** and three diastereomers in the case of **54**.

In order to access segments of the fullerenyne framework, **16 b** was subjected to the conditions of the Hay coupling. Instead of the expected cycles, only a brownish film-forming material was isolated; which examination by GPC and ^{13}C NMR revealed as *linear* polymer. Surprisingly, the hydrocarbon ligand of **56** shows only one set of five (broadened) signals, indicating that the separation of the organometallic stereocenters by a butadiyne group is sufficient to render the diastereomers unresolvable by ^{13}C NMR spectroscopy. A similar observation was made for **55**, where the isomers were likewise indistinguishable.

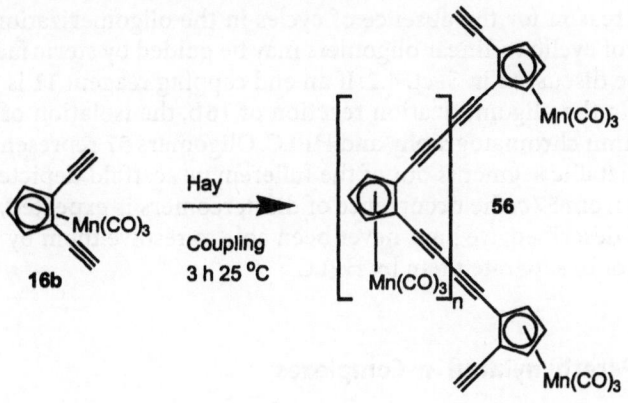

The Oligomer complex **56** is formed from **16b** via Hay Coupling (3 h 25 °C).

11 + 16b → (Hay Coupling)

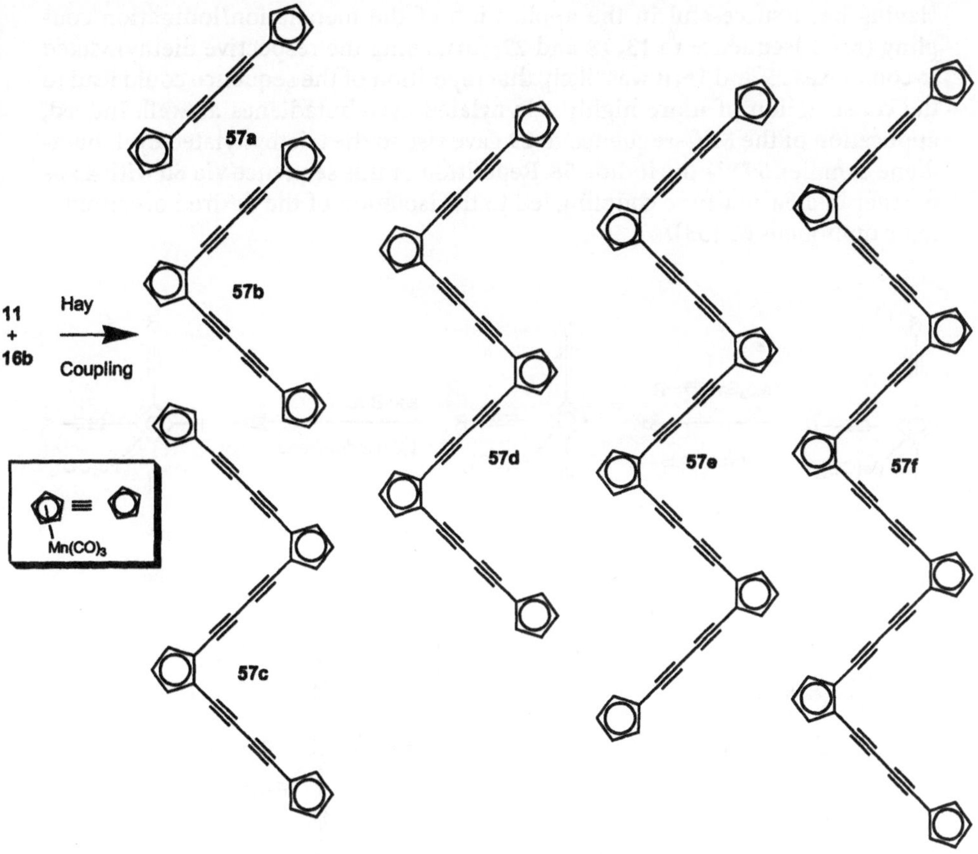

Structures **57a**, **57b**, **57c**, **57d**, **57e**, **57f**

Legend: ⬡ (Mn(CO)₃) = ⬠

What is the reason for the absence of cycles in the oligomerization of **16b**? The formation of cyclic vs. linear oligomers may be guided by steric factors, and the issue will be discussed in Sect. 4.2. If an end capping reagent **11** is provided as an additive in the oligomerization reaction of **16b**, the isolation of **57a–f** is feasible by column chromatography and HPLC. Oligomers **57** represent the first linear organometallic segments out of the fullerenyne scaffold depicted earlier. Again, starting from **57c**, the occurrence of diastereomers is expected, but as in the other cases described, we have never been able to resolve them by ^{13}C NMR spectroscopy nor to separate them by HPLC.

3
Star-Shaped Perethynylated π-Complexes

3.1
Stepwise Construction of Tetraethynylated Cyclobutadienes

Having been successful in the application of the metallation/iodination/coupling (MIC-)sequence to **13, 18** and **27**, furnishing the respective diethynylated π-complexes **29** and **16**, it was likely that repetition of the sequence could lead to the construction of more highly ethynylated cyclobutadienes as well. Indeed, application of the MIC-sequence to **29** gave rise to the triethynylated cyclobutadiene complex **59** via the iodide **58**. Repetition of this sequence via **60** with **45** as partner in a Farina-type coupling led to the isolation of the desired organometallic dumbbells **61** [33].

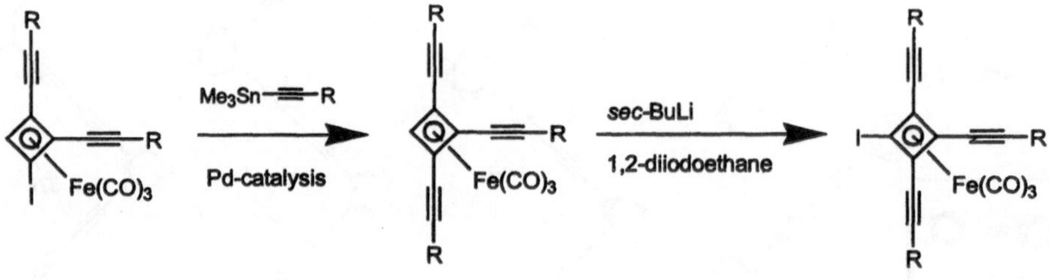

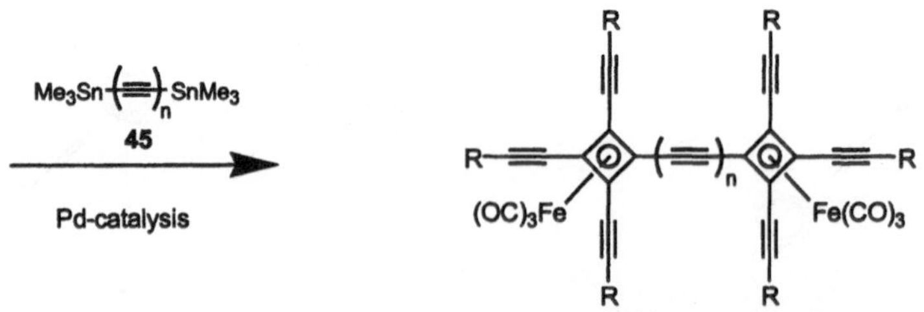

n = 1, 77%, mp 117 °C
n = 2, 66%, mp 172 °C

61

3.2
The One-Pot Procedure to Perethynylated π-Complexes

The synthetic path to **61** gives a good deal of control over the regiochemistry and is useful for the stepwise construction of oligomeric cyclobutadiene complexes, but is paid for by its multi-step character, reducing the yields of the ultimate product and increasing the synthetic effort by the necessary purification steps. The question arose whether the peralkynylation of cyclobutadiene complexes could not be achieved in one step, in the spirit of Vollhardt's hexaethynylation of benzene [2]. And indeed, reaction of **62** with stannylated alkynes [34] under standard Farina-coupling conditions, gave rise to the formation of the corresponding cyclobutadienes **63** in a higher than 80 % isolated yield [35].

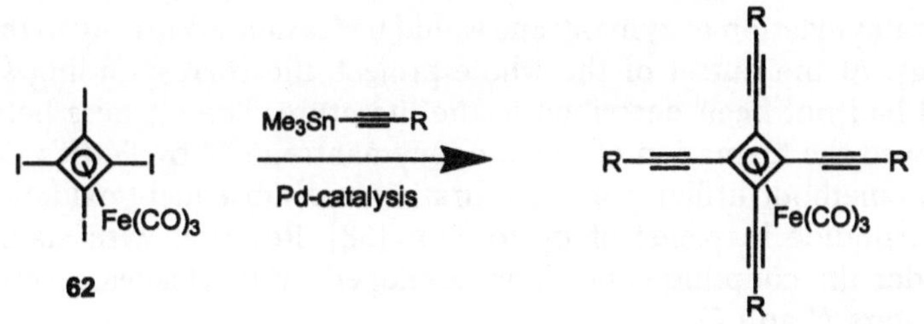

63 a	R = SiMe$_3$	83%
b	Ph	84%
c	Me	81%
d	Octyl	40%
e	(H)	(79%)

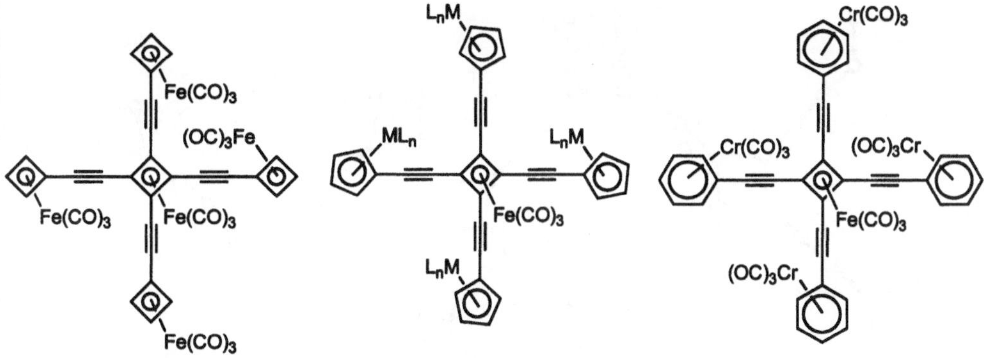

64

65 ML$_n$ = Mn(CO)$_3$
66 ML$_n$ = FeCp

67

But not only organic stannylacetylenes coupled to **62**; organometallic stannanes also gave rise to the formation of pentametallic complexes **64–67** in fair to good yields, and the cross-shaped complex **64** already is a sizable segment out of the proposed organometallic all-carbon network. It was possible to obtain X-ray analyses for **64** and **65**, both of which show unexpected, yet interesting, crystal packing behavior, insofar as the faces of the cymantrene or the cyclobutadiene-tricarbonyliron units are involved in parallel π-stacking. Thus some solid state face-to-face "dimer" or "polymer" formation in **64** and **66**, respectively, do occur [36]. The sandwich-type dimeric character of **64** is shown in Fig. 3.

Not only simple tin-substituted alkynes, but stannylated 1,3-diynes and 1,3,5-triynes react equally well with **62** under palladium catalysis to yield to corresponding expanded stars **68** and **69**, carrying elongated alkyne arms. In the case of the hexatriyne coupling, the yield of the coupling product drops to 16%, probably due to the fact that **69**, exhibiting a large and unprotected π-electron face, is not "inert" towards the reactive Pd catalyst and undergoes uncontrolled crosslinking. This leads to large amounts of intractable material, stuck on top of the chromatography column, when isolating **69**.

Given the close chemical similarity between **9** and **25**, the question arose whether peralkynylation of cymantrene would be feasible according to the same methodology. At the outset of the whole project, the corresponding starting material **70** had not been described in the literature. The likeness between **9** and **25** allowed the formation of pentaiodocymantrene **70** by Pettit's [34] and Winter's [37] method, utilizing permercuration and subsequent oxidation with potassium triiodide in yields of up to 40% [38]. Reaction with stannylated alkynes under the coupling conditions developed by Beletskaya accessed the five-armed stars **71** and **72**.

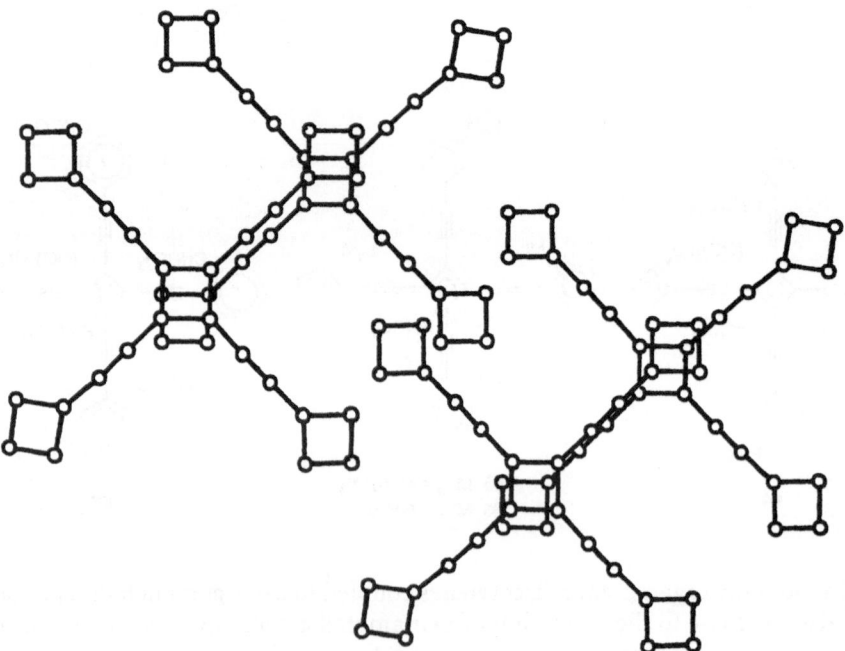

Fig. 3. Ball- and stick-representation of the sandwich structure of **64**

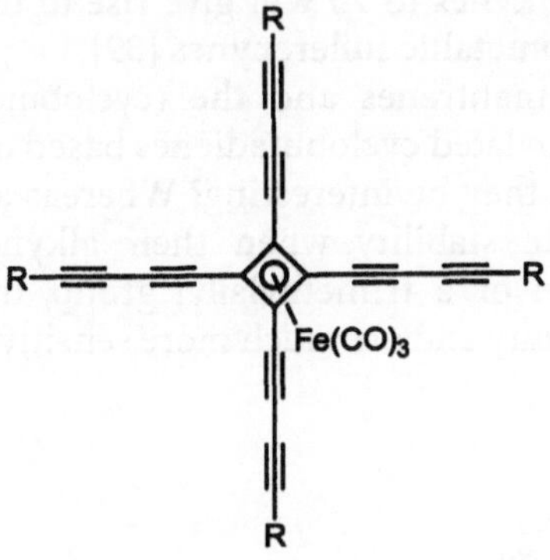
R
R
R
R
Fe(CO)3

68

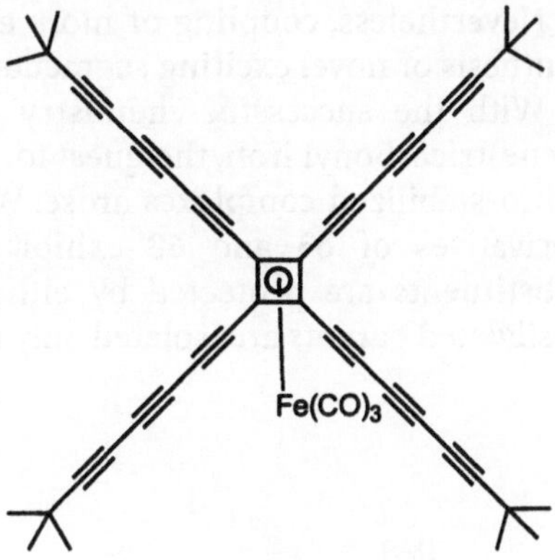
Fe(CO)3

69

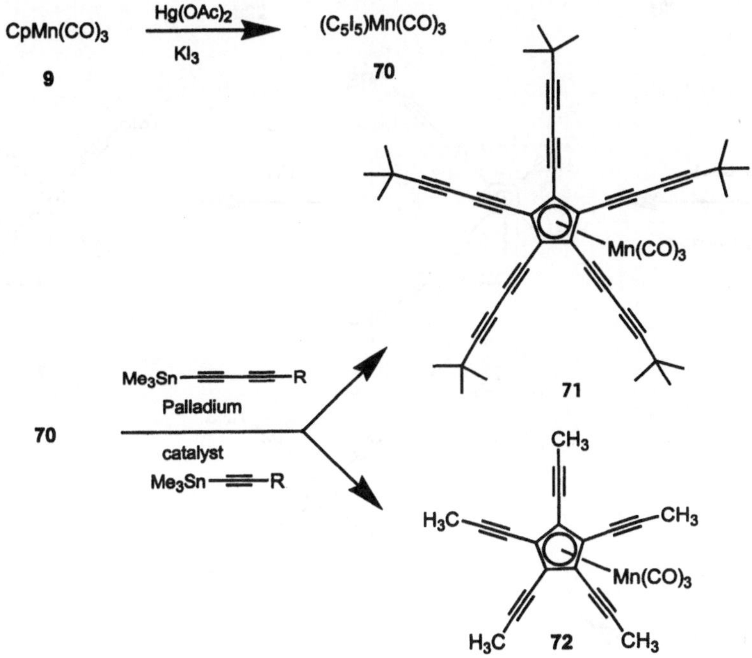
CpMn(CO)3
9
Hg(OAc)2
KI3
(C5I5)Mn(CO)3
70
Mn(CO)3
71
70
Me3Sn—≡—≡—R
Palladium
catalyst
Me3Sn—≡—R
CH3
H3C
CH3
Mn(CO)3
H3C
72
CH3

Whereas in the case of **72** the coupling yield is 38% (82% per coupling step), in the case of the butadiynes the yield of **71** drops to 11% (64% per step). The generally lower yield of the cymantrenes in these couplings has to do with the enhanced steric crowding in **70** as compared to **62**, a fivefold instead of a fourfold coupling, and the presence of a larger π-face, which is more prone to react with the active Pd-species in an unwanted fashion.

Nevertheless, coupling of more elaborate alkynes to **70** will give rise to the synthesis of novel exciting segments of organometallic fullerenynes [39].

With the successful chemistry of the cymantrenes and the (cyclobuta-diene)tricarbonyl iron, the quest for tetraethynylated cyclobutadienes based on CpCo-stabilized complexes arose. Why would they be interesting? Whereas all derivatives of **63** and **68** exhibit reasonable stability when their alkynyl substituents are protected by either an alkyl or a trimethylsilyl group, the desilylated parents are isolated only with difficulty and are much more sensitive.

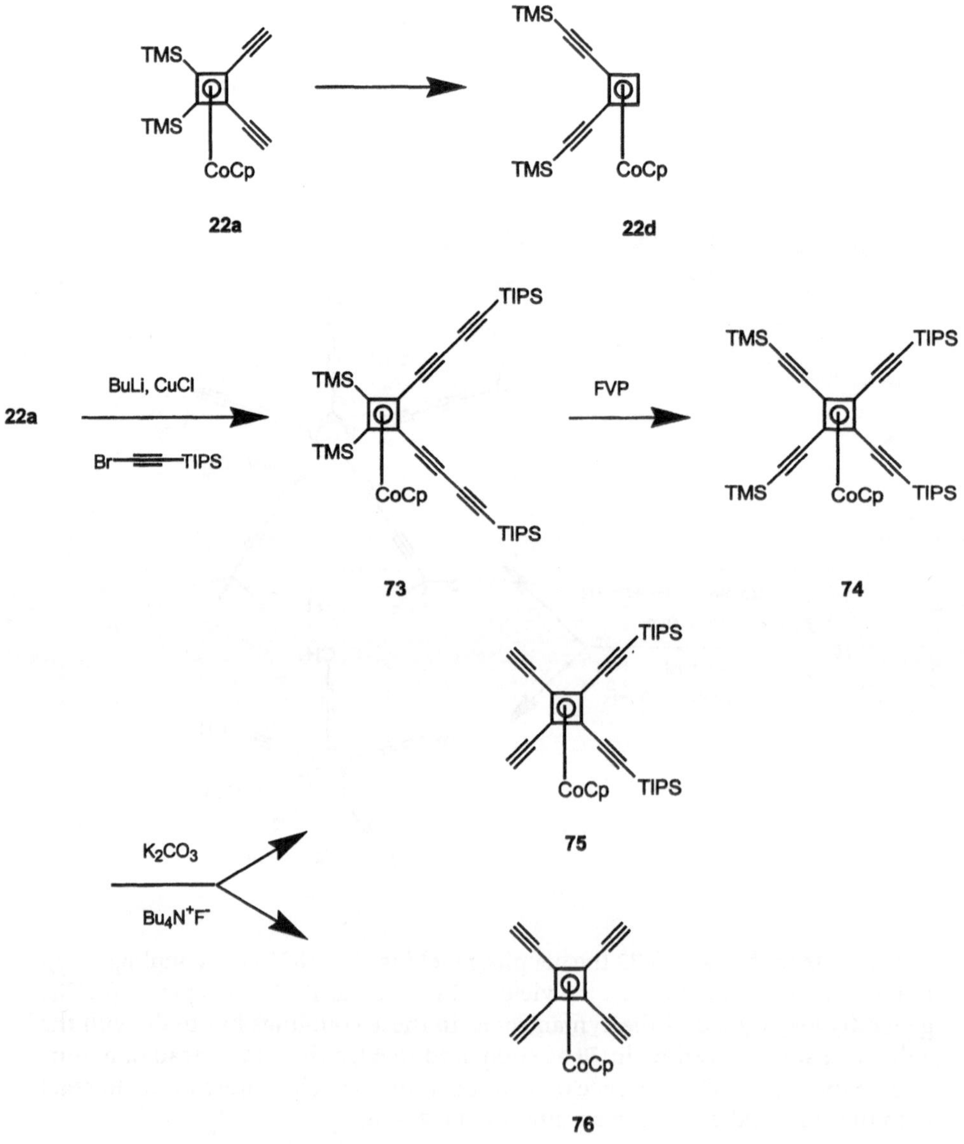

The CpCo complexes, on the other hand, should be more stable due to the presence of the robust and bulky Cp-shield. Unfortunately, however (tetraiodocyclobutadiene)CpCo is not available, and there is no obvious synthetic path to make it. But maybe another way to produce CpCo-stabilized tetraethynylated cyclobutadiene complexe exists: It is known, that 22a undergoes a rearrangement to 22d when subjected to the conditions of flash vacuum pyrolysis at elevated temperatures [24]. The driving force behind this rearrangement is twofold: first, the steric strain between the two adjacent TMS groups is removed in 22d and second, the TMS groups in 22d are not bound to an sp^2-hybridized center but to an sp-hybridized one, which is a more favorable situation from a thermodynamic point of view.

While this rearrangement (22a → 22d) is fascinating from a mechanistic point (it is not completely clear if it is a Bergman rearrangement [40] or an electrocyclic process), it seems, prima facie at least, not particularly valuable for synthetic purposes. It may be possible to extend this rearrangement by adding a suitable substituent to the alkyne groups. It was interesting to see if this substituent would behave analogously to the TMS group upon flash vacuum pyrolysis. To this end, the bisbutadiynylated cyclobutadiene complex 73 was prepared by a copper-catalyzed Cadiot-Chodkiewicz-coupling [16b], involving brominated TIPS-acetylene and 22a.

At 535 °C FVP proceeded as advertised to furnish 74 in 83% yield as the sole product. Despite the presence of the two butadiyne groups, 73 is stable enough to be distilled at 10^{-4} Torr by using a simple heat gun [41]. The product, 74 (rock stable itself) can be converted into the semi- 75 or completely deprotected (tetraethynylcyclobutadiene)cyclopentadienylcobalt 76 with potassium carbonate or tetrabutylammonium fluoride, respectively. With this surprising rearrangement, the regioselective synthesis of tetraethynylated CpCo-stabilized cyclobutadienes was achieved. The substituent pattern in the product is such, that two ortho-alkyne groups carry TMS and the other two TIPS groups. The semi-deprotected species 75 should then provide access to novel peralkynylated oligomers and cycles.

3.3
Stability Considerations

All of the ethynylated cyclobutadienes are completely stable and can be easily manipulated under ambient conditions, as long as the alkyne arms carry substituents other than H. For the deprotected alkynylated cyclobutadiene complexes, obtainable by treatment of the silylated precursors with potassium carbonate in methanol or tetrabutylammonium fluoride in THF, the stability is strongly dependent upon the number of alkyne substitutents on the cyclobutadiene core and the nature of the stabilizing fragment. In the tricarbonyliron series, 27b, 27c, 29b, and 28b are isolable at ambient temperature and can be purified by sublimation or distillation under reduced pressure. The corresponding tetraethynylated complex 63e, however, is not stable under ambient conditions as a pure substance but can be stored as a dilute solution in dichloromethane. It can be isolated at 0 °C and kept for short periods of time with only slow decomposition though. In comparison, Diederich's tetraethynylethylene [3] shows similar behavior. The tetrabutadiynylated cyclobutadiene **68 b**, the least stable in the series, is formed by *in situ* deprotection in deuterated methanol and has to be characterized in this medium immediately: adding potassium carbonate to the yellow solution of **68a** (R = TMS) instantaneously turns the reaction mixture dark.

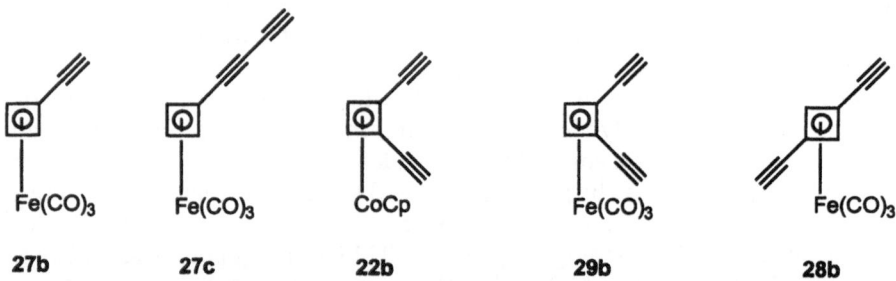

27b   **27c**   **22b**   **29b**   **28b**

Stable, isolable at ambient temperature, can be purified by sublimation under reduced pressure. The CpCo-ligated cyclobutadienes are more stable than the tricarbonyliron ones.

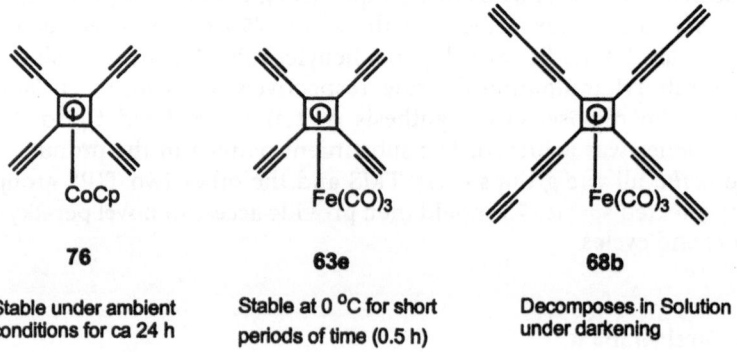

76   **63e**   **68b**

Stable under ambient   Stable at 0 °C for short   Decomposes in Solution
conditions for ca 24 h   periods of time (0.5 h)   under darkening

The CpCo-stabilized ethynylated cyclobutadienes are considerably more robust, and the parent **76** can be isolated as a yellow crystalline material, stable at ambient temperature for several hours. At 0 °C **76** decomposes in the course of several days, which is indicated by darkening of the formerly brillant-yellow needles. The stability of **76** made in X-ray analysis feasible and the bond angles/distances obtained are in good agreement with reported values for ethynylated cyclobutadiene complexes already described [35, 36].

With the enhanced stability as compared to that of **63 e, 76** should be an ideal starting material for the construction of larger organometallic molecular objects of dendrimeric, cyclic or polymeric nature.

4
Novel Organometallic Dehydroannulenes [42]

4.1
Syntheses

While the synthesis of cross-shaped and oligomeric derivatives of **9, 23,** and **25** is straightforward, utilizing standard synthetic techniques, it would be of considerable interest to prepare cyclic structures with similar ease. All of these would be members of the hitherto unknown family of organometallic fused dehydroannulenes, juxtaposing a complexed cyclobutadiene or cyclopentadienyl ring with the central dehydroannulene core (**A**). Besides their esthetic beauty and the fact that they would represent segments of the organometallic tetragonal net or the fullerenyne family (Fig. 1), these organometallic dehydroannulenes would be of considerable interest for the study of electron transfer phenomena in organometallic systems.

Such organometallic dehydroannulenes **A** are expected to be formed as mixtures of diastereomers differing only with regard to the relative position of the organometallic fragments (above or below the plane of the large hydrocarbon ligand) with respect to each other. In respective diastereomers, only the spatial

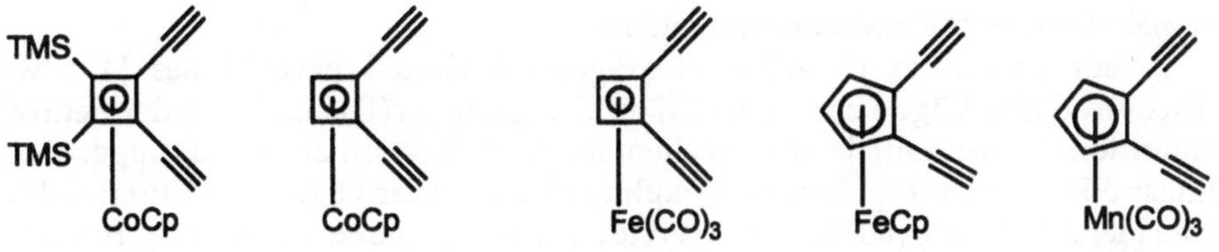

TMS ... CoCp	CoCp	Fe(CO)$_3$	FeCp	Mn(CO)$_3$
22a	**22b**	**29b**	**77**	**16b**
cycles	no cycles	no cycles	cycles	no cycles

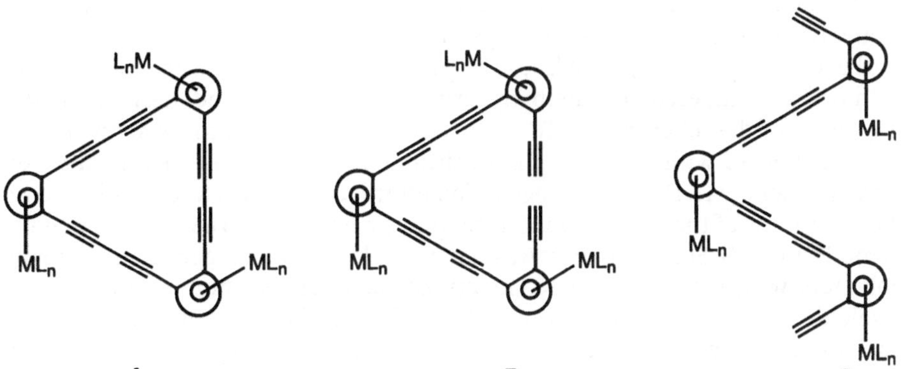

A   B   C

distribution, and as a consequence the distance of the organometallic fragments from each other, varies, whereas the *electronic* situation (as shown later by UV/vis spectroscopy) will be undisturbed. If partially oxidize, through-space vs. through-bond effects in intramolecular electron transfer phenomena can be studied in these systems.

The first synthetic approaches to organometallic dehydroannulenes were not particularly successful, because neither the submission of **22b** nor of **29b** led to the formation of the desired cyclooligomers. In the first case only an insoluble, probably polymeric, yellow material was isolated, whereas in the case of **29b** decomposition under rapid darkening was observed. Surprised by this behavior, it was decided to subject the much less reactive, sterically more encumbered, bistrimethylsilyl-substituted **22a** to the conditions of the Hay coupling.

Surprisingly, workup and subsequent chromatography gave rise to the isolation of *trans*- and *cis*- **78** in 10.1 and 4.4% yield, respectively. While separation of *trans*- and *cis*- **78** was achieved by column chromatography; for the next higher homologe, **79**, preparative HPLC had to be employed. Two of the four isomers of **79** were not completely separable, but they could be enriched, allowing a conclusive assignment of their structures. Further elution made the pentamer **80** accessible in small amounts as a mixture of diastereomers. Due to the low yield, however, preparative diastereomer separation was not attempted. As in the case of **79**, the occurrence of four diastereomers in a ratio of 1:5:5:5 would be expected; three of these were evidenced by analytical HPLC, but the peak of the minor isomer could not be detected and was presumably hidden under the signal of one of the main diastereomers.

In our endeavors to make all-oxygen-substituted pericyclynes [43], we discovered that Vögtle coupling [18], utilizing copper(II) acetate under neutral conditions in acetonitrile, was excellent for the formation of rings as opposed to linear oligomers and polymers. Coupling of **22a** under Vögtle-conditions led to considerably improved yields of cycles when compared to the Hay reaction (Fig. 4). Interestingly enough, ring size and diastereomer distribution in the product mixture were independent of the coupling method used, indicating that the formation/distribution of **78–80** must be determined mainly by statistics.

In favor of this hypothesis is the fact that in both coupling variants the observed diastereomer distribution is roughly in accord with a simple statistical model, excluding any direct interaction between the relatively bulky (by the trimethylsilyl groups and the Cp rings) monomer units.

Hay coupling of **16b** gave a linear polymer **56** (vide supra) while the sterically more encumbered 1,2-diethynylferrocene **77** forms the desired closed fragments of the fullerenynes. To synthesize **77**, **81** was used as a starting material and reacted with $Ph_3P=CHCl$, furnishing **82** in 71% yield as a mixture of stereoisomers [44]. Dehydrohalogenation afforded **77** which, when subjected to the conditions of the classic Eglinton coupling, gave access to **83** and **84**, but only in modest yields. Surprisingly, it was not possible to detect the formation of any polymers, which would have been identified, but copious amounts of black and insoluble material appeared, probably oxidation products of the ferrocene nucleus. Again, the observed ratio of the diastereomers **83/84** (3.5:1) suggested that the distribution obeys statistics and that there is no special steric interac-

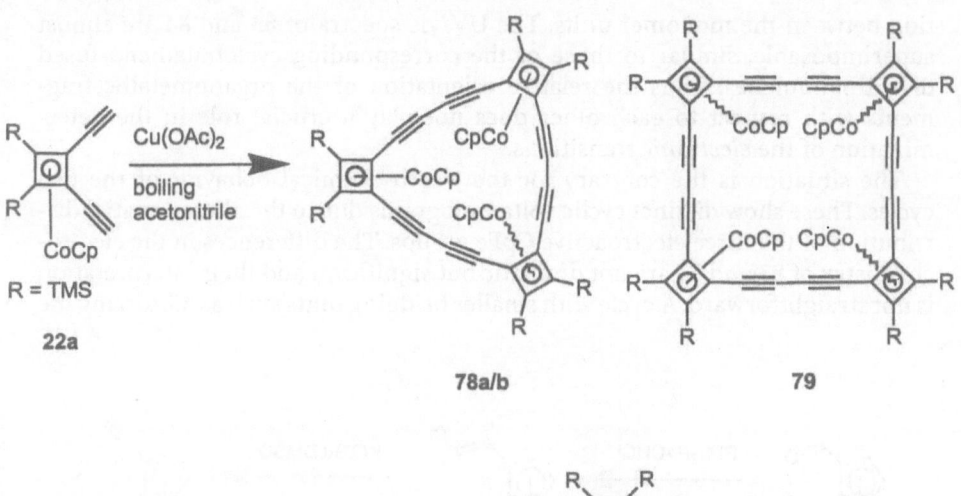

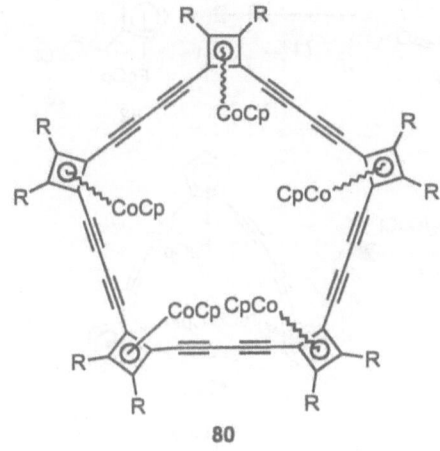

Statistic Ratio	2	1	1	2	4	1
Yield (Hay-conditions)	10.1%	4.4%	4.0%	combined 17.8%		4.7%
Yield (Voegtle-conditions)	16.6%	7.9%	6.6%	combined 28.6%		7.5%

*diastereomer separation not attempted

pentamers 13%*

Fig. 4. Yields and isomer distribution **78** and **79**

tion between the monomer units. The UV/vis spectra of **83** and **84** are almost superimposable, similar to those of the corresponding cyclobutadieno-fused dehydroannulenes. Thus the relative orientation of the organometallic fragments with respect to each other does not play a crucial role in the determination of the *electronic* transitions.

The situation is the contrary for the electrochemical behavior of the two cycles. These show distinct cyclic voltammograms due to the altered spatial distribution of the three electroactive CpFe groups. The differences in the electrochemistry of **83** and **84** are not dramatic but significant and their interpretation is not straightforward. A cycle with smaller bridging units such as **85**, forcing the

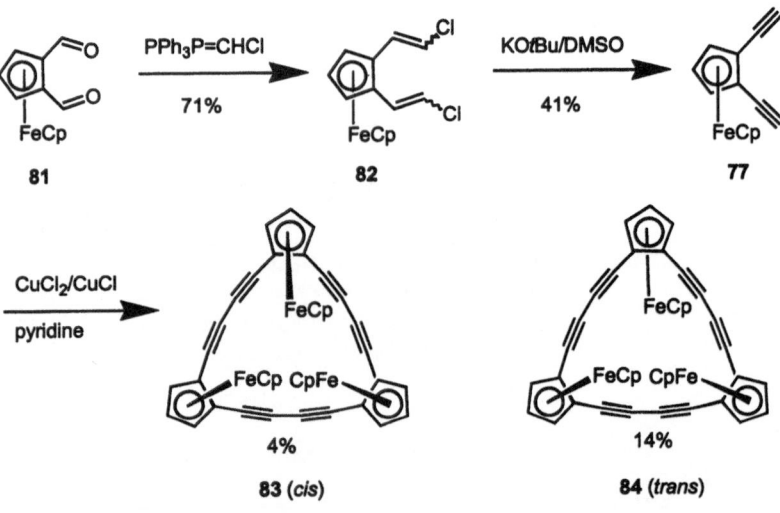

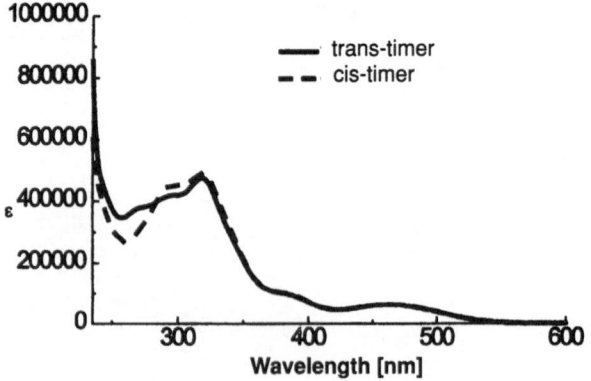

Fig. 5. UV/cis spectra of 83 and 84

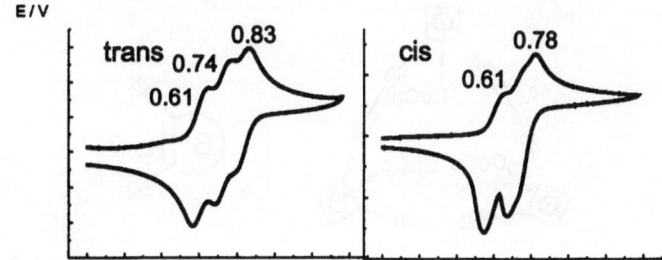

Fig. 6. Cyclic voltammograms of **83** and **84**

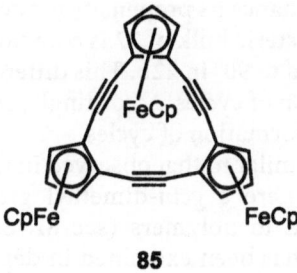

85

electroactive groups into close proximity, should enhance the effect and lead to an larger split, making, the interpretation of the cyclic voltammograms more straightforward.

4.2
Coupling Propensities of Diethynylated π-Complexes

What determines the propensity of cyclization of the different diethynylated complexes? Looking carefully at the monomers, there are three variables which could influence the formation of cycles as opposed to linear polymers. One is the angle α between the two alkyne arms. The second is the parameter ℓ, which indicates how far the two alkyne groups are apart at their origin, while the third is the parameter $\mathscr{B}$, the bulk of the monomer.

Whereas α and ℓ are easily determined, $\mathscr{B}$ is more of a qualitative parameter, and ℓ is constant for all monomers examined. From accumulated experience, the formation of cycles becomes more predominant either with increasing bulk of the organometallic monomer under consideration or with decreasing angle α, or a combination of both. This seems to be true for **22a, b** and **29b**. Only **22a**, the most bulky monomer in this series, but not **22b** or **29b**, give cycles. It could be argued that cycles formed from **22b** or **29b** are too unstable to be isolated. This cannot be true in the case of the cycles derived from **22b**, which were generated independently by desilylation of **78**. The resulting cycle **86** is stable in air up to

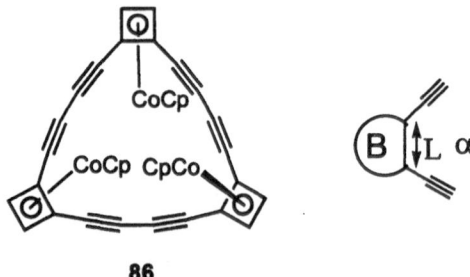

86

180 °C and does not show any sign of decomposition. So the increased steric bulk of 22 a does indeed enhance its propensity for cyclization. In the case of the ferrocene cycles **83/84**, the steric bulk of **77** is comparable to that of **22b**, except that in **77** α is 72° compared to 90° in **22b**. This difference seems to be sufficient for promoting the formation of cycles. Accordingly, **16b** is not sterically encumbered enough to force the formation of cycles.

This general picture is similar to that observed in the ring-closing propensity of open chain alkanes, where a *gem*-dimethyl group greatly improves the yield of cycles as opposed to polymers (see **87**, **88**). This is known as the Thorpe–Ingold effect and has been examined in depth [45]. The steric bulk of the substituents diminishes the conformational space available for the open pre-

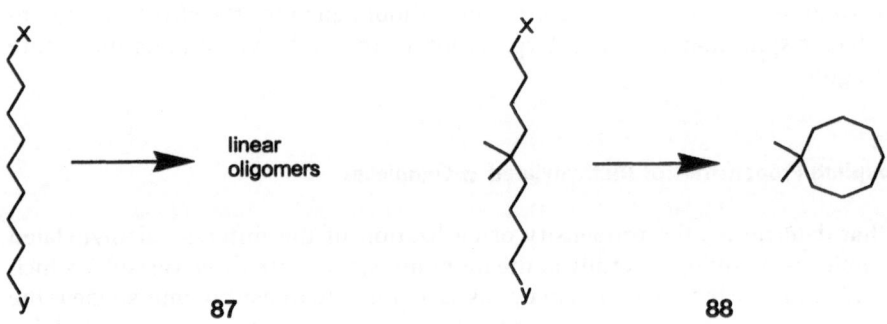

87   **88**

cursor, in such a way that conformations which lead to cycles are greatly preferred. Distortion of bond angles (which is the older interpretation of this effect) does not seem to play any role at all. A similar situation arises for the open oligomers of diethynylated π-complexes, which can either orient themselves to form precursors to cycles **B** or open polymers **C** (see p 153). Despite the fact that the single monomeric units are very rigid, an open oligomer such as **B** or **C** still would be very flexible. This is similar to a polymethylene chain: although every methylene group itself is very rigid, the chain is conformationally unrestricted and floppy.

The increase in steric demand in the open oligomeric precursors of the di-ethynylated π-complexes facilitates the formation of **B**, thus explaining the observed behavior.

5
Conclusions

With the body of work outlined here, it has been shown that multiply ethynyl-ated π-complexes can be accessed quite readily by utilizing Stille- and Hay-type couplings, conventional functional group transformations or – in a more exotic fashion – flash vacuum pyrolysis. Most of the tricarbonyl iron- and manganese-ligated species described here are moderately sensitive and can be handled under ambient conditions for several hours, whereas the CpCo stabilized systems are brutally robust, so that "fire and sword" chemistry may be used for their preparation and transformation. This robustness makes these compounds valuable as building blocks in material science and for the construction of novel mesoscopic organometallic structures and objects. Future developments, utiliz-ing the potential of **22**, **23**, and **74–76** will lead to the synthesis of larger and more elaborate segments from the organometallic networks shown in Fig. 1.

Acknowledgments. The author expresses his gratitude to the coworkers and collaborators who made this article possible (Dr. Jutta E. C. Wiegelmann, Dr. Markus W. Altmann, Dr. Marc Brake, Sepas Setayesh, and Gaby Roidl) by their skill and dedication. He is grateful to Prof. Dr. Klaus Müllen, Priv.-Doz. Dr. Volker Enkelmann, and Prof. Dr. Dieter Neher for encouraging dis-cussion and support. He is indebted to Gaby Roidl for her help with typing the manuscript. Financial support of our work by the Stiftung Volkswagenwerk, the Deutsche Forschungs-gemeinschaft (DFG), and the Fonds der Chemischen Industrie is gratefully acknowledged.

6
References

1. Hirsch A (1994) The Chemistry of the Fullerenes; Thieme Organic Chemistry Mono-graph Series, Stuttgart. Hammond GS, Kuck VJ (eds) (1992) Fullerenes; ACS Symposium Series 481. Acc Chem Res (1992) March issue. Krätschmer W, Lamb LD, Fostiropoulos K, Huffman DR (1990) Nature 347:354
2. Diercks R, Armstrong JC, Boese R, Vollhardt KPC (1986) Angew Chem 98:270; Angew Chem Int Ed Engl 25: 268. Boese R, Green JR, Mittendorf J, Mohler DL, Vollhardt KPC (1992) Angew Chem 104: 1643; Angew Chem Int Ed Engl 31:1643
3. Hopf H, Kreutzer M, Jones PG (1991) Chem Ber 124:1471. Hauptmann H (1975) Tetra-hedron Lett 1931. Vollhardt KPC, Winn LS (1985) Tetrahedron Lett 709. Review.: Tykwinski RR, Diederich F (1997) Liebigs Ann: 649
4. Review: Rubin Y (1997) Chem Eur J 3:1009. Arce MJ, Viado AL, An Y-Z, Khan SI, Rubin Y (1996) J Am Chem Soc 118: 3775. Arce MJ, Viado AL, Khan SI, Rubin Y (1996) Organo-metallics 15:4340. Jux N, Holczer K, Rubin Y (1996) Angew Chem 108:2031; Angew Chem Int Ed Engl 35:1986. Anthony JE, Khan SI, Rubin Y (1997) Tetrahedron Lett 38:3499
5. Scott LT, Cooney MJ (1995) Pericyclynes. In Diederich F, Stang PJ (eds) Modern Acetyl-ene Chemistry, VCH, New York. Kozhusov SI, Haumann T, Boese R, de Meijere A (1993) Angew Chem 105:426. Angew Chem Int Ed Engl 32:401. de Meijere A, Kozhushov SI, Puls C, Haumann T, Boese R, Cooney MJ, Scott LT (1994) Angew Chem 106:934. Angew

Chem Int Ed Engl 33:869. de Meijere A, Kozhusov S, Haumann T, Boese R, Puls C, Cooney MJ, Scott LT (1995) Chemistry 1:124. Review: Scott LT, Cooney MJ (1995) Pericyclynes. In Diederich F, Stang PJ (eds) Modern Acetylene Chemistry. Verlag Chemie, New York

6. Review: Diederich F, Rubin Y (1992) Angew Chem 104:1123. Angew Chem Int Ed Engl 31: 1101. Diederich F (1995) Oligoacetylenes. In Diederich F, Stang PJ (eds) Modern Acetylene Chemistry, Verlag Chemie, New York

7. Rubin Y, Parker TC, Khan SI, Holliman CL, McElvany SW (1996) J Am Chem Soc 118:4308. Schwarz H (1993) Angew Chem 105: 1475; Angew Chem Int Ed Engl 32:1412

8. Scott LT, Bratcher MS, Hagen S (1996) J Am Chem Soc 118:8743. Scott LT (1996) Pure Appl Chem 68:291. Forkey DM, Attar S, Noll BC, Koerner R, Olmstead MM, Balch AL (1997) J Am Chem Soc 119:5766. Rabideau PW, Sygula A (1996) Acc Chem Res 29:235. Faust R (1995) Angew Chem 107:1559; Angew Chem Int Ed Engl 34:1429

9. Morgenroth F, Reuther E, Müllen K (1997) Angew Chem 109:647; Angew Chem Int Ed Engl 36:629. Müller M, Petersen J, Strohmeier R, Günther C, Karl N, Müllen K (1996) Angew Chem 108:947; Angew Chem Int Ed Engl 35:886. Zander M (1995) Polycyclische Aromaten-Kohlenwasserstoffe und Fullerene. Teubner Stuttgart

10. Baughman RH, Eckhardt H, Kertész M (1987) J Chem Phys 87:6687

11a. Haley MM, Brand SC, Pak JJ (1997) Angew Chem 109:864; Angew Chem Int Ed Engl 36:836

11b. Bradshaw JD, Guo L, Tessier CA, Youngs WJ (1996) Organometallics 15:2582. Zhang D, McConville DB, Tessier CA, Youngs WJ (1997) Organometallics 16:824

11c. Brady M, Weng W, Gladysz JA (1994) J Chem Soc Chem Commun 2655. Bartik T, Bartik B, Brady M, Dembinski R, Gladysz JA (1996) Angew Chem 108:467; Angew Chem Int Ed Engl 35:414. Brady M, Weng W, Zhou Y, Seyler JW, Amoroso AJ, Arif AM, Böhme M, Frenking G, Gladysz JA (1997) J Am Chem Soc 119:775

11d. Mongin O, Gossauer A (1997) Tetrahedron 53:6835. Mongin O, Gossauer A (1996) Tetrahedron Lett 37:3825

11e. Müller TJJ, Lindner HJ (1996) Chem Ber 129:607

12. Tobe Y, Kubota K, Naemura K (1997) J Org Chem 62:3430. Tovar JD, Jux N, Jarrosson T, Khan SI, Rubin Y (1997) J Org Chem 62:3432. Sondheimer F (1972) Acc Chem Res 5:81

13. Review: Diederich F (1994) Nature 369:199. Bunz UHF (1994) Angew Chem 106:1127; Angew Chem Int Ed Engl 33:1073. Gleiter R, Kratz D (1993) Angew Chem 105:884; Angew Chem Int Ed Engl 32:842

14. Burdett JK, Mortara AK (1997) Chem Mater 9:812

15. Molecular Electronics: Tolbert LM, Zhao XD (1997) J Am Chem Soc 119:3253. Tolbert LM, Zhao XD, Ding YZ, Bottomley LA (1995) J Am Chem Soc 117:12891. Bumm CA, Arnold JJ, Cygan MT, Dunbar TD, Burgin TP, Jones L, Allara DL, Tour JM, Weiss PS (1996) Science 275:1705. Pearson DL, Jones L, Schumm JS, Tour JM (1997) Synth Met 84:303. Tour JM (1996) Chem Rev 96:537. Taus SJ, Miedema R, Geerlings LJ, Dekker C, Wu J, Wegner G (1997) Synth Met 84:733. Lincoln P, Tuite E, Norden BJ (1997) Am Chem Soc 119:1454. Harriman A, Ziessel RJ (1996) Chem Soc Chem Communs 1707. Mujica V, Kemp M, Roitberg A, Ratner MJ (1996) Chem Phys 104:7296. Coat F, Lapinte C (1996) Organometallics 15:477. Zhou Q, Swager TM (1995) J Am Chem Soc 117:12593. Turro NJ (1995) Pure Appl Chem 67:199. Wegner G (1981) Angew Chem 93:352; Angew Chem Int Ed Engl 20:361

16a. Glaser C (1869) Chem Ber 2:422. Hay AS (1960) J Org Chem 27:3320. Kevelam HJ, Jong KL, Meinders HC, Challa G (1975) Makromol Chem 176:1369

16b. Eastmond R, Walton DRM (1972) Tetrahedron 28:4591. Eastmond R; Johnson TR, Walton DRM (1972) ibido 4601. Johnson TR, Walton DRM (1972) ibido 5221. Bohlmann F, Herbst P, Gleining H (1961) Chem Ber 94:948. Cadiot P, Chodciewicz W (1969) Coupling of Acetylenes. In Viehe H-G (ed) Chemistry of Acetylenes. Dekker, New York

17. Eglinton G, McRae W (1963) Adv Org Chem 4:225

18. Vögtle F, Berscheid R (1992) Synthesis 58

19. Cassar I (1975) J Organomet Chem 93:253. Dieck HA, Heck RF (1975) ibid 93:259.
     Sonoshigara K, Tohda Y, Hagihara N (1975) Tetrahedron Lett 4467. Alami M, Ferri F,
     Linstrumelle G (1993) Tetrahedron Lett 6403
20 a. Stille JK (1986) Angew Chem 98:504; Angew Chem Int Ed Engl 25:508
20 b. Farina V, Krishnan B (1991) J Am Chem Soc 113:9585
20 c. Beletskaya IP (1983) J Organomet Chem 250:551
21. Bunel EE, Valle L, Jones NL, Caroll PJ, Gonzalez M, Munoz N, Manriquez JM (1988)
     Organometallics 7:789
22. LoSterzo C, Stille JK (1990) Organometallics 9:687. LoSterzo C, Miller MM, Stille JK
     (1989) Organometallics 8:2331
23 a. Bunz UHF, Enkelmann V, Beer F (1995) Organometallics 14:2490
23 b. Crocco GL, Gladysz JA (1988) Chem Ber 121:375
23 c. Loim NM, Kondratenko MA, Solokov VI (1994) J Org Chem 59:7485
23 d. Setayesh S, Bunz UHF (1996) Organometallics 15:5470
24. Fritch JR, Vollhardt KPC (1979) J Am Chem Soc 100:1239. Fritch JR, Vollhardt KPC
     (1982) Organometallics 1:590
25. Bunz U (1993) Organometallics 12:3594 and cited references
26. Appel R (1991) Dissertation Technische Hochschule Darmstadt
27. Wiegelmann JEC, Bunz UHF (1993) Organometallics 12:3792
28. Wiegelmann JEC, Bunz UHF, Schiel P (1994) Organometallics 13:4649
29. Wiegelmann JEC, Bunz UHF, Enkelmann V (1995) Chem Ber 128:1055
30. Altmann M, Bunz UHF (1994) Macromol Rapid Commun 15:785
31 a. Altmann M, Enkelmann V, Beer F, Bunz UHF (1996) Organometallics 15:394
31 b. For the determination of conjugation lengths see Grimme J, Kreyenschmidt M, Uckert F,
     Müllen K, Scherf U (1995) Adv Mater 7:292
31 c. Schumm JS; Pearson DL, Tour JM (1994) Angew Chem 106:1445; Angew Chem Int Ed
     Engl 33:1360
32. Altmann M, Enkelmann V, Beer F, Bunz UHF (1996) Chem Ber 129:269
33. Bunz UHF, Wiegelmann-Kreiter JEC (1996) Chem Ber 129:785. Wiegelmann-Kreiter
     JEC, Bunz UHF (1995) Organometallics 14:4449
34. Amiet G, Nicholas K, Pettit R (1970) J Chem Soc, Chem Commun 161
35. Bunz UHF, Enkelmann V (1993) Angew Chem 105:1712; Angew Chem Int Ed Engl
     32:1653
36. Bunz UHF, Enkelmann V (1994) Organometallics 13:3823
37. Winter CH, Han Y, Heeg MJ (1992) Organometallics 11:3169. Winter CH, Han Y, Ostran-
     der RL, Rheingold AL (1993) Angew Chem 105:1247; Angew Chem Int Ed Engl 32:1161
38. Bunz UHF, Enkelmann V, Räder J (1993) Organometallics 12:4745
39. Baughman RH, Galvao DS, Cui C, Wang Y, Tomanek D (1993) Chem Phys Lett 204:8
40. Bergman RG (1973) Acc Chem Res 6:25
41. Altmann M, Roidl G, Enkelmann V, Bunz UHF (1997) Angew Chem 109:1133; Angew
     Chem Int Ed Engl 36:1107
42. Altmann M, Friedrich J, Beer F, Reuter R, Enkelmann V, Bunz UHF (1997) J Am Chem Soc
     119:1472
43. Brake M, Enkelmann V, Bunz UHF (1996) J Org Chem 61:1190
44. Bunz UHF (1995) J Organometal Chem 494:C8
45. Beesley RM, Ingold CK, Thorpe JF (1915) J Chem Soc 107:1080. Schleyer PvR (1961)
     J Am Chem Soc 83:1368. Milstein S, Cohan LA (1972) J Am Chem Soc 94:9166, 9175.
     Lightstone FC, Bruice TC (1994) J Am Chem Soc 116:10 789

Oligo- and Polyarylenes, Oligo- and Polyarylenevinylenes

Ullrich Scherf

Max-Planck-Institut für Polymerforschung Mainz, Ackermannweg 10, D-55128 Mainz, Germany. *E-mail: scherf@mpip-mainz.mpg.de*

An important challenge in the design of novel conjugated polymers is the synthesis of materials with tailor-made solid-state electronic properties. This section outlines the synthesis of the most significant classes of poly(*para*-phenylenevinylene)s (PPVs), poly(*para*-phenylene)s (PPPs), and related structures. Furthermore, this review demonstrates that the chromophoric and electronic properties of conjugated π-systems are sensitive to their molecular and supramolecular architecture.

Insoluble poly(*para*-phenylene) has already been prepared by several synthetic procedures. The introduction of solubilizing substituents leads to soluble polymers. Unfortunately, the resulting mutual distortion of adjacent subunits minimizes the electronic interaction of the π-systems. It is possible to overcome this shortcoming by introducing solubilizing groups that cause minimal mutual distortion of the building blocks. A strategy for accomplishing this is the simultaneous bridging *and* solubilization of the PPP-backbone. PPP-type ladder polymers (LPPP) exhibit a highly efficient photo- and electroluminescence, and can be used in organic-materials-based light-emitting diodes (LEDs) and also as emitters in organic solid-state lasers. Hyperbranched polyphenylenes and polyphenylene dendrimers represent a novel class of soluble phenylene derivatives. The highly branched structures of these materials guarantee their complete solubility.

Poly(*para*-phenylenevinylene)s (PPVs) represent one of the most intensively investigated classes of π-conjugated materials. Many synthetic procedures to generate unsubstituted and substituted PPVs have been developed. They include *1,6-polymerizations* of 1,4-xylylene intermediates as well as several *polycondensation* methods. Parallel to the polymer syntheses, several series of PPV oligomers (OPVs) have been synthesized and characterized. Such model oligomers of different molecular size allow for a study of the dependence of electronic and optical properties on the length of the conjugated π-system.

Keywords: Poly(*para*-phenylene)s, Poly(*para*-phenylenevinylene)s, Oligomers, Photoluminescence, Electroluminescence, Light emitting diodes.

1
Introduction

Conjugated oligomers and polymers can now claim a considerable and uninterrupted degree of attraction over a period of several decades [1]. In the initial years, research concentrated on the synthesis of the first representatives of the new substance-class of conjugated materials (polyacetylene, poly(*para*-phenylene), poly(*para*-phenylenevinylene), polythiophene, polypyrrole, polyaniline). The resulting oligomers and polymers were characterized in most cases by their insolubility and infusibility, properties that were a considerable hindrance to their structural characterization and processing. The majority of such compounds possessed no fully defined structure, and their physical properties were influenced by structural defects. Moreover, it was often difficult to distinguish between neutral molecules and "doped" species, produced as a consequence of oxidation or reduction.

The last few years have now brought about a new qualitative development, as a consequence of considerable advances in the available synthetic methods. In the 1970s and 1980s purely synthetic aspects were in the forefront, whereas in the following years the effective physical function of conjugated polymers has progressively become the main topic of research. However, in order to be able to draw a significant, definite correlation between the unique π-conjugated structure and a specific physical property (e.g., electrical or photoconductivity, electroluminescence, photovoltaic effect), crucial new demands must be made of the materials being investigated. These include the following:

(1) First, the polymers should be as free of defects as possible, in order to exclude the possibility that their physical function is influenced by structural

defects. Reproducibility in the synthesis of the polymers with regard to their properties is included by definition.

(2) Second, the materials used must be processable, so that they can be worked into the desired form (e.g., thin films or layers, fibers). Processability can be achieved, for example, by rendering the polymers soluble. An established strategy involves the introduction of solubilizing groups (alkyl-, alkoxy- or aryl substituents) [2]. Another important procedure is processing of the soluble precursors of essentially insoluble polymers (precursor route) [3,4]. The precursors are brought into the necessary form for processing and then converted in the solid state (most often thermally) to the corresponding conjugated polymers. Other strategies involve the use of solubilizing coun- terions for doped species (polyaniline) [5] or processing via the interme- diate formation of soluble charge-transfer complexes [6].

The requirements outlined above represent a considerable challenge for poly- mer synthesis. This article describes an appealing development, based on two central substance-classes of conjugated polymers, poly(*para*-phenylene)s and poly(*para*-phenylenevinylene)s.

2
Oligo- and Polyarylenes

2.1
Oligo- and Poly(*para*-phenylene)s

Oxidative Condensation of Aromatic Hydrocarbons

Poly(*para*-phenylene)s PPPs and other polyarylenes represent structure-classes of conjugated polyhydrocarbons which are currently under intensive investiga- tion [7]. This, in turn, is the result of important advances that have been made in the chemistry of aromatic compounds in recent years.

The parent, unsubstituted poly(*para*-phenylene) (**PPP 1**), is an insoluble and intractable material, available by a variety of synthetic methods [8,9]. The lack of solubility and fusibility prevent both the unequivocal characterization and the processing of **PPP 1**. Moreover, the intractability of unsubstituted **PPP** mate- rials has thwarted any serious commercial development of the polymer.

The first attempts to generate poly(*para*-phenylene) (1) were undertaken in the 1960s. Kovacic et al. [8] reported that the oxidative treatment of benzene with copper(II) chloride in the presence of strong Lewis acids (aluminium trich- loride) led to a condensation of the aromatic rings. During the reaction, radical cations are formed as reactive intermediates, which subsequently attack neutral benzene molecules. The reaction proceeds to an equilibrium, in which the oligophenyl cations are no longer able to attack uncharged benzene rings. The maximum degree of condensation is ca. 10–12. The benzene subunits are preferentially connected in the 1,4-position; however, cross-linking and oxidative condensation to more highly condensed, aromatic hydrocarbon building blocks occur as side reactions. Adapting the initial procedures of Kovacic et al., several other substituted benzene derivatives and other aromatic hydrocarbon monomers were coupled to oligo- and polyarylenes.

Katsuya et al. [10] reported the oxidative coupling (copper(II) chloride, aluminum chloride) of 2,5-dimethoxybenzene to poly(2,5-dimethoxy-1,4-phenylene) (2). The polymer obtained is only soluble in concentrated sulfuric acid, and is fusible at 320 °C. Ueda et al. [11] described the coupling of the same monomer with iron(III) chloride/aluminum chloride. However, the polymers obtained were not fully *para*-linked.

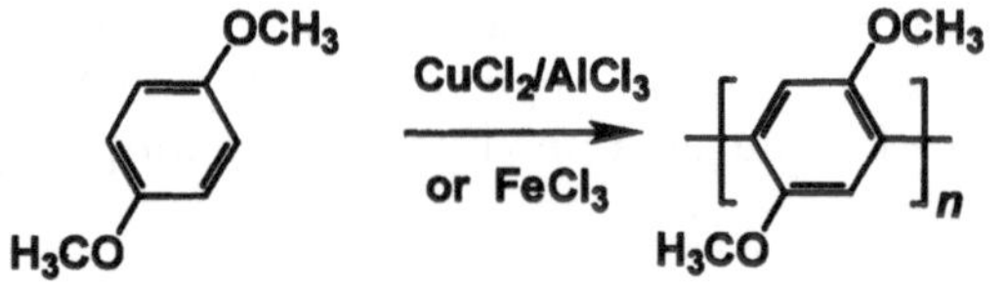

Yoshino et al. [12] prepared 9,9-disubstituted poly(fluorene)s (3), in which the solubilizing substituents are introduced in the form of a di-*n*-hexylmethylene bridge, that spans the neighboring rings in pairs and enforces a planar arrangement. The soluble and fusible poly(9,9-di-*n*-hexylfluorene-2,7-diyl)s (3) are obtained by oxidative coupling of 9,9-di-*n*-hexylfluorene with iron(III) chloride and are characterized by a value of M_n up to a maximum of 5,000. The absorption maximum, λ_{max}, of 3 is centered at about 388 nm. As a result of the partial flattening of the PPP backbone to a "stepladder" polymer, the long-wavelength absorption maximum is shifted bathochromically by about 50 nm relative to that of the parent PPP structure, 1.

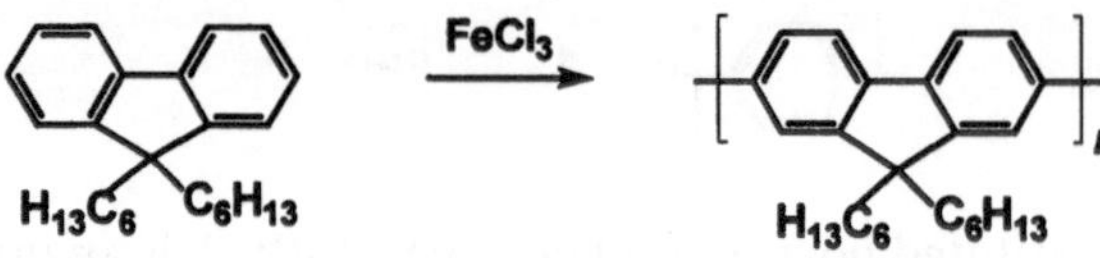

An unsatisfactory aspect of this synthesis is the quite low degree of polymerization: a maximum of 20 aromatic rings. Moreover, in addition to the predominant 2,7-coupling of the building blocks, other types of coupling can occur leading to structural defects.

Transition Metal-Mediated Coupling Reactions

Numerous entries to the preparation of structurally well-defined PPPs have evolved based on a variety of synthetic principles. The availability of newer, more effective methods for aryl-aryl coupling has been an important driving force for the development of new synthetic strategies for poly(*para*-phenylene)s and other polyarylenes. In particular, Pd(0)-catalyzed aryl-aryl coupling developed by Suzuki [13] (arylboronic acid plus aryl halide or tosylate) and nickel(0)-catalyzed or -mediated coupling according to Yamamoto [14] (aryl halide or tosylate plus aryl halide or tosylate) have been employed most successfully.

Kaeriyama et al. [15] reported the Ni(0)-catalyzed coupling of 1,4-dibromo-2-methoxycarbonylbenzene to poly(2-methoxycarbonyl-1,4-phenylene) (4) as a processable **PPP** precursor. The aromatic polyester **PPP** precursor, 4, is then saponified to carboxylated **PPP** (5) and thermally decarboxylated to 1 with CuO catalysts. However, the reaction conditions of the final step are quite drastic and cannot be carried out satisfactorily in the solid state (film).

The strategy of Kaeriyama represents a so-called precursor route, and was developed in order to overcome the shortcomings (insolubility, lack of processability) of previous **PPP** syntheses. Accordingly, the condensation is performed with solubilized monomers, and a soluble polymeric intermediate is formed. This intermediate is then converted to **PPP** (or another polyarylene) in a final reaction step, preferentially carried out in the solid state, allowing the formation of homogeneous **PPP** films or layers.

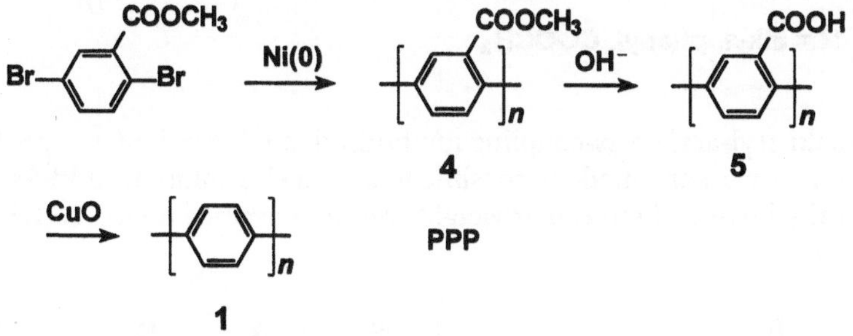

A second, very fruitful synthetic principle for structurally homogeneous, processable **PPP** derivatives involves the preparation of soluble **PPP**s by the introduction of solubilizing side groups. The pioneering work here was carried out at the end of the 80s by Schlüter, Wegner et al. [16, 17] who prepared soluble poly(2,5-dialkyl-1,4-phenylene)s (6) for the first time.

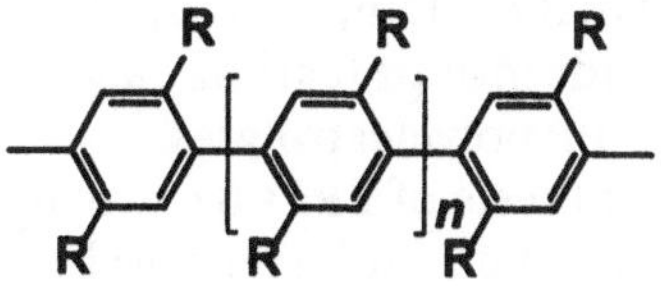

R = alkyl, alkoxy

6

Schlüter et al. [16] were the first to describe coupling of aromatic compounds containing aryl-magnesium halide and aryl halide functions, catalyzed by Ni(0) compounds. Here, the authors adapted the principle of attaching solublizing side chains (in the 2- and 5- positions) and arrived at soluble and processable products. They obtained (soluble) oligo(*para*-phenylene)s with maximum degrees of polymerization of 8–10. The products are characterized by an exclusive 1,4-linking of the benzene rings in the main chain. However, the average molecular weights were quite low.

Several authors further developed the method of Ni(0)-mediated couplings to generate several **PPP** derivatives [14, 18, 19]. They described homocouplings of various 1,4-dihalobenzene (and other dihaloarene) derivatives by means of nickel(II) chloride/triphenylphosphane/zinc or nickel(0)/cyclooctadiene complex. Ni(0)-catalyzed homocouplings of 2-substituted 1,4-phenylene-bis(triflate)s have been reported by Percec et al. [20] and used to prepare substituted poly(*para*-phenylene)s 7 containing alkyl, aryl or ester substituents in the 2- and 3-positions of the 1,4-oligophenylene skeleton. The apparently broad scope of this method of preparation is due, in particular, to the ease of preparation of the bis(triflate) monomers starting from the corresponding hydroquinone derivatives.

$$F_3CSO_2O-\underset{}{\bigcirc}-OSO_2CF_3 \quad \xrightarrow{Ni(0)} \quad 7$$

R = alkyl, phenyl, COOCH$_3$

The Suzuki aryl-aryl cross-coupling method, adapted to polymers by Schlüter, Wegner and co-workers, made it possible to synthesize solubilized **PPPs 6** with a dramatically increased molecular weight (number average up to 100 1,4-phenylene units) [17].

$$Br-\underset{}{\bigcirc}-B(OH)_2 \quad \longrightarrow \quad 6$$

R = alkyl, alkoxy

Soluble **PPPs 6** are known today that contain, not only alkyl substituents, but also alkoxy groups, as well as ionic side groups (carboxy and sulfonic acid functions) [21], which are able to form **PPP** polyelectrolytes.

Scherf et al. [22] reported the synthesis of **PPPs 8a**, which are composed of chiral cyclophane subunits, by means of a Suzuki-type aryl-aryl cross coupling of the corresponding diboronic acid and dibromo derivatives. The monomers containing cyclic $-O-C_{10}H_{20}-O-$loops were resolved into the pure enantiomers by preparative high pressure liquid chromatography on chiral stationary phases, and used to generate the corresponding stereoregular iso- and syndiotactic **PPP** derivatives **8b** and **8c**. In this context, the isotactic derivative **8b** is of special interest because of the chirality of its main chain[22].

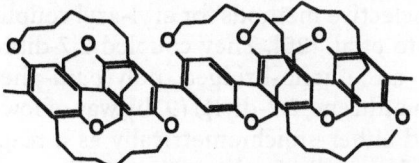

Atactic PPP-Derivative 8a

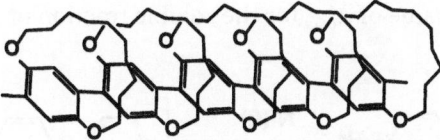

Isotactic PPP-Derivative 8b

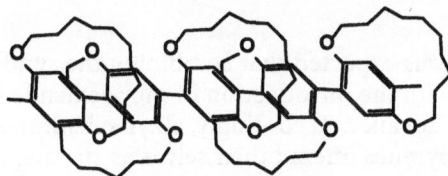

Syndiotactic PPP-Derivative 8c

The electronic properties are adversely affected by changes in conformation due to the substituents. The substituents in the 2- and 5-positions produce a marked mutual twisting of the aromatic subunits, which results in a drastic reduction in the conjugative interaction. In unsubstituted **PPP 1**, there is a 23° twist between adjacent phenylene units [23]. Since the π-overlap is a function of the cosine of the twist angle, a fair amount of conjugative interaction remains even at 23°. If substituents are placed along the **PPP**-backbone, (e.g. at the 2- and 5-positions), the solubility is enhanced, as discussed, but the π-overlap is reduced dramatically (twist angles of 60–80° for alkyl substituents in 2,5-positions, depending on the length of the alkyl substituents [24]). Thus, for poly(2,5-dialkyl-1,4-phenylene)s, neglegable absorption can be detected in the wavelength region above 300 nm, which is characteristic for π–π* transitions of delocalized systems.

The results described thus far sketch the synthetic demands for being able to prepare processable, structurally defined **PPPs**, in which the π-conjugation remains fully intact or is even increased compared with that of the parent **PPP** system **1**. The decisive step in the realization of this principle is the preparation of a **PPP** in which the aromatic subunits could be obtained in a planar or only slightly twisted conformation in spite of the introduction of substituents.

One of the first examples was that reported by Yoshino et al. [12] concerning the synthesis of polyfluorenes via oxidative coupling of fluorene derivatives as decribed above.

It was logical to combine this principle for the preparation of "stepladder" structures possessing a minimized twist between consecutive phenylene units with new, efficient and selective methods for aryl-aryl couplings. A first attempt at this was by Yamamoto et al. [25]. They coupled 2,7-dibromo-9,10-dihydro-phenanthrene to give an ethano-bridged poly(*para*-phenylene) derivative (poly(9,10-dihydrophenanthrene-2,7-diyl)) (**9**) by way of low-valent nickel complexes, which were used either stoichiometrically as a reagent (Ni(COD)$_2$) or were generated electrochemically in the reaction mixture. As a result of the insufficient solubilization of the ethano substituents only the oligomer fraction with $M_n < 1000$ is soluble, the polymeric products precipitating out as an insoluble powder. The value of λ_{max} for the soluble fraction of **9** is about 360 nm.

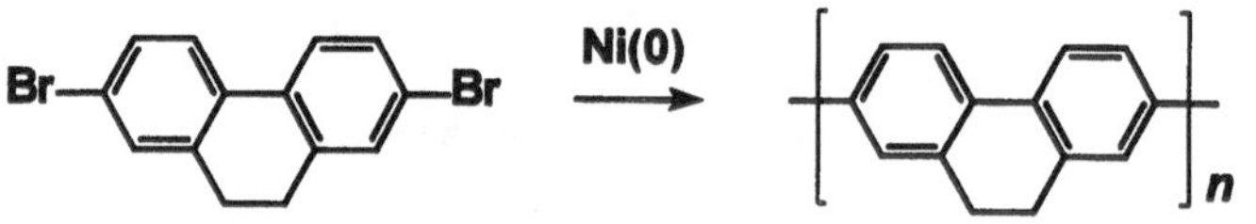

Building on this, it was expected that combining the synthetic procedure of Yamamoto et al. [25] with the introduction of more extended solubilizing substituents would be an advance. Accordingly, alkyl-substituted dihydrophenanthrenes or tetrahydropyrenes offered themselves as starting monomers for the preparation of soluble "stepladder" **PPP**s of this type. 2,7-Dibromo-4,9-dialkyl-4,5,9,10-tetrahydropyrenes (**10**) represent suitable starting monomers for the realization of this synthetic route. These difunctionalized tetrahydropyrene monomers were first prepared by Müllen et al. and reacted in a Yamamoto-type coupling.

Reaction of the dibromide **10** with Ni(COD)$_2$ in DMF/toluene gave a poly(4,9-dialkyl-4,5,9,10-tetrahydropyrene-2,7-diyl) (**PTHP 11**) [26] as a new, completely soluble type of **PPP** derivative with a "stepladder" structure, in which each pair of neighboring aromatic rings is doubly bridged with ethano linkages. The solubilizing alkyl substituents are attached at such positions on the periphery of the molecule that they cannot cause twisting of the main chain. **PTHP 11** possesses a relatively high number-average molecular weight, up to $M_n = 20,000$, which corresponds to the coupling of 46 **THP** units.

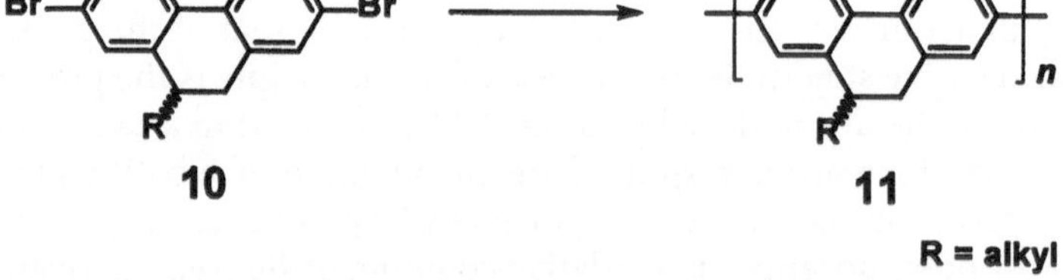

R = alkyl

In the first coupling experiments Müllen et al. used the monomer **10** as a *cis/trans* diastereomeric mixture. When the diastereoisomers are separated by fractional crystallization or chromatography in **10a/b** at the stage of the dibromo monomers, stereoregular **PTHPs 11** are accessible (*cis*- or *trans*-polymer).

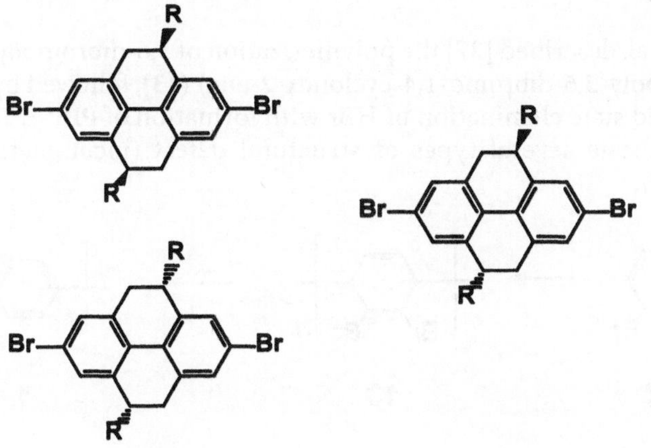

10a

cis - diastereomer

(pair of enantiomers)

10b

trans - diastereomer

PTHP 11 possesses a long-wavelength absorption λ_{max} of 385 nm, almost identical with the value for the "stepladder" polyfluorenes **3** of Yoshino et al. [12]. Thus, two independent proofs exist for the correctness of the "stepladder" concept: The introduction of solubilizing groups combined with simultaneous bridging of the subunits to guarantee the highest possible degree of conjugative interaction.

In solution **PTHP 11** possesses an intense blue photoluminescence (PL) with a quite small Stokes shift between absorption and emission (λ_{max} absorption: 385 nm; λ_{max} emission: 425 nm). In the solid state the PL undergoes a slight bathochromic shift to λ_{max} ca. 457 nm, probably as a result of aggregation. The luminescence characteristic of **PTHP 11** suggests that its suitability as an active component in light-emitting diodes (LED) based on organic polymers should be investigated. Such investigations showed the appearance of a quite intense blue-green electroluminescence (EL) with a quantum yield of 0.1–0.15 % (single layer construction ITO/**PTHP 11**/Ca). Blue polymer-based LEDs represent an attractive target, as blue-emitting LEDs based on inorganic semiconductor materials are not easily accessible. As a result of their band gap energy of 2.7–3.2 eV, **PPP** derivatives are particularly suitable as blue emitters.

In contrast to materials of low molecular mass, polymeric emitters possess the advantage that they can be easily worked into transparent films with a low degree of scattering. In addition, they show a higher morphological stability than that of vapor-deposited low molecular weight compounds (low tendency toward recrystallization).

Other Routes to Poly(para-phenylene)s

Beside the oxidative and transition-metal-catalysed condensation reactions discussed above, several other syntheses were developed to generate **PPP** and **PPP** derivatives.

Marvel et al. described [27] the polymerization of 5,6-dibromocyclohexa-1,3-diene 12 to poly(5,6-dibromo-1,4-cyclohex-2-ene) (13), followed by a thermally induced, solid state elemination of HBr with formation of **PPP** 1. The products, however, indicate several types of structural defect (incomplete cyclization, crosslinking).

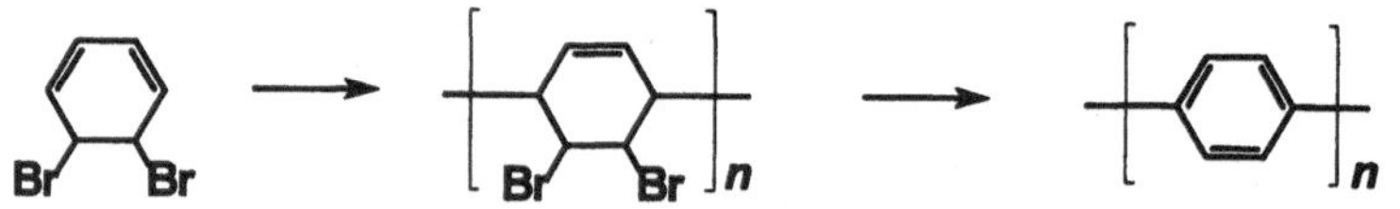

Later on, Ballard et al. [3, 28] developed an improved precursor route, starting from 5,6-diacetoxycyclohexa-1,3-diene (14), the so-called ICI route. The soluble precursor polymer, 15, is then aromatized thermally to **PPP** 1 via elimination of two molecules of acetic acid per structural unit. Unfortunately, the polymerization of the monomer does not proceed as a uniform 1,4-polymerization: beside the regular 1,4-linkages ca. 10% of 1,2-linkages are formed as a result of a 1,2-polymerization of the monomer.

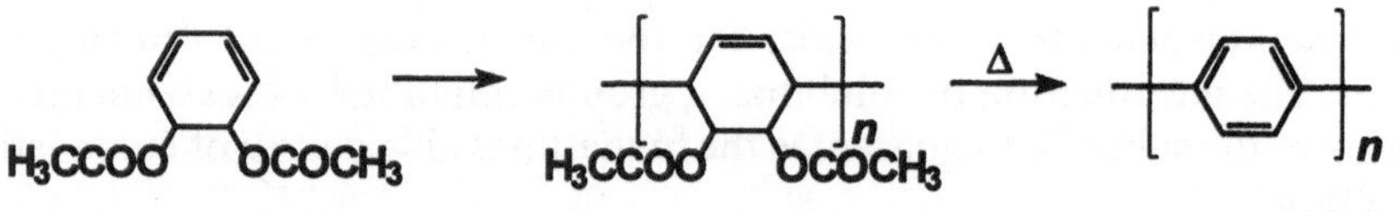

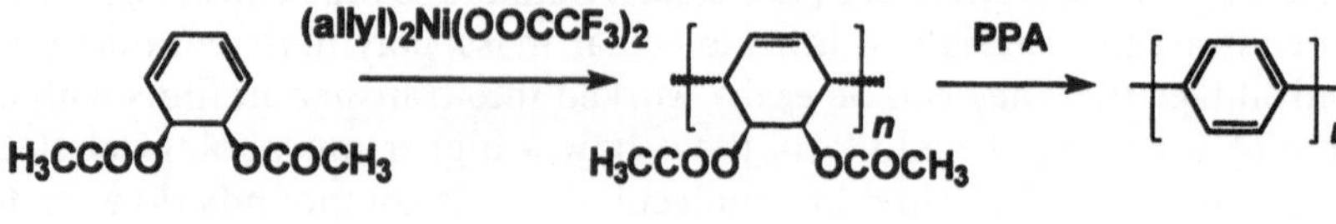

In 1992/1994, Grubbs et al. [29] and MacDiarmid et al. [30] described an improved precursor route to high molecular weight, structurally regular **PPP** 1, by transition metal-catalyzed polymerization, of the cyclohexa-1,3-diene derivative 14 to a stereoregular precursor polymer 16. The final step of the reaction sequence is the thermal, acid-catalyzed elimination of acetic acid, to convert 16 into **PPP** 1. They obtained unsupported **PPP** films of a definite structure, which were, however, badly contaminated with large amounts of polyphosphoric acid, the acidic reagent employed. Nevertheless, in this work, a reliable value for the long wavelength absorption maximum λ_{max} of **PPP 1** could be obtained (about 336 nm). This value is of utmost importance in the interpretation of the optical properties of substituted **PPPs**.

Beside the polymerization routes of 1,3-cyclohexadiene derivatives repetitive Diels-Alder polyadditions were widely used to prepare arylated **PPPs**. Stille et al. developed a set of suitable monomers (1,4-diethynylbenzene and 1,4-phenylene-bis(triphenylcyclopentadienone) derivatives) to generate phenylated **PPPs** (e.g. **17**) with molecular weights of 20,000 – 100,000 [31]. Unfortunately, the repetitive polyadditon does not proceed regioselectively: polymers containing *para*- as well as *meta*-phenylene units within the main chain skeleton are formed.

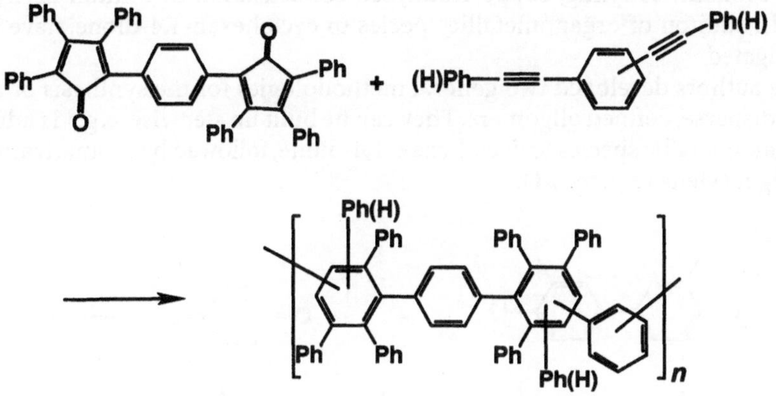

17

Recently, Tour et al. [32] described attempts to prepare **PPP** derivatives via a Bergman cyclization, starting from substituted enediynes, e.g. poly(2-phenyl-1,4-phenylene) (**18**) from 1-phenyl-hex-3-en-1,5-diyne or the structurally related poly(2-phenyl-1,4-naphthalene) (**19**) from 1-phenylethynyl-2-ethynylbenzene.

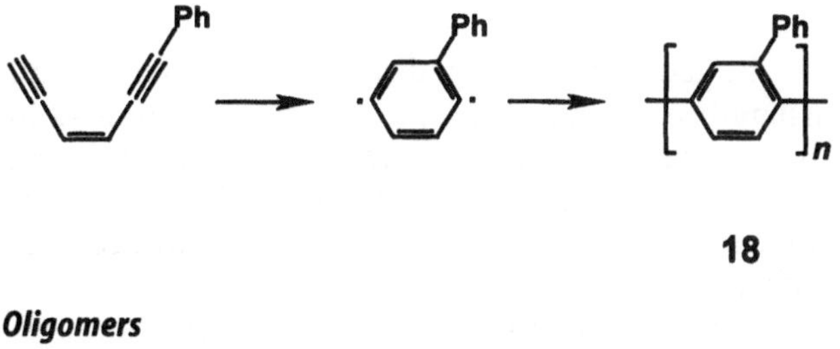

18 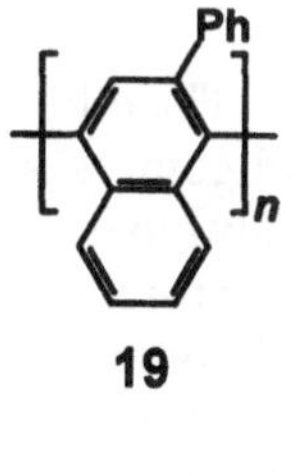　　　**19**

Oligomers

In parallel with polymer synthesis, many activities have been directed towards soluble, well-defined oligomers. Aside from purely synthetic considerations, access to oligomers is, important for the optimization of polymer generation and for the understanding of structure/property relations in the class of **PPP** and other polyarylene materials, e.g. physical properties as a function of chain length. As mentioned above, the problem with oligophenylenes is their low solubilty. The absolute solubility of **PPP** oligomers in conventional organic solvents decreases dramatically with increasing chain length. For example, in the case of octaphenylene in toluene at 25 °C, it reaches the negligibly small value of less than 10 ng/l. In view of this limitation, all attempts to synthesize and characterize longer-chain unsubstituted **PPP** oligomers by direct coupling are not advisable.

The first series of soluble oligo(*para*-phenylene)s **OPVs** (**20**) were generated by Kern and Wirth [33] and shortly thereafter by Heitz and Ulrich [34]. They introduced alkyl substituents (methyls) in each repeat unit and synthesized oligomers (**20**) up to the hexamer. Various synthetic methods, like copper-cata-lyzed Ullmann coupling, copper-catalyzed condensation of lithium aryls, and twofold additon of organometallic species to cyclohexan-1,4-dione, have been investigated.

The authors developed two general methodologies for the synthesis of these monodisperse, defined oligomers. They can be built up stepwise, e.g. via additon of organometallic species to cyclohexan-1,4-dione, followed by aromatization to the oligoarylene (e.g. for **21**).

The other strategy involves the generation of oligodisperse mixtures of several oligomers of different chain length, e.g. copper-catalyzed coupling of mono- and dilithioaryls, followed by a chromatographic resolution of the oligodisperse mixtures into monodisperse components of defined chain length. In this way, oligomers **20** up to a dodecaphenylene derivative have been isolated by prepara-tive thin-layer chromatography.

Rehahn et al. [35] recently presented the synthesis of constitutionally homoge-neous oligophenylenes, **22a/b**, with 2,5-alkyl substituents located on the central aromatic ring, generated via the cross-coupling reaction of Suzuki. They are exclusively linked in the *para*-positions and composed of 3–15 benzene rings.

All of the oligomers described above are characterized by the presence of solubilizing alkyl groups, resulting in increased solubility. However, the elec-

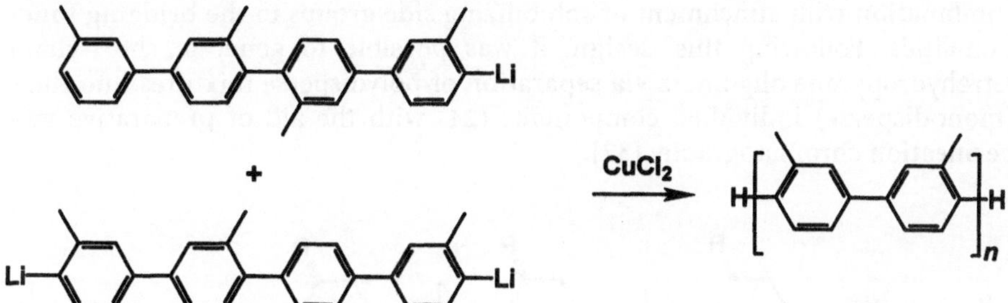

20

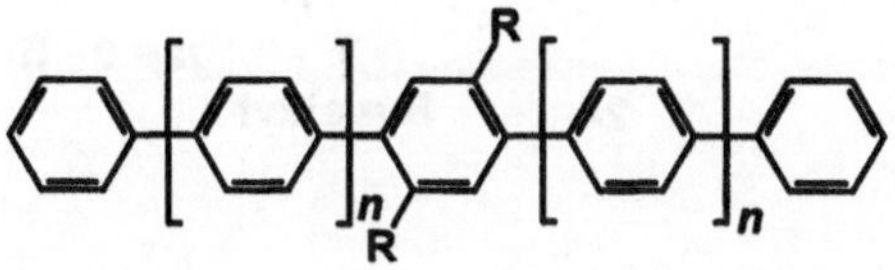

22a

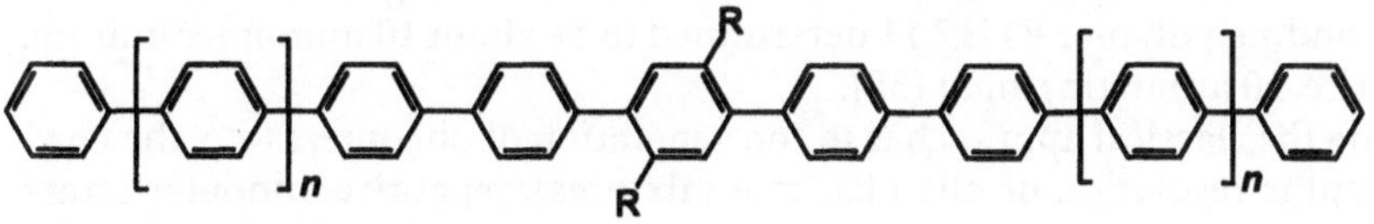

22b

tronic properties of the π-system are disturbed by the mutual distortion of the phenylene units induced by the substituents. Compared to the parent **PPP** system with its 23° twist between adjacent building blocks, the substituted derivatives display torsion angles of 60–80°, minimizing the conjugative interaction within the conjugated backbone. One means of overcome this is substitution exclusively at the terminal rings, as was done by Lüttke et al. [36]. They generated oligophenylenes **23** with *tert*-butyl substituents at the terminal 3- and 5-positions, using Grignard coupling as the key step. However, longer oligophenylenes **23** are not attainable with this approach, since the compounds become insoluble when they reach chain lengths of more then 7 aromatic building blocks.

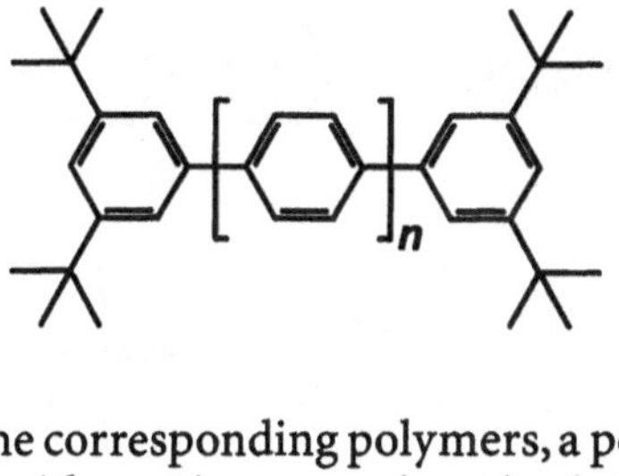

23

As described for the corresponding polymers, a powerful strategy for arriving at soluble oligomers with maximum conjugative interaction is incorporation of the **PPP** backbone into a stepladder – or ladder (see Sect. 2.2.) – framework, in

combination with attachment of solubilizing side groups to the bridging functionalities. Following this design, it was possible to generate short-chain
tetrahydropyrene oligomers via separation of polydisperse mixtures into their
(monodisperse) individual components (**24**) with the aid of preparative gel-
permeation chromatography [37].

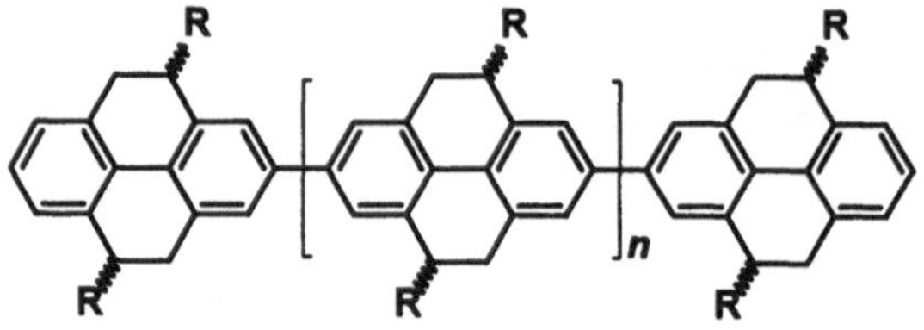

24   R = alkyl

n = 0 - 8

With such a series of oligomers **24**, the convergence of optical properties with
increasing chain length can be followed, and the conjugation length in the
corresponding polymer **PTHP 11** determined to be about 10 monomer building
blocks (i. e. 20 aromatic rings) [38].

Beside the classical approaches to the generation of oligomers (e. g. the chromatographic resolution of oligodisperse mixtures), repetitive, modular strategies for generating extended oligomeric structures have become important in
recent years. Such a strategy involves the repetition of directed protection/
coupling/deprotection steps in a convergent process, in order to minimize the
number of reaction steps necessary to generate the extended oligomers. In such

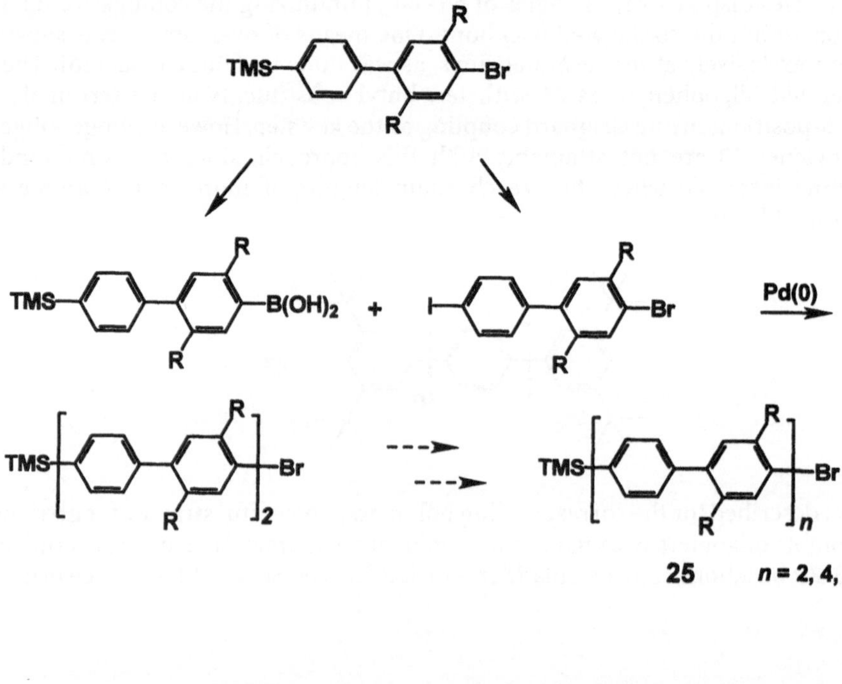

25   *n* = 2, 4, 8

a process, half-protected monomers are sequentially converted to dimers, tetra-mers, octamers, etc. Tour et al. [39] employed such a strategy for the preparation of linear poly(*para*-phenyleneethynylene)s **PPEs**.Using a related strategy, Schlüter et al. [40] described the synthesis of monodisperse oligophenylene rods (**25**) with up to 16 phenylene rings and with well-defined functional end groups. The synthesis is based on a convergent (exponential) growth using the Suzuki reaction as the coupling step. The basic principle is defined by the significantly faster coupling of iodoraryl functions compared with the corresponding bromoaryls in the aryl-aryl cross-coupling reaction by the method of Suzuki. Therefore, derivatives containing both iodo and bromo functions undergo coupling with an arylboronic acid preferentially at the iodo site, leaving the unreacted bromo site for further derivatization (conversion into a boronic acid function).

2.2
Ladder-Type Oligo- and Poly(*para*-phenylene)s

Polymers

The logical continuation of the "stepladder" strategy outlined above for minimi-zing the mutual distorsion of adjacent main chain phenylene units was the incorporation of the complete **PPP**-parent chromophore into the network of a completely planar ladder polymer. The complete flattening of the conjugated π-system by bridging of all the subunits should then lead to maximum conjuga-tive interaction. As with the **PTHP 11** systems, alkyl or alkoxy side chains should lead to solubilization of the polymers.

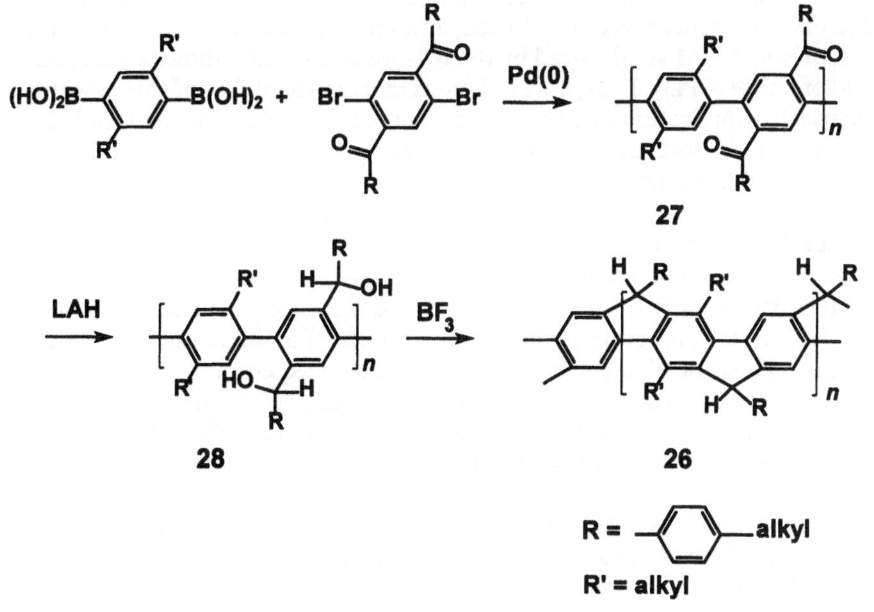

R = ——⟨ ⟩——alkyl

R' = alkyl

This idea was realized impressively in 1991 with the first synthesis of a soluble, conjugated ladder polymer of the **PPP**-type [41]. This **PPP** ladder polymer, **LPPP 26**, was prepared according to a so-called classical route, in which an open-chain, single-stranded precursor polymer was closed to give a double-stranded ladder polymer. The synthetic potential of the so-called classical multi-step sequence has been in doubt for a long time; in the 1980s synchronous routes were strongly favoured as preparative method for ladder polymers.

In a classical multi-step route the main point is to be able to conduct the ring closure quantitatively and regioselectively. In the synthesis of **LPPP**, the precursor polymer **27** is initially prepared by aryl-aryl coupling of an aromatic diboronic acid and an aromatic dibromoketone.

The cyclization to structurally defined, soluble **LPPP** takes place in a two-step sequence, consisting of a reduction of the keto group followed by ring closure of the secondary alcohol groups of **28** in a Friedel-Crafts-type alkylation.

The resulting ladder polymer **LPPP 26** has an average molecular weight of 25,000, which corresponds to the incorporation of 65 phenylene units. No structural defects could be detected using NMR spectroscopy. **LPPP 26** is characterized by unusual electronic and optical properties as a consequence of planarization of the chromophore, the absorption maximum undergoes a marked bathochromic shift to a λ_{max} value of 440–450 nm. In addition, the long-wavelength π-π^* absorption band possesses an unusually sharp absorption edge.

The photoluminescence of **LPPP 26** in solution is blue and very intense (λ_{max} emission: 450–460 nm). The Stokes shift between absorption and emission is extremely small (ca. 150 cm^{-1}), a consequence of the geometric fixation of the chromophore in the ladder structure. The PL quantum yields are high, in comparison with those of many other conjugated polymers: values between 60 and 90% have been measured in solution, and up to 40% in the solid state [42]. In comparison, **PPP 1** synthesized by the ICI-precursor route shows a PL quantum yield of only 4% [43]. Thus, it was obvious that the suitability of this new type of material for application as an active component in blue organic materials-based LEDs should be investigated. The initial result of this investigation was surprising: Although efficient LEDs can be assembled with **LPPP**, the emission in the solid state (film) is yellow (PL and EL). In addition to the primary emission of the **LPPP 26** chromophore in the blue region, the PL and EL spectra show a broad unstructured emission band in the yellow region (ca. 600 nm; Fig. 3) [44]. The relative intensity ratio of these two bands is strongly dependent on the process used for the preparation of the films (solvent, film thickness, preparation of the film). Thus, the blue emission band of the isolated chromophore disappears almost completely on annealing the film at about 150°C. The yellow emission band could then be characterized unequivocally in photophysical experiments as an aggregate emission [45–47]. This result is consistent with:

(1) PL investigations of "site-selective" excitation, according to which the films show additional absorptions at lower frequency than those seen in dilute solution [45];

(2) time-resolved PL studies, demonstrating that the aggregate states are directly populated from initially formed singlet excitons [45, 46];

(3) photovoltaic experiments with **LPPP 26** [47].

The experiments provided proof that weak aggregate absorption can be detected beyond the absorption edge even in the ground state. Time-resolved PL spectroscopy established an unusually short lifetime of about 50 ps for the blue primary emission. This observation can only be explained by a fast relaxation of the initially formed excitons to lower energy aggregate states. The maximum aggregate population is reached several picoseconds after excitation. The lifetime of the latter, ca. 450 ps, is roughly an order of magnitude longer. The photovoltaic experiments furnished a highly efficient sensitivity for **LPPP 26** films in the region of the aggregate absorption/emission bands at 600 nm.

EL experiments showed that the yellow-emitting LEDs prepared from **LPPP 26** exhibit quite remarkable characteristics (single layer construction ITO/**LPPP 26**/Ca; quantum efficiency: ca. 1.0%, applied voltage: 4–6 V [48]). These figures are in the range of the best values described hitherto for polymeric emitters in a single layer arrangement, for example, poly(*para*-phenylenevinylene) **PPV** and **PPV** derivatives. The comparison of **LPPP 26** and **PPV** is also interesting for another reason. In addition to the spontaneous emissions (PL or

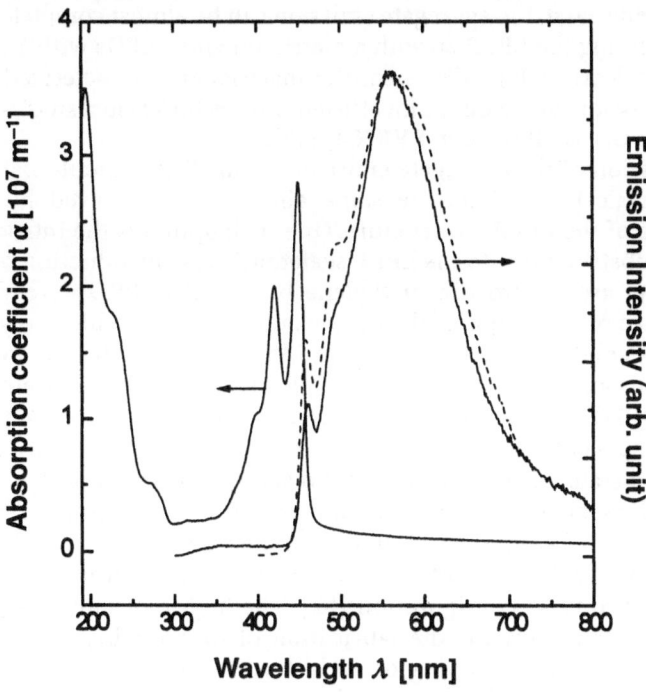

Fig. 1. Absorption and photoluminescence (*dashed line*) spectra of a thin film of **LPPP 26** and electroluminescence (*solid line*) spectrum of an ITO / **LPPP 26** (60 nm) / Al device (from [50])

EL), conjugated polymers are also of interest as materials for optically or electrically pumped stimulated emission. For effects of this type, the ratio of stimulated emission to photoinduced absorption (PA) is of particular interest for conjugated polymers. In this context, the orign of the PA is controversial: the PA can be a result of the formation of either charge-separated "polaron pair"-states or excimers. Initial experiments support the conjecture that **LPPP 26** is significantly superior[49], as the stimulated emission of **LPPP 26** is markedly more intense than that of **PPV** under comparable conditions.

Nevertheless, from the perspective of the strategies aimed at the fabrication of efficient *blue*-emitting LEDs, the results outlined above regarding *yellow* **LPPP** light emitting diodes are unsatisfactory. In order to prepare *blue* LEDs from **LPPP** materials, it is necessary to efficiently mask out or suppress the dominant yellow aggregate emission. Specifically, the preparation of blue-emitting LEDs would make use of the inherent advantage of **LPPP 26** (band gap energy ca. 2.75 eV). With other conjugated polymers, such as **PPV**, that have a smaller band gap energy, blue-emitting LEDs cannot normally be prepared. In that case, only the transition to very short chain, oligomeric chromophores would shift the color of emission to blue. However, the change to oligomers would introduce disadvantages in handling, processing, and stability (re-crystallization tendency of amorphous films, poor film forming properties, low mechanical stability).

Suppression of the aggregate emission can be accomplished by two quite different means. First, the aggregate emission can be almost completely shut off by simply diluting the **LPPP 26** with a matrix polymer. LEDs with 1% **LPPP 26** in poly(9-vinylcarbazole) **PVK** as emitter material are characterized by a pure blue light emission with a quantum efficiency of ca. 0.15% in a single-layer configuration (ITO/1% **LPPP 26** in **PVK**/Ca) [48].

If suppression of the aggregate emission in this first example was based on a purely physical principle, the same aim can be achieved by chemical modification of the **LPPP 26** structure. One such option is the introduction of additional substituents into the **LPPP** skeleton. Thus, introduction of an additional methyl group into the methylene bridge of **LPPP 26** (reaction with $LiCH_3$ in place of reduction of the keto groups) leads to ladder polymers **Me-LPPP 29**. The solid state PL spectra of **Me-LPPP 29** show only very weak aggregate emissions and their solution- and solid state PL spectra are almost identical (Fig. 2 [51]). The suppression of aggregate formation is accompanied by a dramatic increase of the PL quantum efficiency to >90% (solution) and up to 60% (solid state). This enormous solid state PL quantum efficiency favors **Me-LPPP 29** as an emitter for organic materials-based solid state lasers. Leising et al. and Lemmer et al. have demonstrated very recently the high potential of **Me-LPPP** as emitter in solid state lasers both in waveguide and "distributed feedback" configuration [52]. The high molecular weights of **29** (M_n up to 50,000) allow for the fabrication of high quality thick films and stripes (up to 10 μm). Pumping of such devices with blue light results in a laser-like emission even for low energies of the pumping pulses (<2 nJ). The devices are characterized by an enormous stability: lasing was observed over a period of more than 10^7 pulses [52].

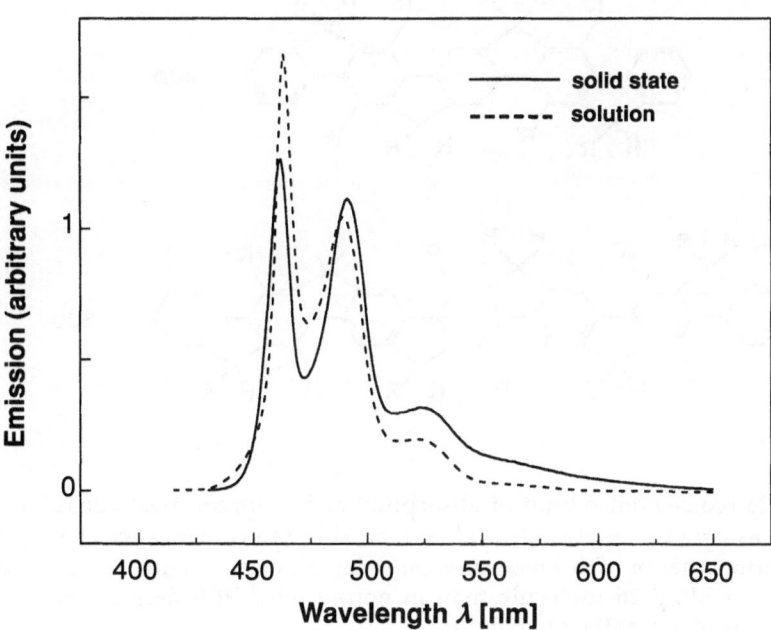

Fig. 2. Photoluminescence spectra of the methyl-substituted Me-LPPP (**29**) (*solid line*: solid state; *dashed line*: solution, methylene chloride)

Oligomers

Besides the polymeric structures described above, there is a considerable interest to generate the corresponding **LPPP** oligomers (**30**) [53]. They were synthesized following a so-called "oligodisperse approach". The two bifunctional (chain forming) monomers for the synthesis of polymer **26** were mixed with a suitable amount of a monofunctional (end capping) monomer, to generate an oligodisperse mixture of molecules of different chain length. These mixtures were resolved into their monodisperse components by means of liquid chromatography, preferably by size exclusion chromatography. Following this strategy, **LPPP** oligomers **30a**, **30b** and **30c** were synthesized containing three, five and seven 1,4-phenylene units within the planar ladder-type main chain.

From UV/Vis investigations of this series of monodisperse, oligomeric model compounds (**30**) the effective conjugation length of the corresponding polymer

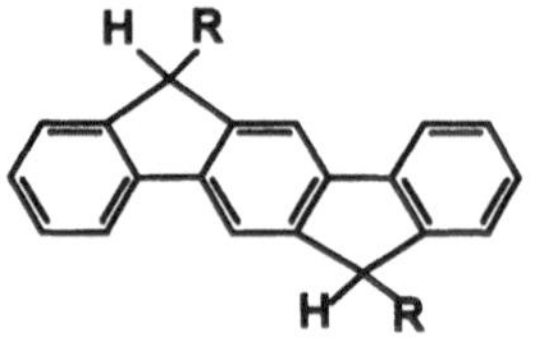

30a

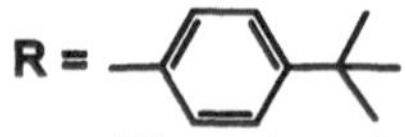

R = —⟨benzene ring⟩—C(CH₃)₃

R' = *n*-hexyl

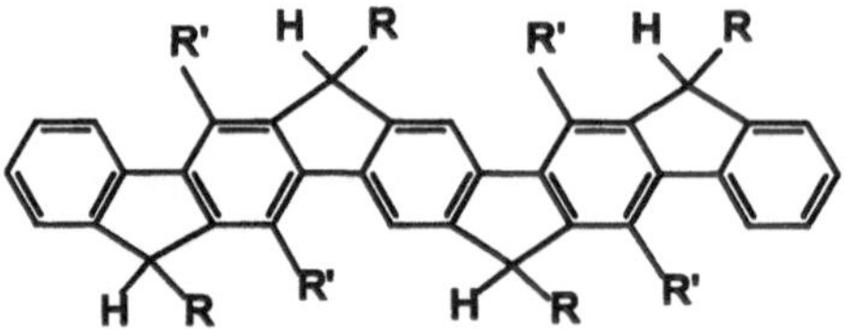

30b

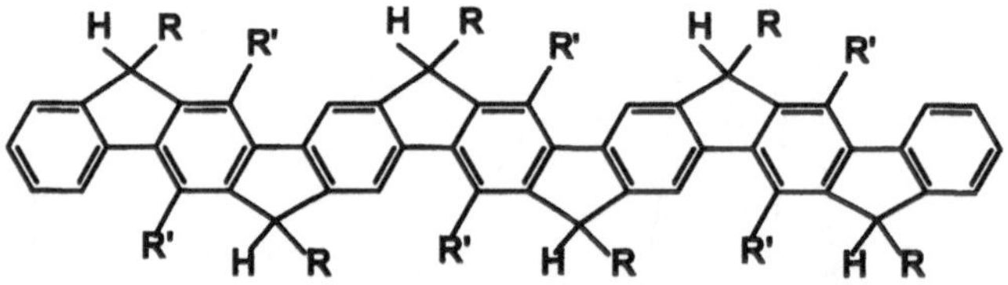

30c

LPPP 26 (convergence limit of absorption and PL properties) could be determined as ca. 11 phenyl rings [38]. Remarkably, the convergence of the optical
absorption energy with increasing chain length occurs much more rapidly in
the planar **LPPP 26** molecule than in non-bridged **PPP** derivatives or in the
partially bridged **PTHP 11** structure.

2.3
Phenylene-Type Dendrimers and Hyperbranched Polymers

Hexa(oligophenyl)benzenes

A main problem concerning the solid state electronic properties of **PPPs** is its
tendency to form aggregates. One synthetic strategy for overcoming this
problem is the transition from linear, one-dimensional systems to branched or
dendritic, two-dimensional ones.

Hexa(oligophenyl)benzenes (e.g. **31** or **33**) present one possible approach to
the realization of this aim. Two efficient synthetic routes have been elaborated
for the preparation of hexa(terphenyl)- and hexa(quaterphenyl)benzene. The
first, involving palladium-catalyzed trimerization of diarylacetylenes [54] as
the key step, was demonstrated by the synthesis of a hexakis-alkylated
hexa(terphenyl)benzene derivative **31** from the corresponding bis(terphenyl)
acetylene (**32**). The peripheral *tert*-alkyl substituents serve to solubilize the
molecule.

The second synthetic route consists of the coupling of hexa(4-iodophenyl)benzene (34) with an alkylated oligophenylboronic acid to produce a hexa(oligophenyl)benzene by extending the aromatic chain [52]. This route is illustrated by the reaction of hexa(4-iodophenyl)benzene (34) with an alkylated terphenyl boronic acid with formation of the hexa(quaterphenyl)benzene derivative 33. Once again, the aliphatic substituents serve to guarantee sufficient solubility.

Higher hexaphenylbenzene homologues were also prepared following a repetitive Diels-Alder procedure. The synthesis of a bis(hexaphenyl)benzene was first described by Ogliaruso et al. [55].

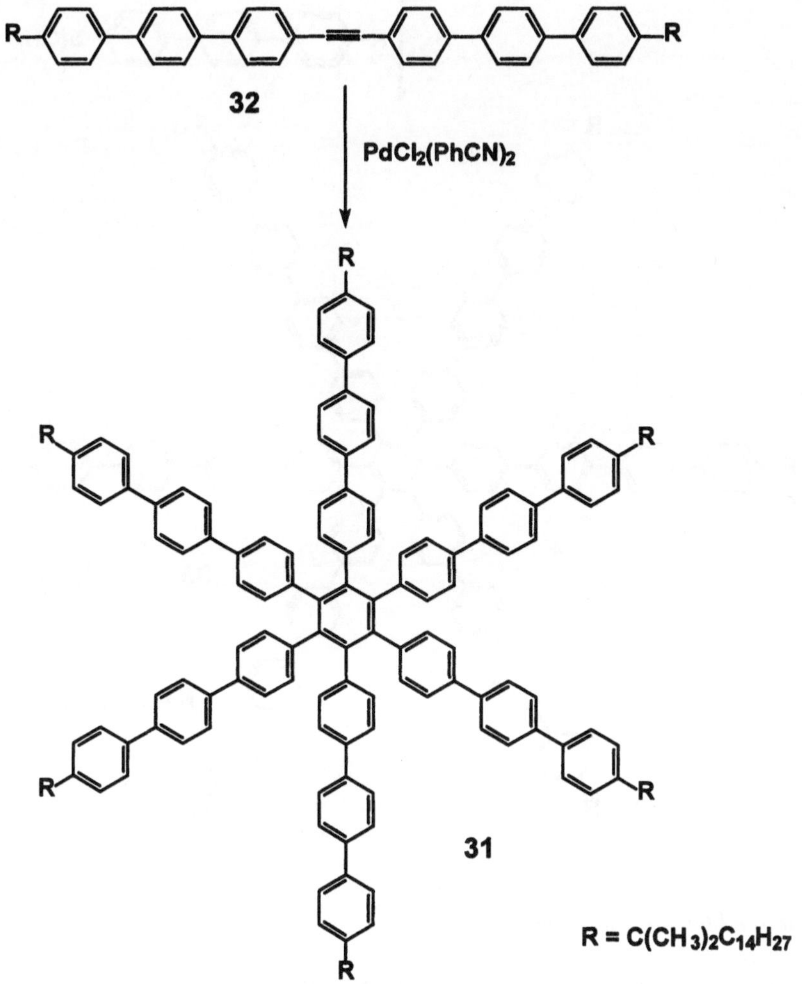

R = C(CH$_3$)$_2$C$_{14}$H$_{27}$

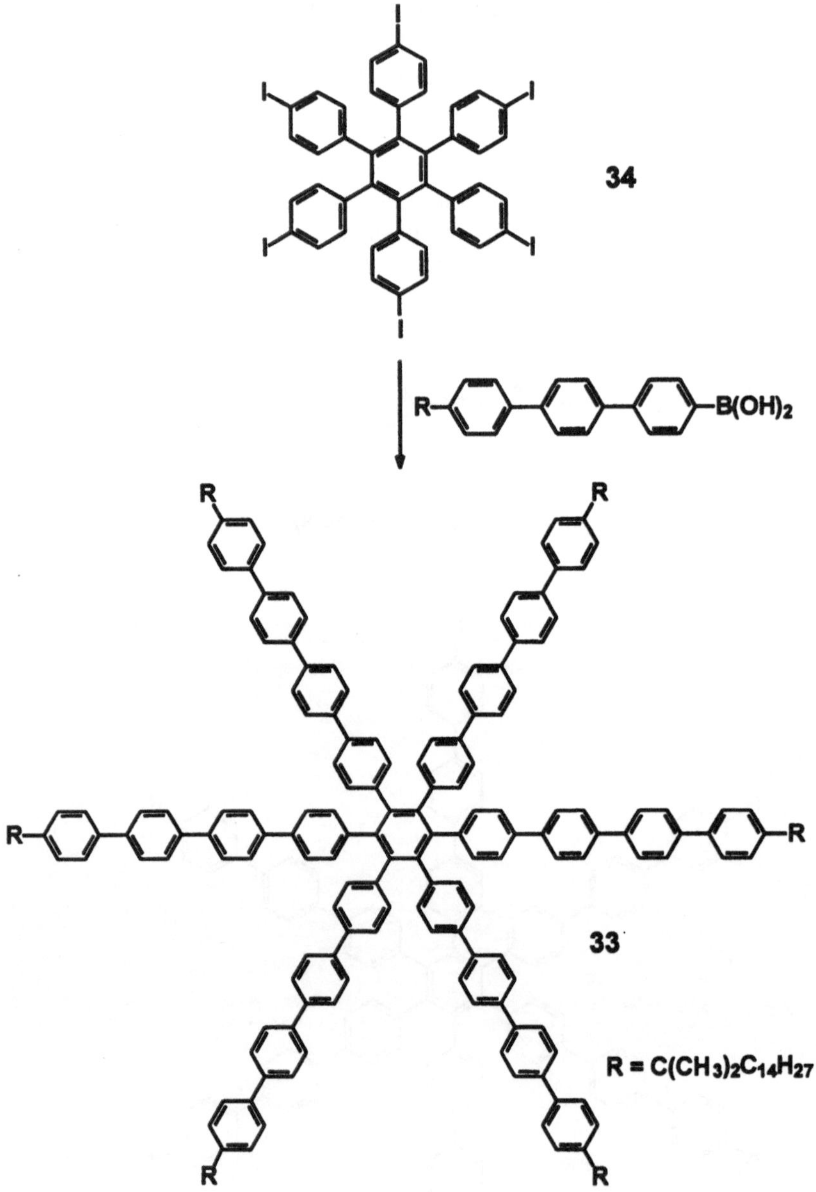

9,9-Spirobifluorene Derivatives

Structurally related star-shaped molecules with oligophenyl arms, derived from
9,9-spirobifluorene as the central unit, have been synthesized by Tour et al. [56]
and Salbeck et al. [57] and proposed as potential emitter materials for blue LEDs.
The authors employed the synthesis of 9,9-spirobifluorenes (35) which had been

developed by Clarksen and Gomberg [58] (additon of biphenyl-2-yl-magnesium iodide to fluorenone and subsequent cyclization with protic acids), and generated 2,2′, 7,7′-tetraarylated 9,9-spirobifluorenes (37).

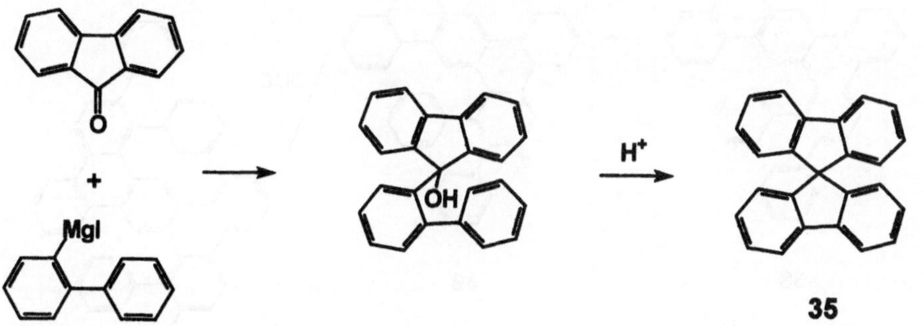

35

For this purpose 9,9-spirobifluorene (35) was tetrabrominated (see Tour et al. [56]) to 36, followed by a Suzuki-type aryl-aryl coupling with various oligoaryl- and oligoheteroarylboronic acids to produce the 2,2′, 7,7′-arylated derivatives (37). The star-shaped molecules (37) are characterized by a drastically increased solubility compared to the corresponding unsubstituted **PPP** oligomers, and by extraordinarily high glass-transition temperatures [57]. Therefore, the efficiently luminescent **PPP** derivatives (37), consisiting of two orthogonally arranged oligophenylene chains, are promising candidates as emitters in organic materials-based light emitting diodes (LEDs).

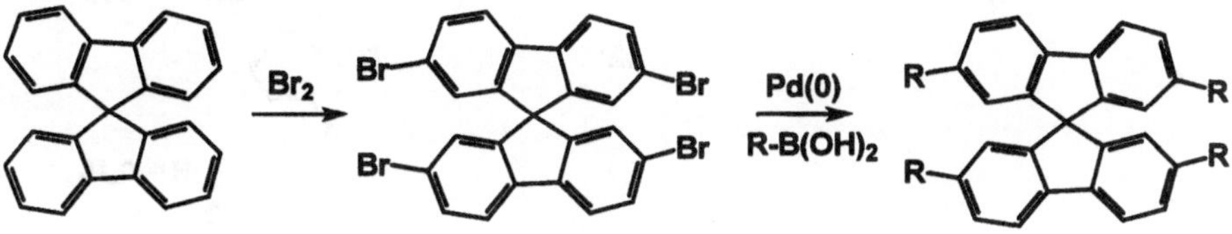

35   **36**   **37**

Branched Polyarylenes and Phenylene-Type Dendrimers

More recently, Müllen et al., have worked out very efficient methods for generating highly arylated, branched oligophenylene- and related oligoarylene derivatives, following different synthetic procedures based on polyaddition reactions in the key step.

A first milestone was the development of a novel intramolecular Diels-Alder cyclization of terphenyl monomers 38 and 41, containing both 4-phenylbuta-dienyl and styryl functions. The formation of the [4+2] cyclization adducts 39 and 42 is followed by a simple aromatization of the cyclohexene moieties [59]. In this way, the phenylated, two-dimensional arylene structures, 40 and 43, based on a tetrabenzoanthracene core and possessing a hitherto unknown topology, were generated.

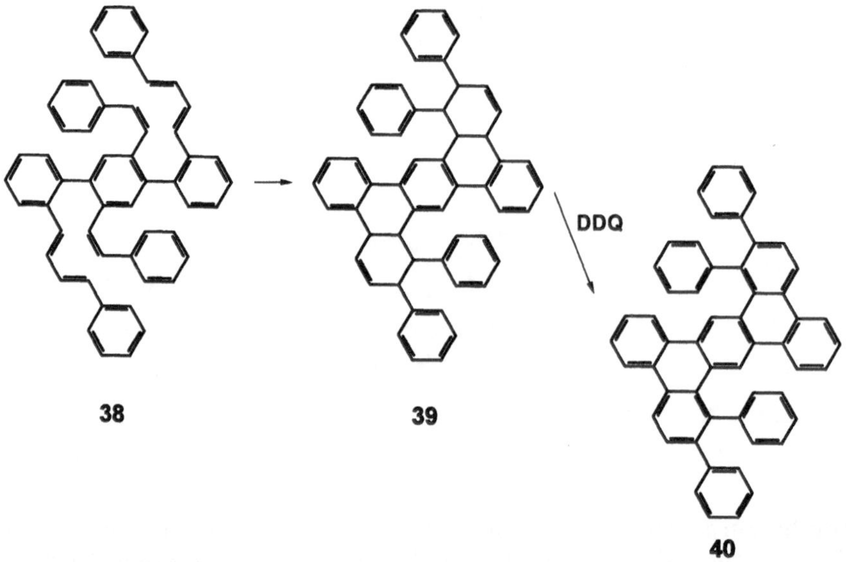

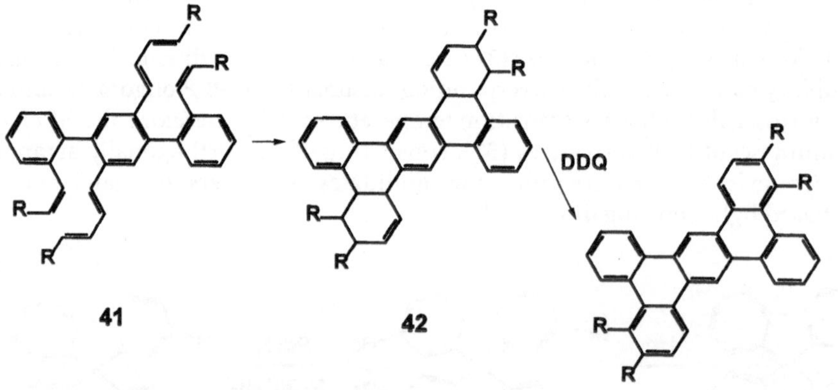

The branched oligoarylenes, **40** and **43**, can undergo further oxidative cyclization with copper(II) chloride or triflate in the presence of aluminum trichloride with formation of large, hitherto unknown polycyclic aromatic hydrocarbons, PAHs, **44** and **45**.

In line with a second novel synthetic principle, the authors further developed the repetitive Diels-Alder procedure, in which monomers containing cyclopentadienone (dienophile) units were reacted with protected/deprotected ethynylene functions (see [31]). In this way, they generated a novel class of highly arylated phenylene dendrimers **46**, starting from a 3,3′,5,5′-tetraethynyl-substituted biphenyl core [60].

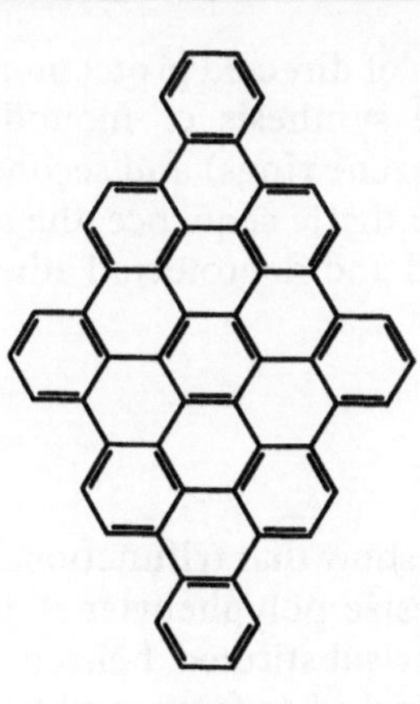

44

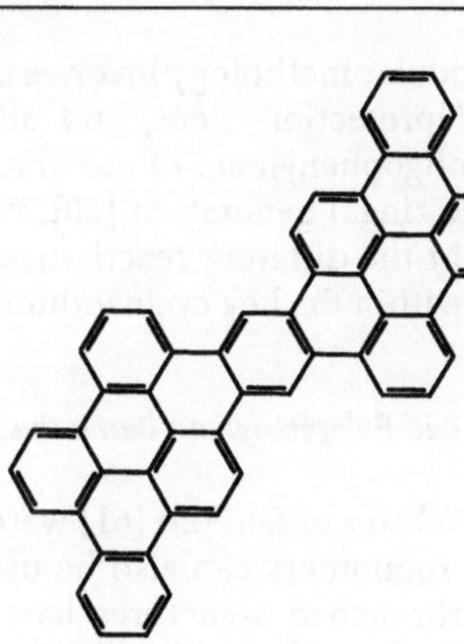

45

46a

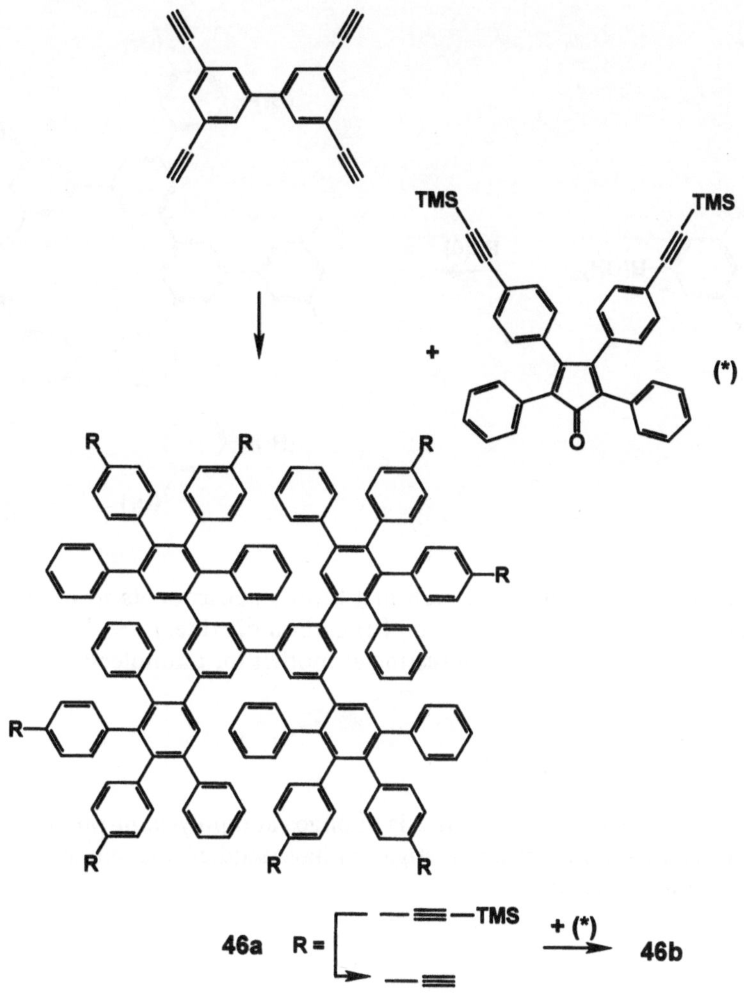

This modular metholology involves the repetition of directed protection/cyclo-addition/deprotection steps, and allows for the synthesis of monodisperse dendritic oligophenylenes of the first (**46a**, 22 benzene rings) and second (**46b**, 62 benzene rings) generation [60]. Within the synthetic sequence, the authors made use of the different reactivities of protected and deprotected ethynylene functions within the key cycloadditon step.

Hyperbranched Polyphenylene Derivatives

Kim and Webster of DuPont [61] were the first to show that trifunctional benz-ene-based monomers can also be used to synthesize polyphenylenes, in their case hyperbranched structures based on 1,3,5-trisubstituted benzene rings. They self-condensed 1,3-dibromophenyl-5-boronic acid, to form soluble, hyper-branched **PPP**-type macromolecules, **47**.

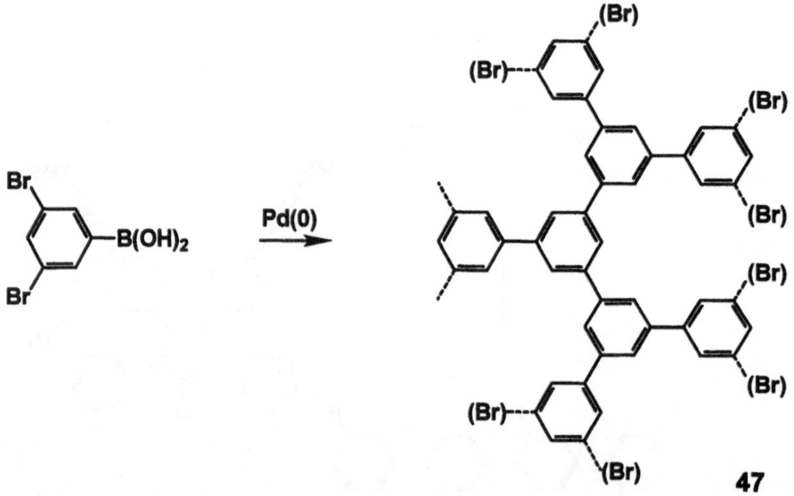

The transformation of the terminal bromo substituents to carboxylic acid functions with: (i) *n*-butyl lithium; (ii) carbon dioxide, provides water soluble derivatives of **47** which are interesting as models for unimolecular micelles.

2.4
Other Oligo- and Polyarylenes

Based on the results of the synthesis of oligo- and poly(*para*-phenylene)s **PPP**s (see Sects. 2.1–2.3) several investigators have studied the generation of other oligo- and polyarylenes.

Poly(meta-phenylene)

Unsubstituted Poly(*meta*-phenylene) is far more soluble than poly(*para*-phenylene) **PPP 1** of comparable molecular weight due to the angular structure of the *meta*-derivative. Poly(*meta*-phenylene) **48** can be prepared using the Yamamoto procedure [20]; ca. 35 % of the material formed is soluble in toluene.

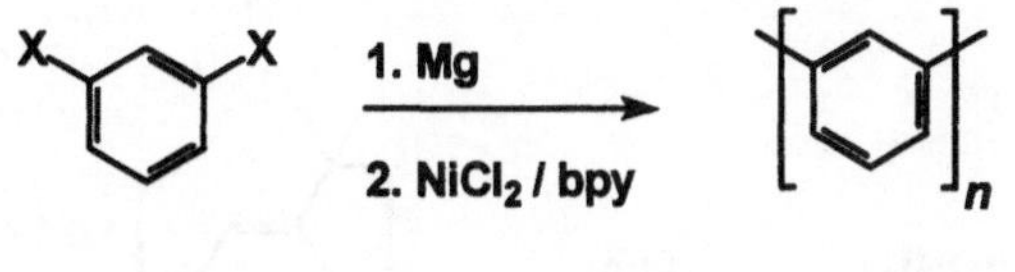

48 X = Cl, Br, I

Staab et al. showed that the intramolecular coupling of terminally halogenated *meta*-linked oligophenylenes can afford cyclic structures, as shown here for the cyclization of the dibromo-sexi(*meta*-phenylene) derivative, **49**, to the cyclic *meta*-phenylene hexamer, **50** [62].

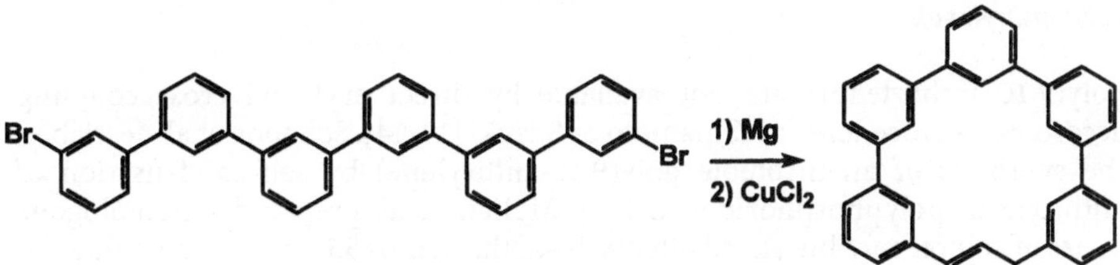

49

50

Poly(naphthylene)s

Insoluble, structurally undefined poly(naphthylene)s can be prepared electrochemically or by using $AlCl_3$ or $AlCl_3/CuCl_2$ according to the Kovacic oxidative procedure, starting from naphthalene [8, 63]. A number of soluble copolyarylenes composed of alternating naphthalene and biphenyl building blocks, e.g. **51**, have been prepared by Percec et al. by oxidative coupling with $FeCl_3$ [64].

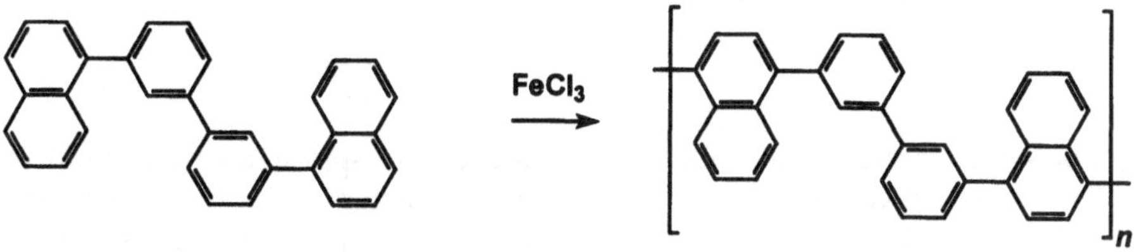

51

As mentioned above, Tour et al. described the synthesis of poly(2-phenyl-1,4-naphthylene) (19) following a Bergman-type cyclization approach [32]. Müllen et al. [65] published the synthesis of fully soluble poly(3,7-di-*tert*-butyl-1,5-naphthylene) (52) starting from 1-bromo-3,7-di-*tert*-butyl-naphthalene-5-boronic acid according to Suzuki. This polynaphthylene derivative is characterized by a regular chemical structure (1,5-coupling) and a surprisingly high degree of polymerization: ca. 25–28.

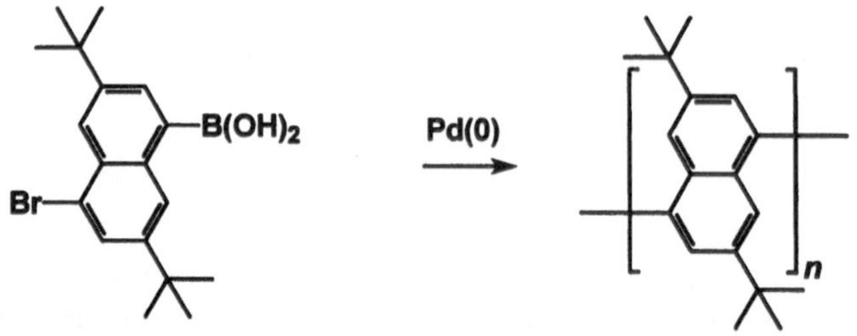

52

Poly(anthrylene)s

Poly(9,10-anthrylene)s are not available by direct aryl-aryl cross coupling according to procedures of Yamamoto or Suzuki [13, 14]. Schopov et al. described the synthesis of an insoluble poly(9,10-anthrylene) by self-condensation of anthrone in polyphosphoric acid [66]. Müllen et al. prepared a homologous series of soluble *tert*-butylated oligo(9,10-anthrylene)s 53 up to the heptamer by reacting alkyl substituted lithio- and dilithioanthrylenes with anthracene- and bianthryl-based quinones, followed by reduction of the polyalcohols formed with hypophosphoric acid [67].

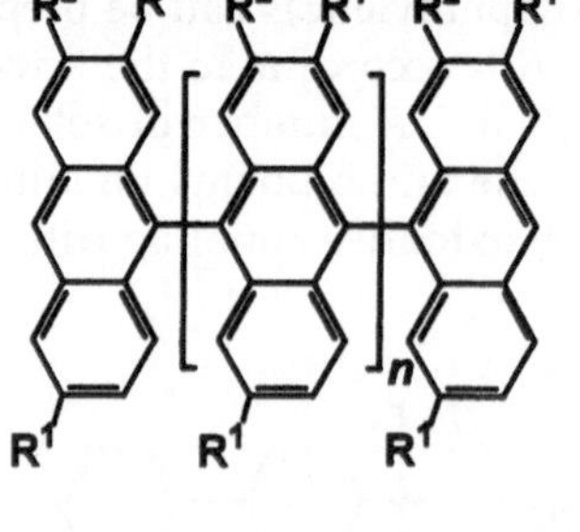

53

$R^1, R^2 = H, C(CH_3)_3, C_6H_{13}$

The soluble products are able to form charged high-spin states after chemical and electrochemical oxidation. The high-spin character is the result of the lack of conjugative interaction between the highly distorted, orthogonally arranged aromatic subunits (decoupled π-systems) [68].

Poly(phenanthrene)s

Hörhold et al. have synthesized soluble poly(2,9-phenanthrylene) derivatives (55) starting from bis(4-alkoxyphenyl)-substituted poly(1,3-phenylenevinylene)s (54) [69]. The polymer-analogous cyclization proceeds intramolecularly under mild conditions (FeCl$_3$ x6H$_2$O) and was shown to be approximately complete. Starting from the corresponding *para*-phenylenevinylene precursors (56), they could isolate alternating copolyarylenes (57) composed of 1,4-phenylene and 9,10-phenanthrylene moieties. The latter example has a highly distorted structure, the optical properties of the polymers are similar to those of isolated phenanthrene chromophores.

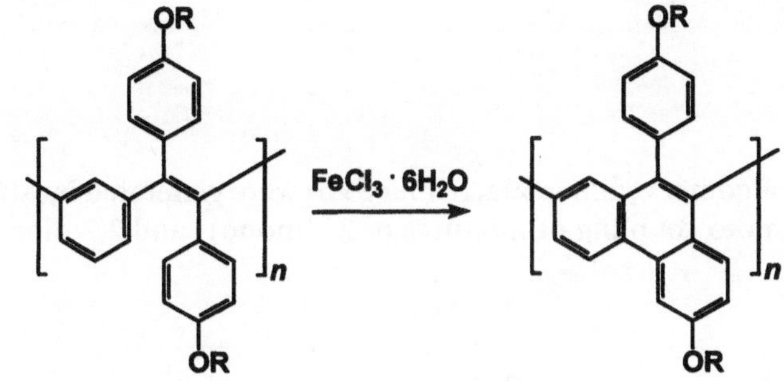

54   **55**   R = C$_6$H$_5$, CH$_3$

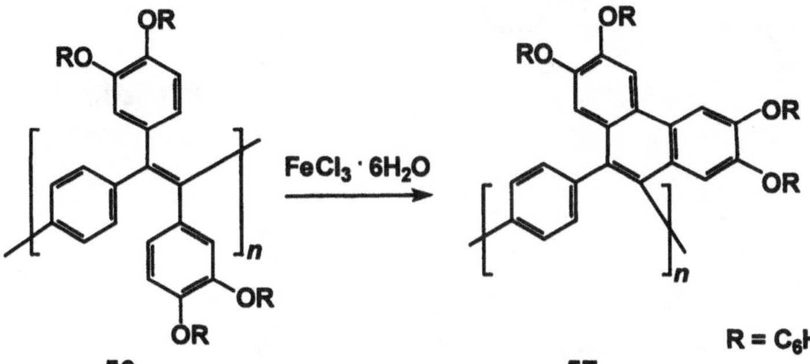

56   **57**   R = C$_6$H$_5$, CH$_3$

Poly(perylene)s

Soluble poly(1-butylperylene) (58) was prepared in very high yields by Anton and Müllen [70] who used the procedure of Taylor [71], which involves the oxidative coupling of bis-Grignard reagents with *cis*-1,4-dichloro-2-butene as an oxidant. The products contain 4,9- and 4,10-perylenylene moieties, are fully soluble and possess average degrees of polymerization of ca. 22.

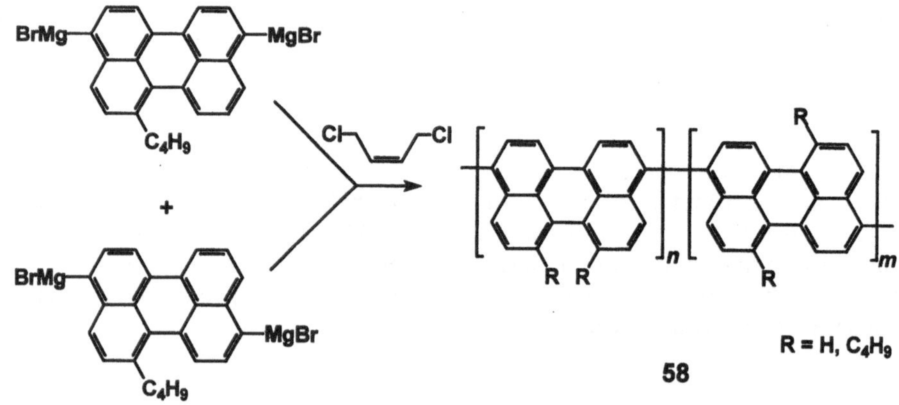

58

R = H, C₄H₉

Oligo(pyrene)s

Two pyrene-based oligomers (dimer **59 a**, trimer **59 b**) were generated by Müllen et al. [72] via Yamamoto coupling of mixtures of 2- (mono)- and 2,7-dibromo-pyrene.

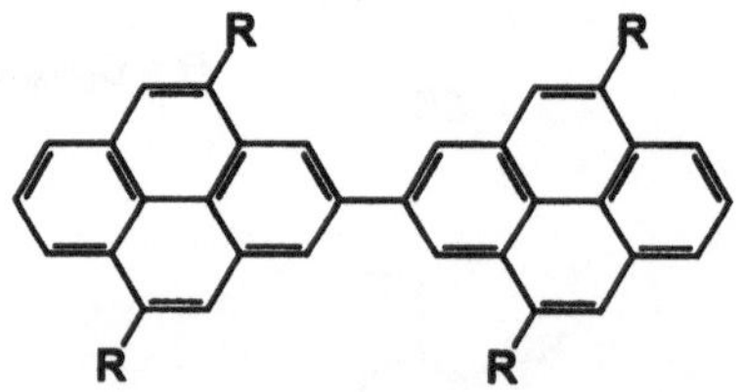

59a

R = C₈H₁₇

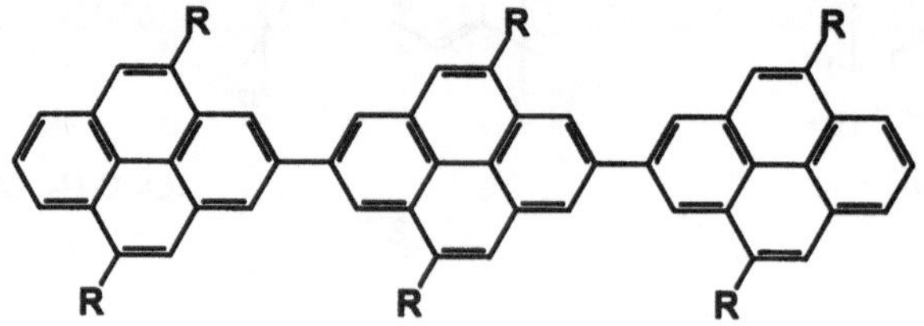

59b

The optical absorption spectra and the first reduction potentials are virtually independent of the number of pyrene units present in the molecule, as a result of the specific stereoelectronic situation. Since the orbital coefficients of the bridgehead centers are almost zero, the rings are electronically decoupled. Thus, oligopyrenes differ significantly from oligo(*para*-phenylene)s (OPVs).

3
Oligo- and Polyarylenevinylenes

Beside oligo- and polyarylenes, oligo- and polyarylenevinylenes represent one of the most intensively investigated classes of π-conjugated hydrocarbon molecules. From a structural point of view, oligo- and polyarylenevinylenes can be regarded as alternating copolymers composed of unsubstituted/substituted arylene and vinylene moieties, and, therefore, as a hybrid between polyarylenes on one hand, and polyacetylene(s), on the other hand.

Oligo- and polyarylenevinylenes are available via a variety of different, often very powerful, synthetic approaches; the next section will give a brief overview of the most common strategies to generate this type of structures. In this context, emphasis will be placed on the synthesis of *structurally defined* and *well-characterized* materials.

3.1
Poly(*para*-phenylenevinylene)s via Polymerization Methods

The Wessling Procedure

Among the various means by which poly(*para*-phenylenevinylene)s, especially unsubstituted poly(*para*-phenylenevinylene) (**PPV 60**) may be synthesized, the so-called Wessling method [73] has been very intensively employed. The method was developed at Dow Chemical Corp. in 1968 by Wessling and Zimmerman, and represents a polymerization route via a water-soluble, processible polyelectrolyte precursor. At present, the Wessling procedure is one of the most convenient methods of obtaining high quality films and fibers of high molecular weight **PPV 60**, a conjugated polymer which is itself an insoluble and intractable material.

PPV

60

The general process involves polymerization of 1,4-bis(dialkylsulfoniomethyl)benzene dihalides (**61**) by addition of base. The immediately formed polyelectrolyte (**62**) is then converted thermally to the final **PPV** derivative (**63**). The process was first developed for the synthesis of unsubstituted **PPV 60**. The mechanism of the Wessling process is still not fully clear. First, Hörhold et al. [74] showed by means of UV/Vis spectroscopy that a transient *para*-xylylene intermediate (**64**) is formed after the initial elimination of HX and dialkyl sulfide, and then polymerizes to yield the polyelectrolyte **62** as a suitable precursor for thermal conversion to **PPV**. There is some evidence for a radical chain propagation mechanism for the polymerization of **64**, as shown by Karasz et al. [75] in experiments with radical trapping agents.

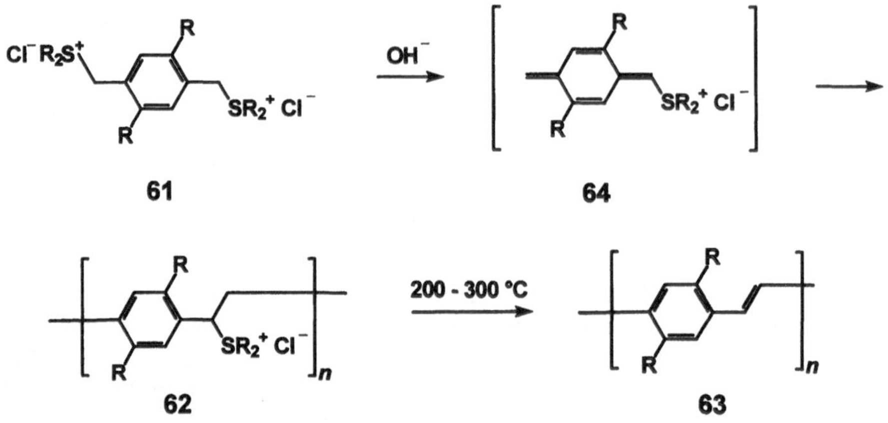

The Wessling method is applicable to a variety of substituted derivatives, however, the delicately balanced series of equilibria producing **64**, the species that actually polymerizes, is greatly affected by the substituents attached to the monomer species, **61**. Alkoxy, alkyl, aryl and halogen ring substituents R (2-, 5- and 2,5-position) do not greatly influence the ability to generate and polymerize the *para*-xylylene intermediates (**64**) [76, 77]. Strong acceptors, such as nitro or cyano substituents, reduce the concentration of **64** in the reaction mixture, possibly due to the formation of resonance-stabilized ylide species, which do not undergo fast elimination to **64** [77]. In contrast to the high molecular weight precursors formed, for example, in the case of alkoxy- and alkyl-substituted derivatives (DP up to 5.000), the 2,5-biscyano-substituted monomer produces only low molecular weight polyelectrolytes (DP of ca. 8). The "push-pull"-substituted 2-methoxy-5-cyano derivative gives a somewhat higher molecular weight material with a DP of ca. 150 [78].

However, not all types of *para*-xylylene analogues are able to polymerize under the Wessling conditions to give polyelectrolyte precursors, as shown by failure to polymerize the anthracene homologue, **65** [77].

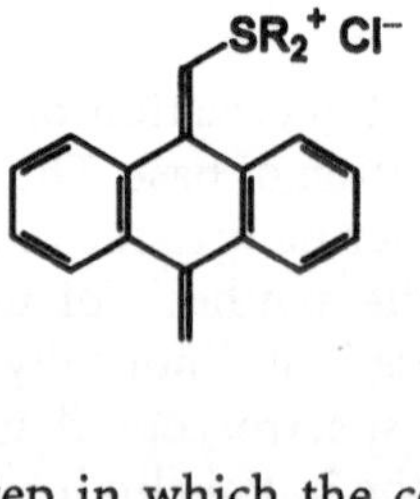

The final elimination step in which the conjugated **PPV** derivative (**63**) is generated from the sulfonium polyelectrolyte precursor polymer (**62**) was reported by Wessling and Zimmerman to be heating in vacuo to 200–300 °C. The target **PPV** derivative is formed with elimination of dialkyl sulfide and hydrogen halide; the process can be easily monitored by UV/Vis spectroscopy. **PPVs**

display the typical absorption spectrum of an one-dimensional π-conjugated polymer, with an intense, long-wavelength π–π^*-transition; in unsubstituted PPV this transition has an absorption energy of ca. 2.39 eV (ca. 520 nm). In contrast to poly(*para*-phenylene)s, the absorption properties of PPVs are only slightly affected by substituents in the 2- and 5-positions of the aromatic ring. The additional vinylene group in the PPVs elongates the distance between two aromatic rings and, therefore, drastically reduces steric hindrance, resulting in a minimum mutual twisting of the aromatic subunits. As a consequence, substituents on the 2- and 5-positions of the aromatic rings influence the optical properties of most PPVs only via their electronic effects. Unsubstituted PPV 60 itself is characterized by a yellow photoluminescence, centered at ca. 2.2 eV, both in solution and the solid state, with a solid state PL quantum yield of about 8%. Friend et al. were the first to show the possibility of using PPV 60, prepared via the Wessling procedure as a yellow-green emitting material, in light emitting diodes (LEDs) based on organic materials [79].

Several approaches have been tried in order to facilitate the conversion process and lower the conversion temperatures, e.g. by using appropriate sulfonium groups (tetrahydrothiophene derivatives) and/or different counter ions. With chloride as a counter ion [80], the synthesis of 2,5-dimethoxy-PPV can be performed at room temperature starting from the corresponding dimethylsulfonium polyelectrolyte precursor.

The Dihalide Approach

The polymerization of 1,4-bis(halomethyl)benzenes to PPVs in the presence of a large excess of potassium *t*-butoxide is referred as the Gilch route [81]. The method was first described for the synthesis of unsubstituted PPV 60, but – unfortunately – this route produces the PPV as an intractable, insoluble powder. However, the adaptation of the Gilch route to the polymerization of 1,4-bis(halomethyl)benzenes possessing solubilizing side groups gives access to soluble PPV materials.

Since the products often precipitate during the polymerization, a modification was reported by Swatos et al. [82] involving the use of only about one equivalent of *t*-BuOK. This method, the so-called "chlorine precursor route", first gives a soluble non-conjugated precursor (66) which is then converted thermally in the film or in a high boiling solvent, e.g. cyclohexanone. In the latter case, homogeneous solutions of (soluble) PPV derivatives 63 can be obtained.

In the polymerization reaction, a 1,6-elimination of HCl (or HX) takes place, leading, as described above for the Wessling procedure, to the formation of a reactive 1,4-xylylene intermediate (67), which polymerizes to give the corresponding precursor polymer (66). The method was applied to the synthesis of 2,5-dialkyl- [83], 2,5-dialkoxy- [84, 85], 2,3-diphenyl-substituted [86] PPVs, as well as a broad variety of other soluble PPV derivatives. For potential application in organic materials-based light emitting diodes (LEDs), MEH-PPV, a 2-methoxy-5-(2-ethylhexyloxy)-PPV derivative has been broadly investigated [85]. MEH-PPV exhibits a red-orange photoluminecence with a PL quantum

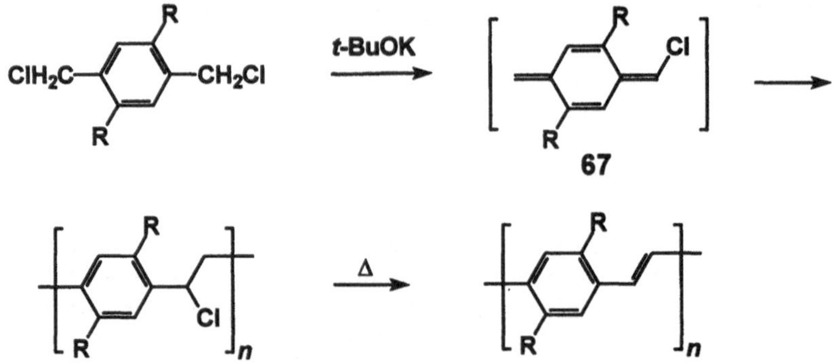

67

66   **63**

yield of about 15 % in the solid state. Heeger et al. reported the fabrication of efficient LEDs using **MEH-PPV** as the light emitting layer [87].

The generation of **PPV** and corresponding derivatives via the dihalide
approach is possible not only in solution reaction, but also – via the gas phase –
in a so-called chemical vapor deposition (CVD) process. In this process, the
vapor of a dichlorinated *para*-xylene (α,α' or α,α) is pyrolyzed at moderately
low pressures (0,1–0,2 torr) to form a chlorinated *para*-xylylene intermediate,
which then condenses and polymerizes on a suitable, cooled substrate. The
coating of the chlorinated precursor polymer can be heated to eliminate HCl, to
form **PPV 60** (or a **PPV** derivative) [88]

The Vanderzande Procedure

In 1995 Vanderzande et al. [89] published a novel, modified Gilch procedure to
unsubstituted **PPV 60** starting from 1-chloromethyl-4-(alkylsulfinyl)methyl-
benzenes (**68**). The initial step, elimination of HCl with NaH as a strong base in
NMP or DMF, leads to the formation of the sulfinyl-substituted 1,4-xylylene

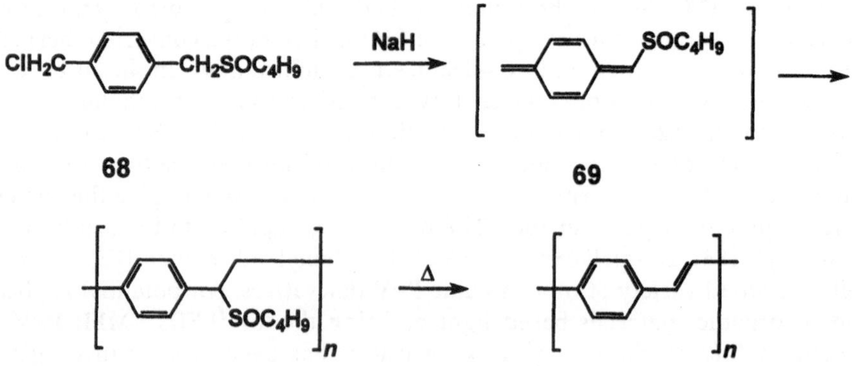

68   **69**

70   **60**

intermediate (69) which polymerizes to a soluble precursor polymer (70). Fortunately, in contrast to the Wessling polyelectrolyte- and the dichloride approaches to the generation of unsubstituted **PPV 60**, the precursor polymers (70) are soluble in organic solvents like chloroform. The non-ionic nature and high stability of the precursor polymers (70) allow for their detailed character- ization (e.g. with GPC), and define an important advantage of the Vanderzande method over the Wessling polyelectrolyte precursor route.

The conversion of **70** to the final **PPV 60** is then carried out thermally at rela- tively low processing temperatures (about 100–150 °C) with elimination of (unstable) alkylsulfinic acid. TGA-mass spectroscopy, FT-IR, UV/Vis and ^{13}C CP/MAS NMR spectroscopy are all consistent with quantitative elimination and formation of **PPV 60**.

Dehydrochlorination of Unsymmetrically Substituted para-Xylylene Dichorides

In contrast to the dichloride approach of Gilch et al., involving the dehydroha- logenation of 1,4-bis(halomethyl)benzenes, Hörhold, Raabe and Scherf devel- oped a novel method to synthesize phenylsubstituted **PPVs 71**, based on the thermally induced or base catalyzed dehydrochlorination of 1-(phenyldich- loro)methyl-4-methyl- [90, 91] or 1-(phenyldichloromethyl)-4-benzylbenzene [92]. The process is very efficient if the polymerization is carried out with or- ganic bases (pyridine, quinoline) in high boiling organic solvents like 1,2-di- chlorobenzene. Again, the mechanism involves the formation of a 1,4-xylylene intermediate, 72. Polymerization of 72 to the non-conjugated precursor polymer (73) and subsequent dehydrochlorination result in the generation of the phenyl- substituted **PPVs 71** (elimination-polymerization-elimination mechanism).

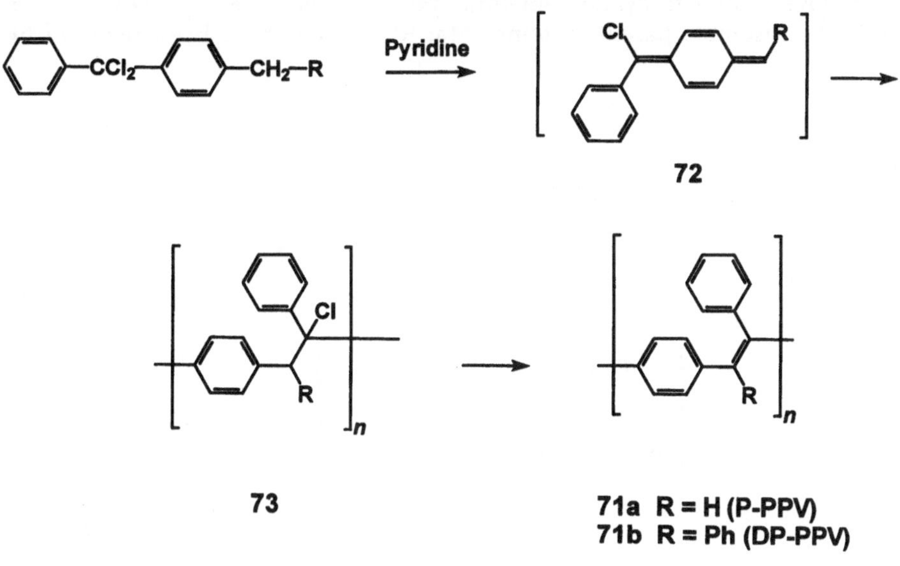

73

71a R = H (P-PPV)
71b R = Ph (DP-PPV)

The yellow colored, soluble products display high molecular weights (M_n > 10.000), the long-wavelength optical absorption energies of the phenylated derivatives (71) are blue shifted relative to the parent **PPV 60** system: **PPV 60**: 2.39 eV (520 nm), **P-PPV 71a** (R=H): 2.80 eV (443 nm), **DP-PPV 71b** (R=Ph): 2.88 eV (430 nm). This finding can be interpreted as the result of mutual distortion of the main chain aromatic rings, caused by the introduction of the sterically demanding phenyl substituents at the vinylene moieties. The results of Hörhold, Raabe and Scherf have been confirmed more recently by Hsieh et al. [93] in the synthesis of **71b**.

3.2
Oligo- and Polyphenylenevinylenes via Polycondensation Methods

The polymerization methods to **PPV** and **PPV** derivatives described in the previous section involve 1,6-polymerization of an immediately formed 1,4-xylylene derivative. Aside frome this polymerization approach, a broad spectrum of polycondensation procedures (step-growth methods) to **PPV** and **PPV** derivatives has been developed. The methods can be classified as follows:

(i)   The carbon skeleton of **PPV** is generated in an olefination reaction (e.g. the Knoevenagel, Wittig-Horner and McMurry reactions) with formation of the olefinic double bond.

(ii) The **PPV** backbone is synthesized via a transition metal-catalyzed aryl-olefin-coupling (e.g. the Heck reaction) with formation of the aryl-vinyl single-bond.

Knoevenagel-Type Polycondensation

Knoevenagel-type condensation of 1,4-xylylene dinitriles and aromatic dialdehydes gives access to cyano-containing poly(*para*-phenylene-cyanovinylene)s (74). The insoluble parent systems (74a: R^1, R^2:=H) have been reported by

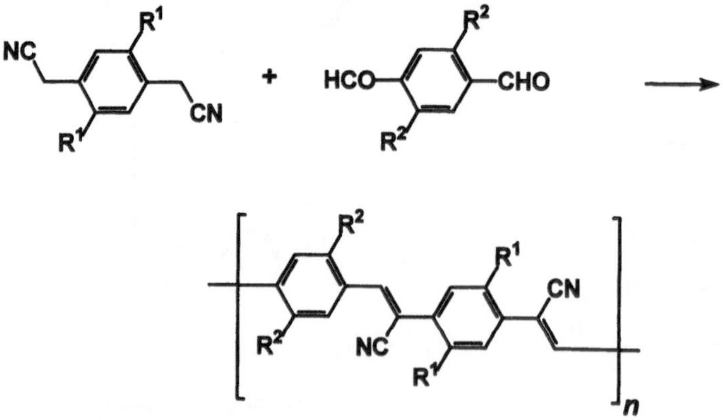

74a (R^1, R^2 = H)

74b (R^1, R^2 = alkyl, alkoxy)

Hörhold et al. and Lenz et al. [94, 95]. The polycondensation provides the cyano-PPVs as insoluble, intractable powders. Holmes et al. [96], and later on Rikken et al. [97], described a new family of soluble, well-characterized 2,5-dialkyl- and 2,5-dialkoxy-substituted poly(*para*-phenylene-cyanovinylene)s (**74b**) synthesized by Knoevenagel condensation-polymerization of the corresponding alkyl- or alkoxy-substituted aromatic monomers. Careful control of the reaction conditions (tetra-*n*-butyl ammonium hydroxide as base) is required to avoid Michael-type addition.

The soluble polymers display average molecular weights M_n of 4000–7000, and can be processed to thin films by spin coating. By judicious choice of the solubilizing substituents, the HOMO-LUMO gap of the resulting polymers (**74b**) and – consequently – the electro-optical properties, i.e. the wavelength of the photo- and electroluminescence emissions, can be controlled; it ranges from the red to the blue. Dialkyl substitution on the aromatic ring was found to increase the emission energy, resulting in a blue shift of the emitted light, whereas alkoxy substitution results in a red shift of the UV/Vis absorption and emission maxima [98]. As the lowest unoccupied orbitals of the per-alkoxy-substituted derivative (R^1, R^2: -alkoxy) of **74b** lie at lower energies than those of **PPV 60**, highly efficient light emitting diodes LEDs have been fabricated on the basis of these two PPV-type materials, with electron injecting electrodes made from stable metals such as aluminium. In a bilayer device,: indium tin oxide ITO/PPV 60/cyano-PPV 74b/Al LEDs were fabricated that possess high (internal) quantum efficiencies of up to 4%. The bilayer setup helps to localize charges at the interface between **PPV** and cyano-**PPV**, increasing the efficiency of recombination.

Wittig-Type and Wittig-Horner-Type Polycondensations

Polymers
A second example of step-growth polycondensations with formation of the olefinic double-bond are Wittig- and Wittig-Horner-type condensations. The Wittig-type polycondensations involve AA/BB-type reactions of aromatic bisaldehydes with bisphosphonium ylides [99, 100] with formation of **PPV** derivatives (**75**) and lead to products of only moderate molecular weight (DP: 10–20).

75 ($R^1, R^2 =$
     H, alkyl, alkoxy)

The Wittig-Horner procedure, starting from bisphosphonate or aromatic bisphosphine oxide monomers, allows for AA/BB-coupling of the PO-activated bismethylene monomers, not only with aromatic dialdehydes but also with aromatic diketones to the corresponding **PPV** derivatives (**76**), and for the self-condensation of AB-type aromatic starting compounds containing both aldehyde/keto and PO-activated methylene functions [101].

76

(R = H, alkyl, alkoxy)

Oligomers

In general, polycondensation-type step-growth polymerizations do not only allow for the generation of polymers, they are also favor formation of the corresponding oligomers. The **PPV** oligomers (**OPVs**) (i) can be generated in a step-by-step approach of repetitive condensations, or (ii) can be isolated from oligodisperse mixtures of oligomers of different chain length. Polymerization methods (e.g. the Wessling procedure) are not suited for the generation of **OPVs**, they are difficult to stop at a specific degree of polymerization.

The first series of unsubstituted **OPVs** was synthesized by Drefahl and Hörhold [102]. They started from 4-bromomethylbenzaldehyde and benzyltriphenylphosphonium chlorides and successively produced the **OPVs 77** with up to 8 phenylenevinylene units via repetitive *in situ* generation of the methylene-triphenylphosphonium functions.

The poor solubility of these materials motivated efforts to increase the solubility of the **OPVs**. The synthetic approach of Müllen et al. was to attach 3,5-di-*t*-butylphenyl substituents at the terminal positions of **OPVs**. The oligomers, **78**, were generated, starting from 3,5-di-*t*-butylbenzyltriphenylphosphonium bromide, in a repeated reaction sequence of 1) Wittig-type coupling with 4-methylbenzaldehyde, 2) NBS bromination of the terminal methyl group and 3) regeneration of the triphenylphosphonium salt functions with triphenylphosphine [103], yielding the phosphonium salt precursor components **79** ($n=0,1$).

The resulting phosphonium salts, **79** ($n=0,1$), containing up to 2 phenylenevinylene moieties, were then coupled with terephthalic dialdeyde, **80** ($m=0$), or 4,4'-stilbene dicarbaldehyde, **80** ($m=1$), to produce the target structures, terminally *t*-butylated **OPVs** (**78**) up to the heptamer ($x=6$). With results based on these model oligomers, the effective conjugation length of unsubstituted **PPV 60**

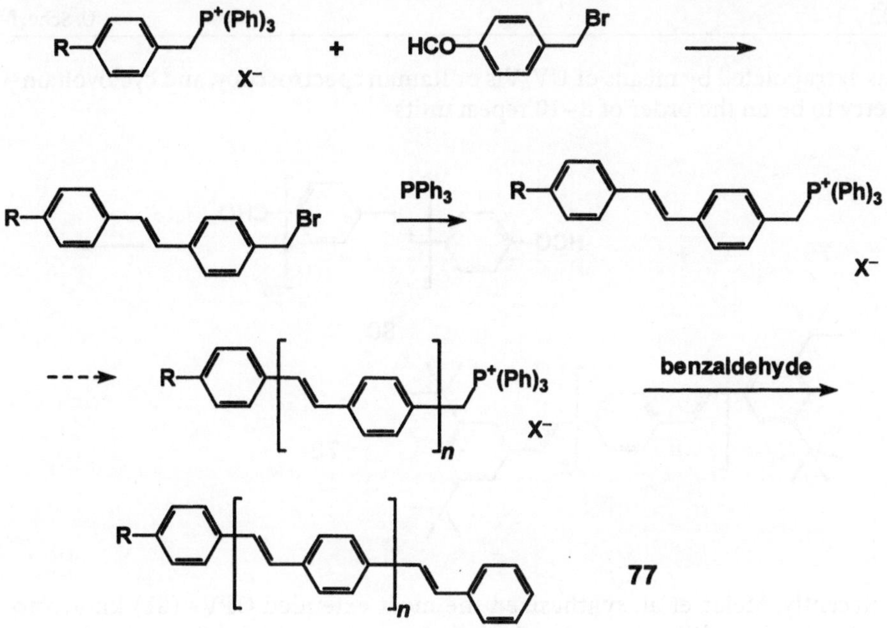

P+(Ph)3
R
X−
HCO
Br
PPh3
benzaldehyde
n

77

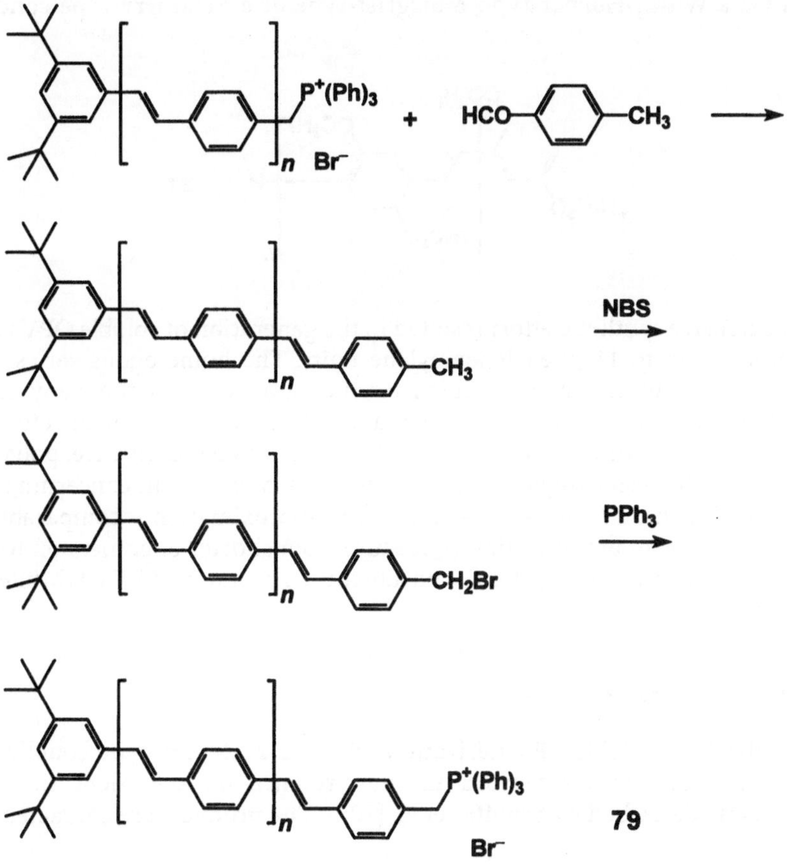

P+(Ph)3
Br−
HCO
CH3
NBS
CH2Br
PPh3
n

79

was extrapolated by means of UV/Vis or Raman spectroscopy, and cyclovoltam-
metry to be on the order of 8–10 repeat units.

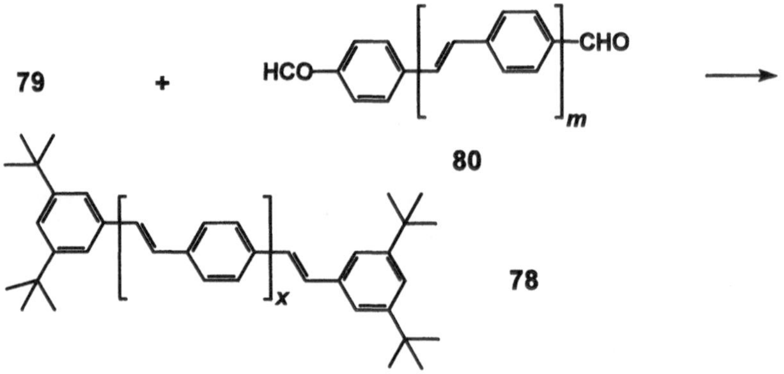

Recently, Meier et al. synthesized the most extended **OPVs** (**81**) known to
date [104]. Four different synthetic pathways were used for the generation of
these **OPVs**; the final step involves the formation of one or two olefinic double
bonds via a Wittig-Horner-type, a Siegrist-type or a McMurry-type condensation.

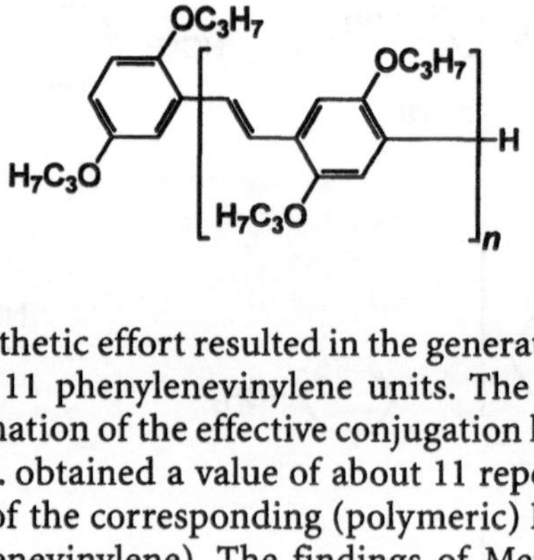

The extensive synthetic effort resulted in the generation of soluble **OPVs** (**81**)
composed of up to 11 phenylenevinylene units. The homologous series now
allows for a determination of the effective conjugation length without any extrapolation. Meier et al. obtained a value of about 11 repeat units for the effective
conjugation length of the corresponding (polymeric) **PPV** derivative, poly(2,5-
dipropyloxy-phenylenevinylene). The findings of Meier et al. concerning the
effective conjugation lengths are – in the first approximation – comparable to
those derived by an extrapolation procedure from shorter oligomers on to the
corresponding polymers, e.g. for the unsubstituted prototype **PPV 60** (Müllen et
al. [103]).

McMurry-Type Polycondensation

The synthesis of soluble **PPV** derivatives (**82**), via McMurry-type coupling of
solubilized aromatic dialdehydes in the presence of low-valent titanium
reagents, was described by Schlüter et al. [105]. The products are characterized

by a *cis/trans* ratio of the olefinic moieties of about 0.4 and an average degree of polymerization of about 30.

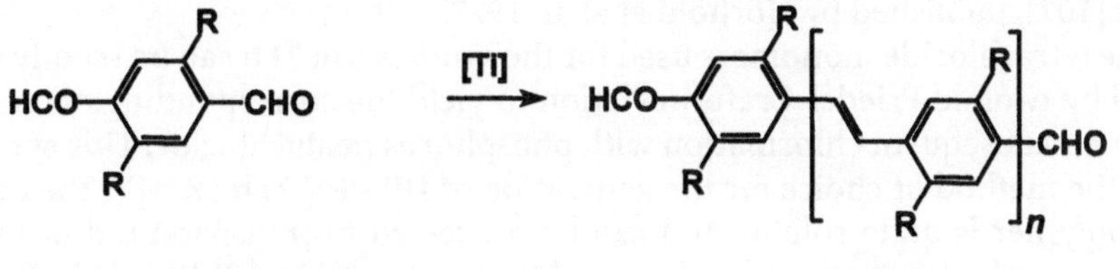

82

Coupling of Bis(diazo) Derivatives

A **PPV** derivative which is twofold phenylsubstituted at the vinylene unit, poly(1,4-phenylene-1,2-diphenylvinylene **DP-PPV**), (**71b**) (see also the discussion of dehydrochlorination of unsymmetrically substituted *para*-xylylene dichlorides in Section 3.1) was first synthesized by Smets et al., using acid-catalyzed elimination of nitrogen from 1,4-bis(diazobenzyl)benzene **83** [106]. The yellow products obtained are fully soluble in common organic solvents (toluene, chloroform, ethylene chloride, DMF, THF).

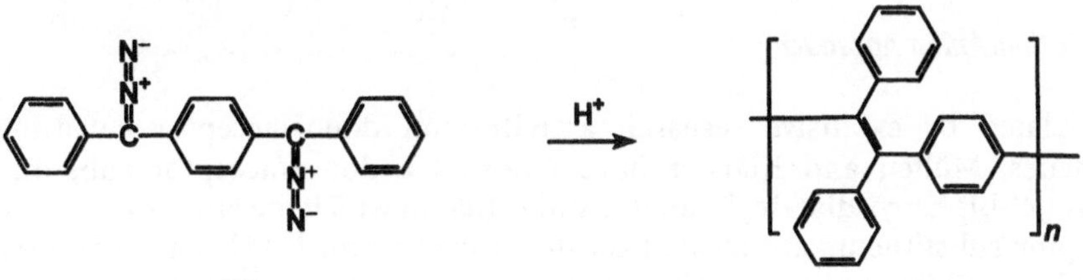

83   **71b**

Reductive Dehalogenation Polycondensation

The class of phenylsubstituted **PPVs** (**71a/b**) has been under very intensive investigation since the 70s, by Hörhold et al. Phenyl substituents at the vinylene positions both solubilize the polymers and stabilize them against the attack of

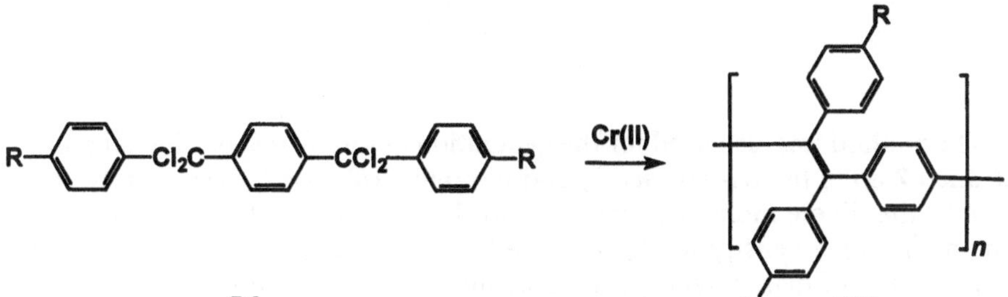

84   **71b**

air and/or light. A very effective method for the synthesis of diphenylated **DP-PPV 71b** is the reductive dechlorination-polycondensation of 1,4-bis(phenyl-dichloromethyl)benzene derivatives (**84**) with chromium(II) acetate as reducing agent [107], published by Hörhold et al. in 1977.

The tetrachloride monomers used for the synthesis of **71b** can be readily prepared by twofold Friedel-Crafts acylation to yield the corresponding diketones and their subsequent chlorination with phosphorus pentachloride. This synthesis is the method of choice for the generation of **DP-PPV 71b** (R=H), the resulting polymer is quite soluble and can be processed to transparent, thin films; and has number-average molecular weights of up to 20,000. **DP-PPV 71b** (R=H) exhibits high photoconductivity and displays a strong yellow photoluminescence [108]. The dehalogenation method of Hörhold et al. allows the introduction of various substituents R (e.g. alkoxy, phenoxy and ester side groups) in the 4- (and 3-) positions of the phenyl moieties. In the solid state, the photoluminescence quantum yield of **71b** (R=OPh) exceeds the highest solution values (about 60%) [109]. Bässler et al. have published detailed electroluminescence studies performed on the **DP-PPV 71b** derivative in which R=OPh. In single layer devices the onset of light emission reflects the transit time of injected holes, whereas in bilayer devices it is determined by the time needed for minority carriers (electrons) to reach the internal interface [110].

The Cation/Anion Approach

Stimulated by extensive research activities on donor/acceptor substituted stilbenes, Müllen and Klärner have reported a donor/acceptor substituted poly(4,4'-biphenyl-diylvinylene) derivative (**85**) in which the NR_2 donor and CN acceptor substituents are located on the vinylene unit [111]. The synthesis is based on a C-C-coupling reaction of *in situ* generated carbanion functions with a (pseudo)cation function, followed by a subsequent elimination of MeSH with formation of the olefinic double bond.

85

The method was also used for the generation of the corresponding oligomers **86** and **87** up to the tri- (**86**, $n=2$) and tetramers (**87**, $n=3$), respectively. In **86** and **87**, the dipole moments increase with increasing length of the oligomers. The high dipole density in **85**, **86** and **87** is a promising starting point for the construction of highly hyperpolarizable materials for potential applications in non-linear optics.

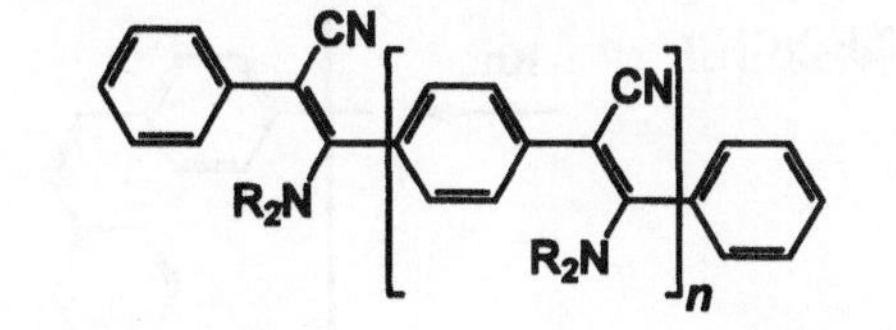

86

R = alkyl

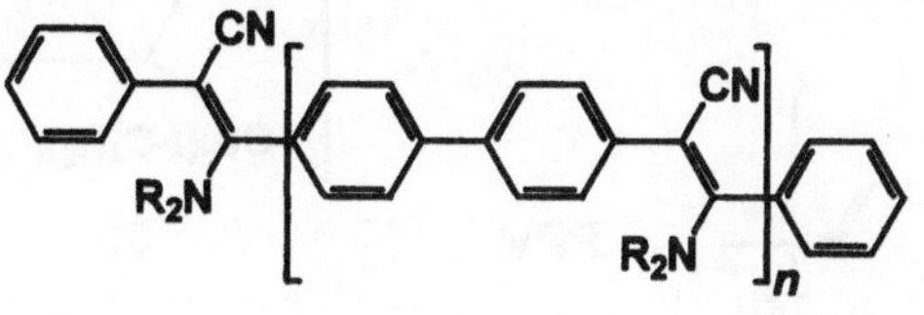

87

Metathesis Polycondensation

The olefin metathesis reaction opens an elegant way to the synthesis of **PPV** derivatives or **PPV 60** precursors in a simple single-step process. A first metathesis route to **PPV 60**, starting from bicyclic monomers possessing a bicyclo[2.2.2]octadiene skeleton, was published by Grubbs et al. in 1992 [112]. The bicyclic monomers were polymerized by ring-opening metathesis polymerization (ROMP) with Schrock-type molybdenum carbene catalysts, to yield well-defined, soluble precursor polymers containing carboxylic ester functions. The non-conjugated precursors were then thermally converted to the conjugated **PPV 60** structure, in analogy to the synthesis of **PPP 1** via the ICI-route [3, 28, 29].

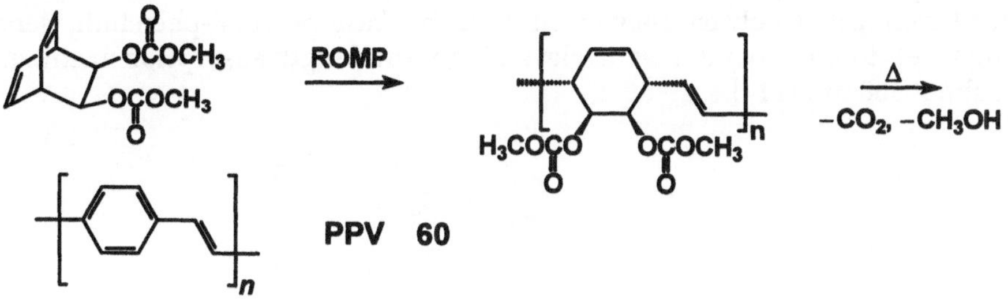

PPV 60

Bazan et al. [113] described the synthesis of a silyloxy-substituted **PPV** precursor (**88**) by ring-opening metathesis polymerization (ROMP) of 9-(t-butyldimethylsilyloxy)-[2.2]paracyclophan-1-ene (**89**) with a Schrock-type molybdenum carbene initiator. The reaction proceeds as a living polymerization: films of the cis-olefinic **PPV** precursor (**88**: M_n: 47000, M_w/M_n: 1, 2) can be converted to the conjugated **PPV 60** by heating to 120–140 °C in the presence of HCl gas. The infrared spectra of these films display only signals of trans-olefinic bonds. Therefore, although native **88** is mostly in the cis-configuration, the elimination step generates **PPV 60** in which the double bonds are exclusively in the trans-configuration.

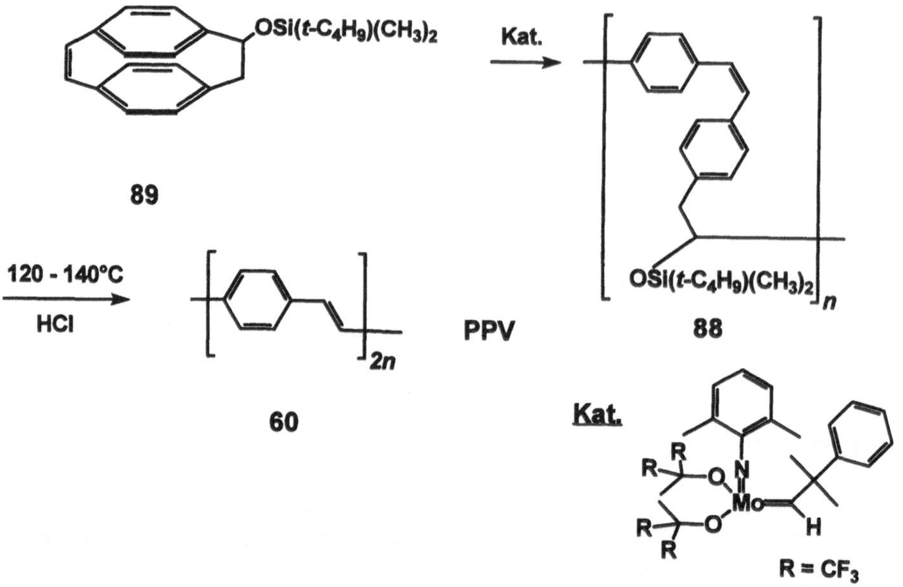

Adapting this method, block copolymers composed of **PPV** and norbornene blocks can also be made available. These copolymers are promising, highly efficient emitters for potential application in organic materials-based LEDs.

The silyloxy-substituted precursor polymers (**88**) can be photocyclized in a polymer-analogous fashion to yield polymeric intermediates (**90**) containing the 3,6-phenanthrylene unit. These intermediates can be converted thermally to yield conjugated polyarylenevinylenes, in this case poly(3,6-phenanthrylene-vinylene) (**91**), a polymer that displays a long-wavelength absorption maximum at about 360 nm [114].

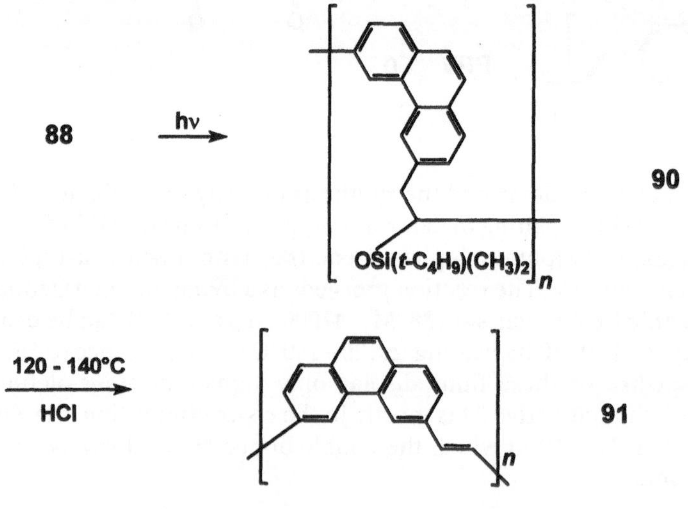

The synthesis of **PPVs** via metathesis polycondensation is also possible following the ADMET (acyclic diene metathesis) procedure. Thorn-Csanyi et al. have described the generation of a 2,5-dialkylated **PPV 63** (R = n-heptyl) via the ADMET process [115]. ADMET polycondensation involves an elimination of ethylene from the divinylbenzene derivative (**92**) catalyzed by a Schrock-type molydenum-carbene complex. The reaction affords low molecular weight products (**63**: R = n-heptyl) with degrees of polymerization of about 10. Metathesis polymerization provides polymers **63** with an all-*trans* configuration of the olefinic bonds, as detected by ^{1}H-NMR spectroscopy. The low molecular weight products can undergo a further metathesis reaction, since the vinyl end groups are much more reactive than the inner vinylene double bonds; this paves the way to the generation of novel block copolymers.

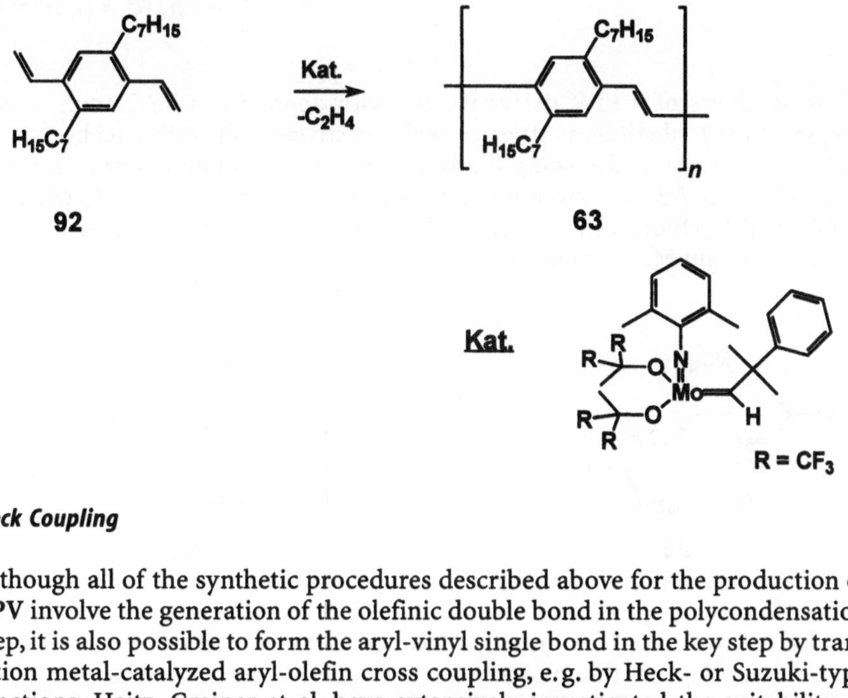

Heck Coupling

Although all of the synthetic procedures described above for the production of **PPV** involve the generation of the olefinic double bond in the polycondensation step, it is also possible to form the aryl-vinyl single bond in the key step by transition metal-catalyzed aryl-olefin cross coupling, e.g. by Heck- or Suzuki-type reactions. Heitz, Greiner et al. have extensively investigated the suitability of Heck-type coupling for the generation of **PPV 60** and **PPV** derivatives, e.g. poly(2,5-dialkoxy-1,4-phenylenevinylene) (**63**) and poly(2-phenyl-1,4-phenylenevinylene) (**93**) [116]. They coupled 1,4-dibromobenzene derivatives (e.g. 2-phenyl-, 2,5-dialkoxy-) with ethylene (and other olefins, e.g. styrene) catalyzed by palladium acetate, $PdCl_2$, Pd(dba)$_2$, Pd/C or Pd(PPh)$_4$, to form conjugated products with M_n up to 5000. The vinylene units of the PPVs are predominantly in the *trans*-configuration. Some side reactions, such as reductive dehalogenations and 1,1-diarylations at the olefin moiety, could be detected. The regioselectivity of Heck-type coupling can be controlled by the reaction conditions, by the leaving group, by the substituents R^1 and R^2 on the arylene component, and by choice of the olefinic monomer. The soluble phenylsubstituted **PPV** derivative, **93**, represents a very suitable emissive material for organic materials-based LEDs [117].

R^3 = H, Ph

63 (R^1, R^2 = alkoxy, R^3 = H)

93 (R^1 = Ph, R^2 = H, R^3 = H)

The synthesis of a **PPV** derivative, **94**, with donor (di-*n*-hexylamino) and acceptor (nitro) substituents attached regioselectively to the **PPV** backbone, was published by Yu et al. following the Heck-type cross coupling approach [118] starting from an AB-type monomer (**95**). The red-orange polymer (**94**), which is soluble in THF, chloroform and 1,2-dichloroethane, displays a number-average molecular weight M_n of about 12,000.

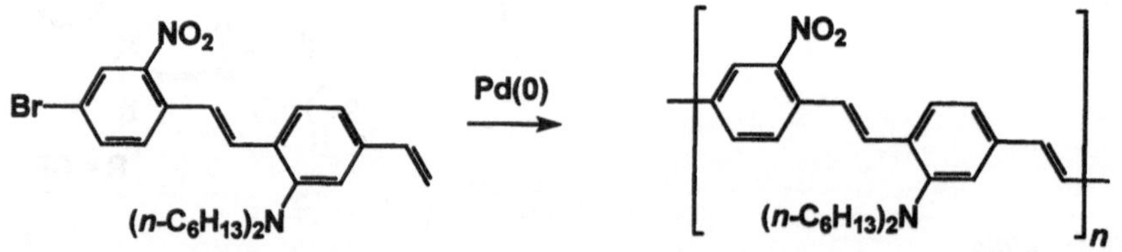

95   **94**

A drawback of the Heck-type reaction is that it is not strictly regioselective [119]. Depending on the substituents >1% of 1,1-diarylation is observed. Soluble 2,5-dialkoxy-**PPV**s **63** or 2-phenyl-**PPV** PPPV **93**, without 1,1-diarylated moieties, were synthesized by Heitz et al. in a Suzuki-type cross coupling of substituted 1,4-phenylenediboronic acids and *trans*–1,2-dibromoethylene, catalyzed by Pd(0) compounds [120]. However, about 3% of biaryl defect structures are observed in the coupling products (M_n up to 12,000), resulting from homocoupling of boronic acid functions.

In 1996, Wegner et al. published the synthesis of poly(oligophenylenevinylene)s (**96**), consisting of biphenylene-, terphenylene- and quinquephenylene moieties as aromatic building blocks, via Suzuki-type aryl-aryl cross coupling of AA/BB-type monomers [121]. By judicious choice of the arylene moieties, the optical properties of the resulting polymers can be tailored within a wide range.

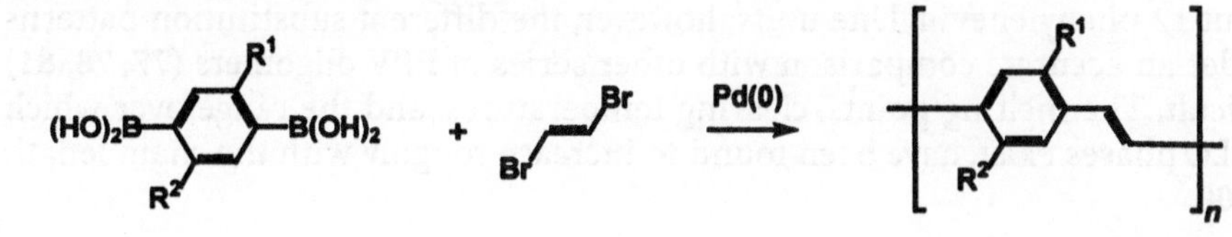

63 (R^1, R^2 = alkoxy)
93 (R^1 = Ph, R^2 = H)

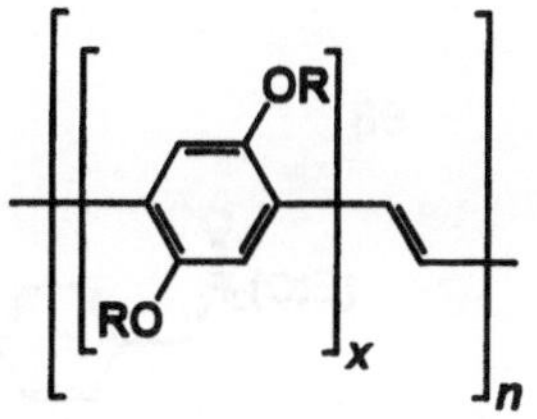

96 (x = 2, 3, 5)

R =

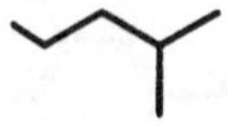

Some of the derivatives **96** show high photoluminescence quantum yields of up to 73%, and can be applied as blue and green polymeric emitters in organic materials-based LEDs [122].

The Orthogonal Approach (Combined Heck- and Wittig-Horner-Type Condensations) to PPV Oligomers

Yu et al. have recently published [123] a novel, modular strategy for the production of a series of **PPV** oligomers of defined chain length. Their elegant, stepwise synthesis represents a so-called "orthogonal" approach that involves the combination of two non-interacting reaction types, here the Wittig-Horner- and the Heck-type couplings, for the generation of the **PPV** skeleton. The reaction sequence allows for sequential, modular growth of the chain length, eliminates the need of protecting groups, and provides a series of **PPV** oligomers in a simple fashion. Starting from a monofunctional iodostilbene derivative (**97**), the two types of bifunctional stilbene monomers, **98** (possessing a vinyl and an aldehyde function) and **99** (possessing an iodo and a phosphonate function) are added alternately, beginning with **98**. In this way, the iodoarene functions are coupled with the vinyls, and the phosphonates with the aldehyde functionalities, respectively. The oligomers (**100**) with n = 1 – 5 (corresponding to trimer, pentamer, up to the undecamer) were characterized by their UV/Vis spectroscopic properties and their ability to form thermotropic liquid crystalline mesophases. The UV/Vis data display a covergence limit of the optical absorption energies at about 12 phenylenevinylene units; however, the different substitution patterns render an accurate comparison with other series of **PPV** oligomers (77, 78, 81) difficult. The melting points, clearing temperatures, and the range over which the LC phases exist, have been found to increase roughly with the chain length of **100**.

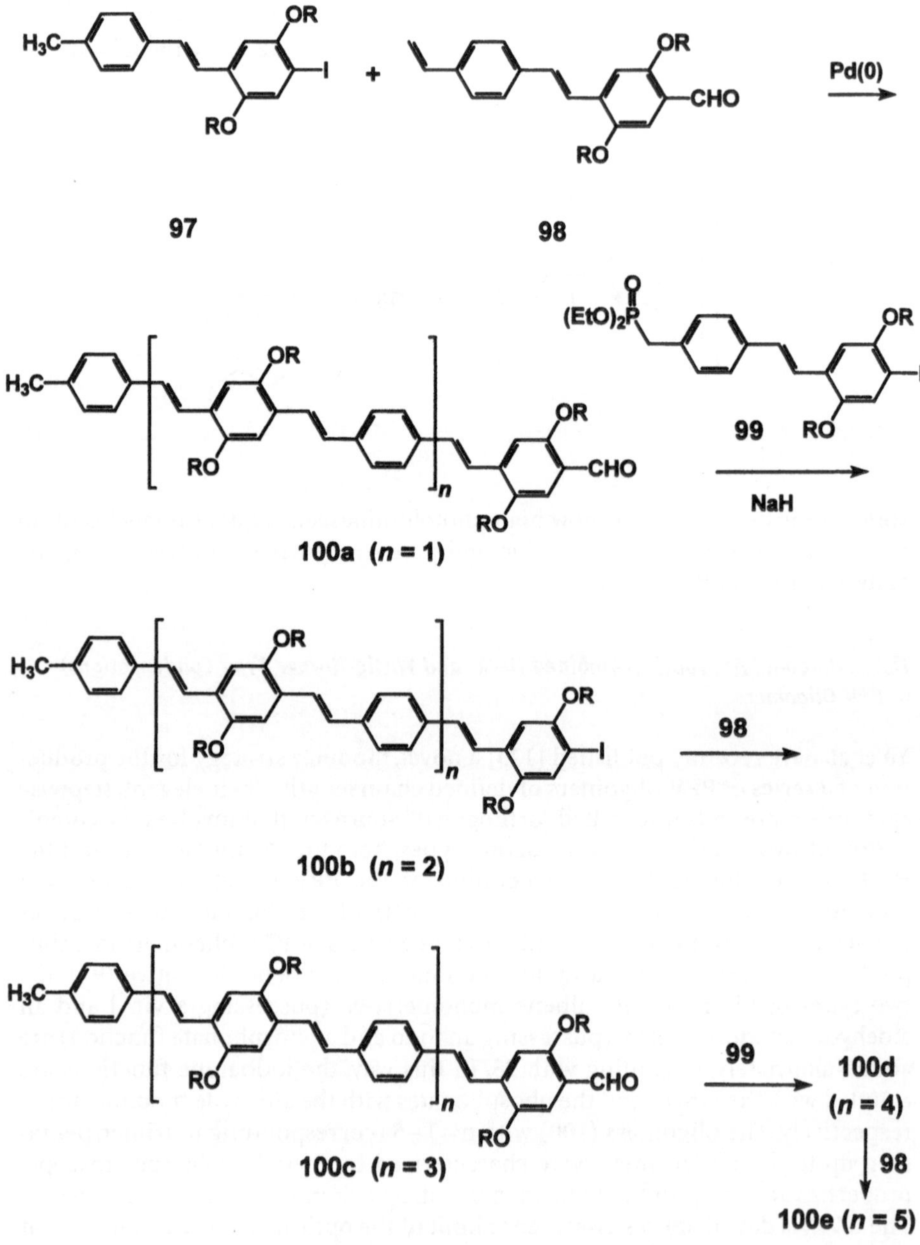

97   **98**

100a (*n* = 1)

100b (*n* = 2)

100c (*n* = 3)

99

98

100d
(*n* = 4)

98

100e (*n* = 5)

3.3
Other Oligo- and Polyarylenevinylenes

Poly- and Oligo(meta-phenylenevinylene)s

meta-Substituted phenylenevinylenes are not accessible via the polymerization approach. Several meta-substituted phenylenevinylenes were generated by poly-condensation methods. For example, Hörhold et al prepared diphenyl-substituted meta-phenylenevinylenes, using the reductive dehalogenation polyconden-sation method [69, 101]. Alkoxy- and phenoxyphenyl-substituted poly(meta-phenylenevinylene)s can be further cyclized to poly(2,9-phenanthrene)s [69].

A series of oligo(meta-phenylenevinylene)s (101) with terminal 3,5-di-t-butylphenyl end groups were prepared via a Wittig-type approach by Müllen et al. [124]. The meta-oligomers (101) display increased solubility when compared to their para-substituted counterparts (78), but - on the other hand - the 1,3-phenylene units act as conjugation barriers. The UV/Vis absorption spectra are, therefore, nearly independent of the chain length of the oligomers (101), and correspond to stilbene as the longest conjugated segment.

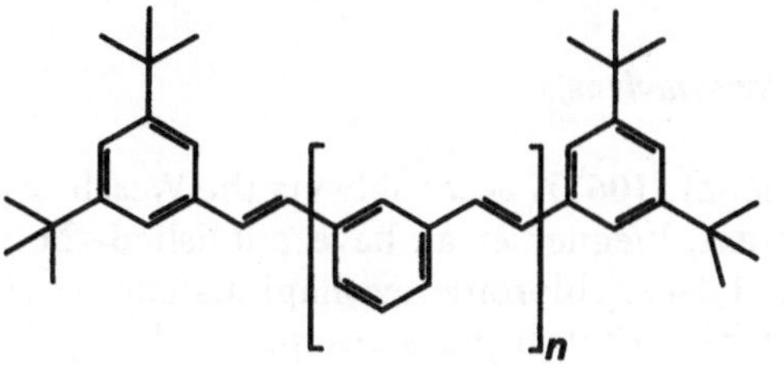

101

Oligo(ortho-phenylenevinylene)s

Oligo(ortho-phenylenevinylene)s (102) constitute an intermediate case between the para- and meta-analogues, 78 and 101. The ortho-topology of 102 allows, on the one hand, an extended π-conjugation, but - on the other hand - the inter-action is inhibited by the non-planar geometry due to steric hindrance between adjacent vinylene units. Using various coupling methods (McMurry-type, Wittig-type, Heck-type), Müllen et al. have generated a serious of ortho-phen-ylenevinylene oligomers (102) up to the hexamer ($n = 5$) [125, 126].

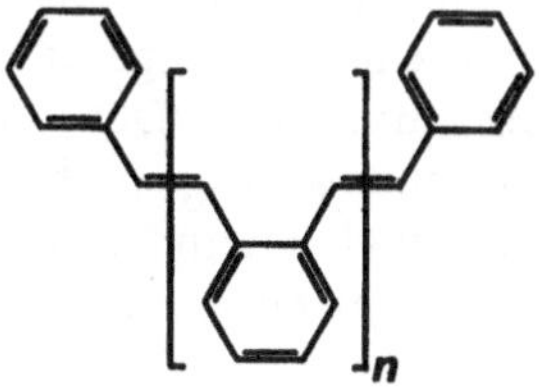

102 (n = 1 - 5)

Higher oligomers and polymers (103) can be generated via Pd(0)-catalyzed Stille-type coupling of 1,2-diiodobenzene or 1,2-bis(2-iodostyryl)benzene (104) with bis(tri-*n*-butylstannyl)ethylene (105) [126].

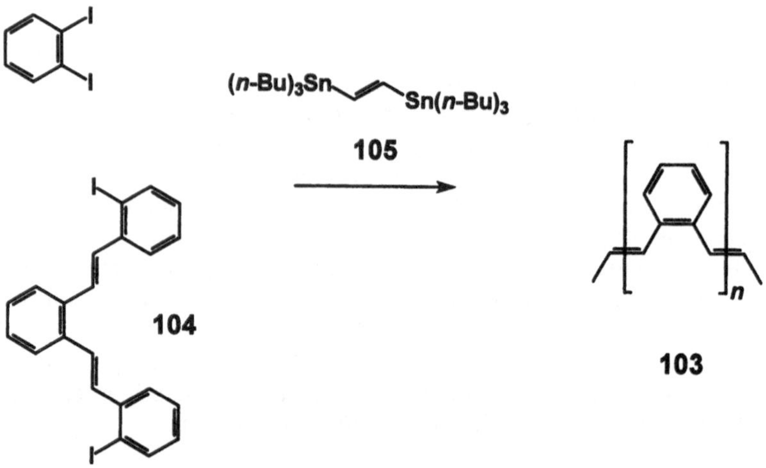

Poly- and Oligo(1,4-naphthylenevinylene)s

Poly(1,4-naphthylenevinylene) (106) is accessible via the Wessling polymerization procedure. Lenz, Karasz, Wegner et al. have published the synthesis of **PNV 106**, starting from 1,4-bis(chloromethyl)naphthalene [127, 128]. The poly(1,4-naphthylenevinylene) (106) displays an optical absorption energy of 2.05 eV, slightly red-shifted by about 0,3 eV relative to the parent **PPV 60**-system, due to the electronic effect of the annelated benzene ring.

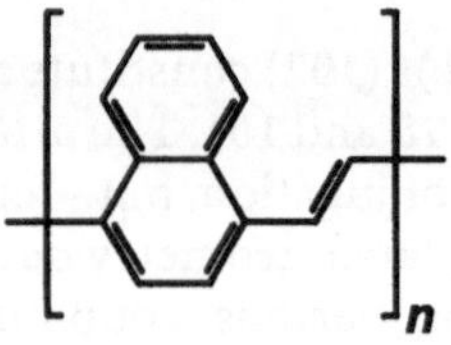

Alkylated poly(1,4-naphthylenevinylene)s, poly(6-hexyl- or 6-undecyl-1,4-naphthylenevinylene) (107) were generated by Grubbs et al. applying the ring-opening metathesis procedure (ROMP), starting from a benzobarrelene monomer (108) [129]. After oxidation of the non-conjugated intermediates (109) with 2,3-dichloro-5,6-dicyano-1,4-benzoquinone (DDQ), the soluble, conjugated polymers (107) are formed with number-average molecular weights M_n of up to 10,000. The alkylated poly(1,4-naphthylenevinylene)s (107) were used as the active, orange-red (λ_{peak}: 620 nm) emitting component in organic materials-based LEDs [130].

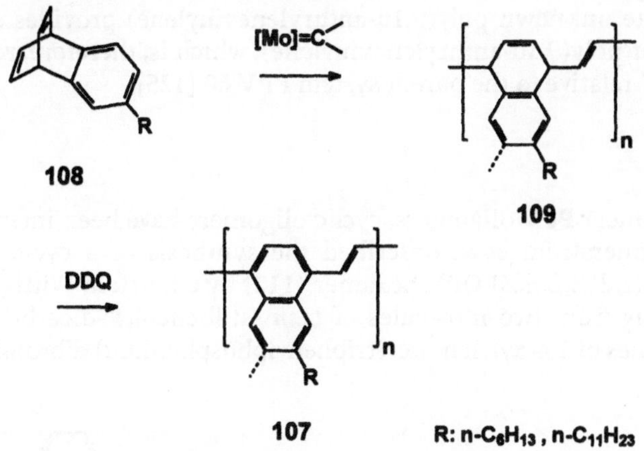

108 → **109** (via [Mo]=C<)

→ (DDQ) **107**

R: n-C$_6$H$_{13}$, n-C$_{11}$H$_{23}$

Oligo(3,7-di-*t*-butyl-1,5-naphthylenevinylene)s (**110**), up to the tetramer (*n* = 2), were generated by Müllen et al. by the combination of Wittig-type and McMurry-type couplings [131].

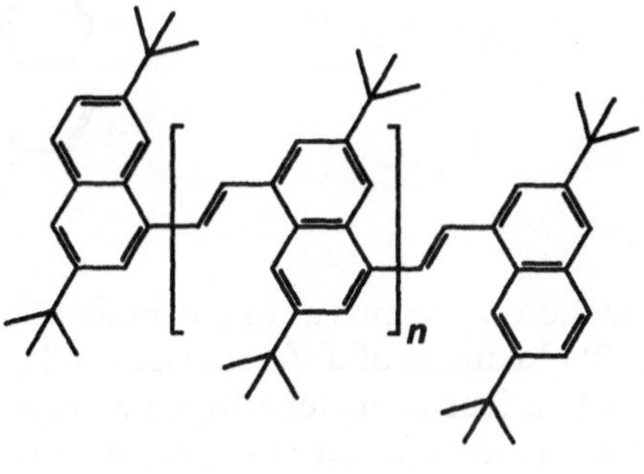

110

Oligo(9,10-anthrylenevinylene)

As mentioned above, poly(9,10-anthrylenevinylene) is not accessible by means of the Wessling polymerization procedure. Well defined oligo(9,10-anthrylene-vinylene)s (**111**) were synthesized via Horner-type coupling [132]. Extrapolation of the optical absorption energies in the **111** series against the corres-

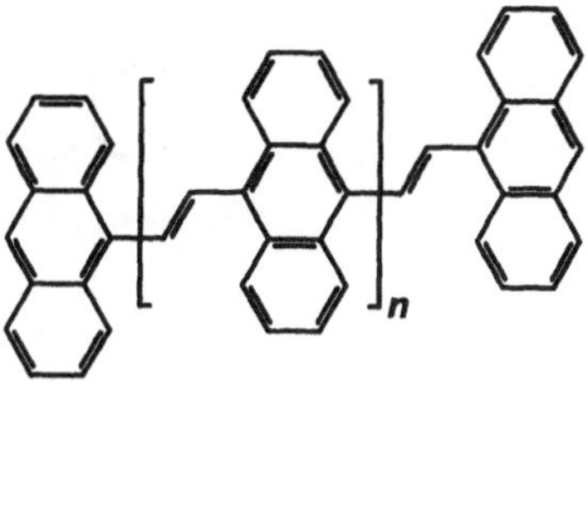

111

ponding, to date unknown poly(9,10-anthrylenevinylene) provides a value of about 2.0 eV for poly(9,10-anthrylenevinylene), which is, therefore, red-shifted by about 0.4 eV relative to the parent system **PPV 60** [125].

Cyclic Oligomers

In addition to linear **PPV** oligomers, cyclic oligomers have been intensively investigated. Wennerström et al. described the synthesis of a cyclic *cis,trans, cis,cis,trans,cis* (*Z,E,Z,Z,E,Z*)-**OPV** hexamer (**112**) by a fourfold Wittig-type olefination, starting from two molecules of *trans*-stilbene-4,4′-dicarboxaldehyde and two molecules of 1,4-xylylene-bis(triphenylphosphonium) dibromide [133].

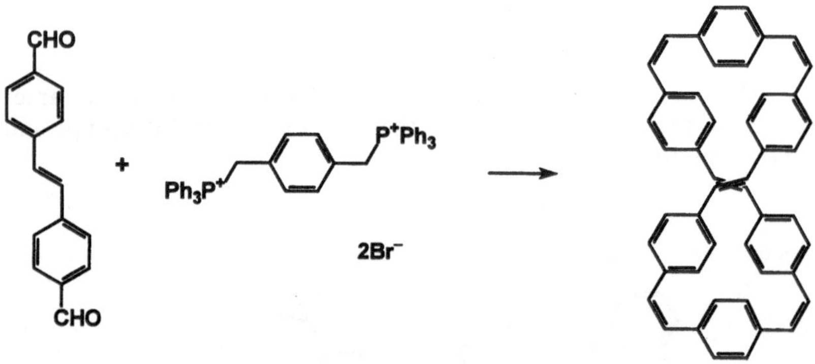

This type of one-pot reaction represents a simple method for the synthesis of a large variety of cyclic **OPV** oligomers of different sizes and composed of different arylene building blocks, such as phenylene, biphenyl-diyl and naphthylene. The compounds are suitable subjects for the study of photoinduced *Z/E*-isomerisation and photocyclization.

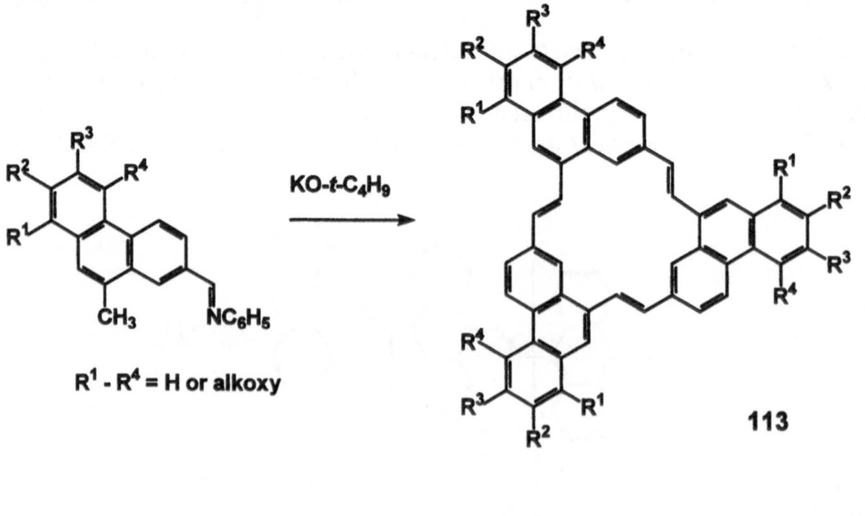

Meier et al. have also contributed to the field of cyclic **PAV** oligomers with the synthesis of cyclic all-*trans* (all-*E*) trimers (e.g. **113**) containing alkoxy-substituted 1,7-naphthylene and 1,9-phenanthrylene building blocks, via a Siegrist-type trimerizing olefination [134, 135]. Suitable substitution at the periphery of the cyclic trimers allows for the formation of stable, thermotropic discotic mesophases [134].

Polyindenofluorenes

In 1996, Scherf et al. described the synthesis of a novel low-band-gap hydro-carbon polymer, poly(3,9-di-*t*-butylindeno[1,2-b]fluorene) (**PIF 114**) [136]. The polymer (**114**), a poly(*para*-phenylene-diphenylvinylene) derivative containing additional aryl-aryl linkages, was generated via reductive coupling of alkylated 6,6,12,12-tetrachloro-6,12-dihydroindeno[1,2-b]fluorene **115** with low valent transition metal agents [e. g. cobalt(0), chromium(0), and nickel(0) compounds]. The UV/Vis spectra of **114**, with a degree of polymerization (DP) of ca. 20, indicate the presence of a conjugated one-dimensional π-system with a long-wavelength absorption maximum, λ_{max}, of about 800 nm (Fig. 3). Particularly noteworthy, however, is the position of the long-wavelength absorption band: it is red shifted far into the NIR region. Compared to **PPV 60**, λ_{max} of **PIF 114** is red shifted by more than 350 nm!. This fact can be interpreted as resulting from the contribution of quinoid states to the electronic ground state [136]. The completely soluble material (**114**) is characterized by high thermooxidative stability, and can be processed to thin films by spin-coating. **PIF 114** displays huge third-order non-linear optical responses [137].

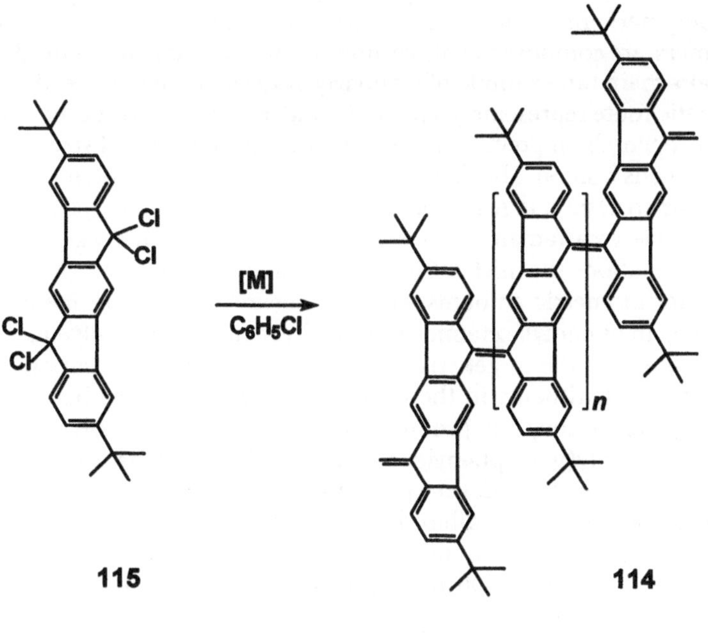

115   **114**

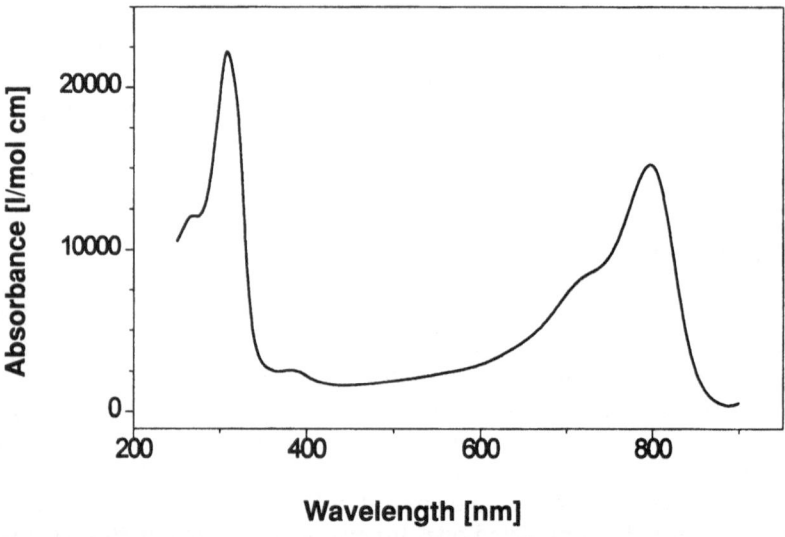

Wavelength [nm]

Fig. 3. UV/Vis absorption spectrum of **PIF 114** (solvent: methylene chloride)

3.4
Ladder-Type Poly(*para*-phenylene-cis-vinylene)s

In 1993, Scherf and Chmil described the first synthesis of a ladder-type
poly(*para*-phenylene-*cis*-vinylene) (**116**) [138]. On the one hand, ladder polymer **116** represents, a planar poly(phenylene) containing additional vinylene
bridges; on the other hand, it is a poly(phenylenevinylene) with aryl-aryl linkages in the polymeric main chain. The target macromolecules, as fully aromatic
ladder polymers, are composed of all-carbon six-membered rings in the double-
stranded main chain (an example of angularly annelated poly(acene)s).

The synthetic route represents a "classical" ladder polymer synthesis: a suitably substituted, open-chain precursor polymer is cyclized to a band structure in a
polymer-analogous fashion. The first step here, formation of the polymeric, open-
chain precursor structure, is AA-type coupling of a 2,5-dibromo-1,4-dibenzoyl-
benzene derivative, by a Yamamoto-type aryl-aryl coupling. The reagent employed for dehalogenation, the nickel(0)/1,5-cyclooctadiene complex (Ni(COD)$_2$),
was used in stoichiometric amounts with co-reagents (2,2'-bipyridine and 1,5-
cyclooctadiene), in dimethylacetamide or dimethylformamide as solvent.

For the synthesis of the target structures, it is absolutely necessary to introduce solubilizing substituents in the positions peripheral to the benzoyl substituents. The primary coupling product, **117**, a poly(2,5-dibenzoyl-1,4-phenyl-
ene) derivative – a poly(*para*-phenylene) with two benzoyl substituents in each
structural unit – is, as expected, very poorly soluble. Highly substituted monomers (2,5-dibromo-1,4-bis(3,4-dihexyloxy-benzoyl)benzene), containing four
solubilizing alkoxy groups per monomer unit, allow the synthesis of polymeric
materials with M_n of about 12,000 and M_w of about 22,000 [139].

The open-chain precursor polymers can be cyclized to the ladder-type structure (116) in a polymer-analogous ring closure. A carbonyl olefination reaction, first described in 1992 by Steliou, Salama, and Yu [140], was used with boron sulfide, generated *in situ*. In the course of the cyclization, the corresponding thioketones are initially formed. In the next step, the C=S group undergoes dimerization ([2 + 2] cycloaddition) with formation of cyclic disulfide bridges. In turn, these intermediates stabilize themselves by elimination of sulfur, to give the conjugated aromatic ladder polymer.

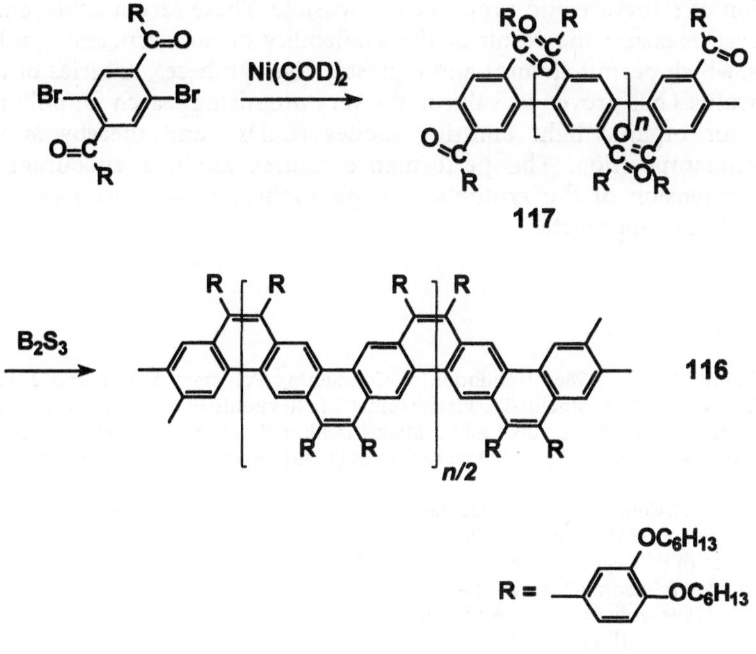

117

116

$$R = \text{[aromatic ring with } OC_6H_{13} \text{ and } OC_6H_{13}]$$

$n \sim 25$

The polymer-analogous cyclization is accompanied by a remarkable change in the absorption properties. The colorless intermediates (117) are converted to a deep-yellow, planar ladder polymer (116); associated with this is a strong bathochromic shift of the long-wavelength absorption maximum. The polymer (116) possesses an absorption band with well defined vibrational fine structure and a sharp absorption edge of the 0–0 transition (437 nm; 2.83 eV). The optical spectrum of 116 is in very good accord with band gap calculations of Bredas et al. [141], who predicted an optical transition energy of 2.86 eV for the conjugated ladder skeleton of 116. The photoluminescence behavior of 116 is characterized by the appearance of a sharp, structured emission band (short-wavelength emission maximum: 484 nm). This new class of ladder-type poly(*para*-phenylene-*cis*-vinylene)s can be used as active material in light emitting diodes [142].

4
Conclusion

The results discussed in this article on the syntheses of new, structurally defined **PPP** and **PPV** derivatives demonstrate the qualitative leap from a purely structurally motivated search for new conjugated molecules to the synthesis of structurally defined, processable materials with properties that are tailor-made for particular applications. Specifically, the combination of the solubilization required for processability *and* complete retention of the conjugated character led to a new generation of **PPP** and **PPV** derivatives, in which a comprehensive correlation of structure and properties is possible. These recent achievements are in large measure the result of the availability of new, efficient synthetic methods, which permit chemo- and regioselective syntheses. A series of these new derivatives has proven its value in the very promising search for polymeric emitters for organic light emitting diodes (LEDs), and thereby attained a predominantposition. The performance figures attained encourage far-reaching extension of the synthetic and photophysical work in a genuinely interdisciplinary approach.

5
References

1. Skotheim TA (ed) (1986) Handbook of Conducting Polymers, Vols 1 and 2, Marcel Dekker, New York; Skotheim TA, Elsenbaumer RL, Reynolds JR (eds) (1998) Handbook of Conducting Polymers, Vols 1 and 2, Marcel Dekker, New York, Basel, Hong Kong
2. Heitz W (1986) Chem.-Ztg. 110:385; Ballauf M (1989) Angew Chem 101:261; Int Ed Engl 28:253
3. Ballard DGH, Courtis A, Shirley IM, Taylor SC (1983) J Chem Soc, Chem Comm 1983:954
4. Wessling RA (1985) J Polym Sci, Polym Symp 72:55
5. Cao Y, Smith P, Heeger AJ (1993) Synth Met 55–57:3514
6. Jenekhe SA, Johnson PO, Agrawal AK (1989) Macromolecules 22:3216
7. Tour JM (1994) Adv Mater 6:190; Schlüter A-D, Wegner G (1993) Acta Polymer 44:59
8. Kovacic P, Jones MB (1987) Chem Rev 87:357
9. Baughman RH, Bredas JL, Chance RR, Elsenbaumer RL, Shacklette LW (1982) Chem Rev 82:209
10. Katsuya M, Teshirogi T, Kuramato N, Kitamura T (1995) J Polym Sci, Polym Chem Ed 23:1259
11. Ueda M, Abe T, Awano H (1992) Macromolecules 25:5125
12. Fukuda M, Sawada K, Yoshino K (1993) J Polym Sci A 31:2465
13. Miyaura M, Yanagi T, Suzuki A (1981) Synth Commun 11:513
14. Kanbara T, Saito N, Yamamoto T, Kubota K (1991) Macromolecules 24:5883; Yamamoto T, Morita A, Miyazaki Y, Maruyama T, Wakayama H, Zhou Z, Nakumura Y, Kanbara T, Sasaki S, Kubota K (1992) Macromolecules 25:1214
15. Chaturvedi V, Tanaka S, Kaeriyama K (1993) Macromolecules 26:2607
16. Rehahn M, Schlüter A-D, Wegner G, Feast WJ (1989) Polymer 30:1054
17. Rehahn M, Schlüter A-D, Wegner G, Feast WJ (1989) Polymer 30:1060
18. Yamamoto T, Hayashi Y, Yamamoto A (1978) Bull Chem Soc Jpn 51:2091; Noll A, Siegfield N, Heitz W (1990) Makromol Chem, Rapid Commun 11:485
19. Ueda M, Ichikawa F (1990) Macromolecules 23:926; Ueda M, Miyaji Y, Ito T (1991) Macromolecules 24:2694, Colon I, Kwiatkowski GT (1990) J Polym Sci A 28:367
20. Percec V, Okita S, Weiss R (1992) Macromolecules 25:1816

21. Vahlenkamp T, Wegner G (1994) Macromol Chem Phys 195:1933; Wallow TI, Novak BMJ (1991) J Am Chem Soc 113:7411; Child AD, Reynolds JR (1994) Macromolecules 27:1975; Rau IU, Rehahn M (1994) Acta Polymer 45:3; Rulkens R, Schultze M, Wegner G (1994) Macromol Rapid Commun 15:669
22. Huber J, Scherf U (1994) Macromol Chem Phys 15:897, Fiesel R, Huber J, Apel U, Enkelmann V, Hentschke R, Scherf U, Cabrera K (1997) Macromol Chem Phys 198:2623
23. Elsenbaumer RL, Shacklette LW (1986) in: Skotheim, TA (ed) Handbook of Conducting Polymers, Vol 1 Marcel Dekker, New York, chap 7
24. Park KC, Dodd LR, Levon K, Kwei TK (1996) Macromolecules 29:7149
25. Saito N, Kanbara T, Sato T, Yamamoto T (1993) Polym Bull 30:285
26. Kreyenschmidt M, Uckert F, Müllen K (1995) Macromolecules 28:4577
27. Marvel CS, Hartzell GE (1959) J Am Chem Soc 81:448
28. Ballard DGH, Curtis A, Shirley IM, Taylor SC (1987) Macromolecules 21:1787
29. Gin DL, Conticello VP, Grubbs RH (1992) J Am Chem Soc 114:3167
30. Gin DL, Avlyanov JK, MacDiarmid AG (1994) Synth Met 66:169
31. Noren GK, Stille JK (1971) Macromol Rev 5:385
32. Tour JM, John JA (1993) Polym Prepr (Am Chem Soc Div Polym Chem) 34(2):372; John JA, Tour JM (1994) J Am Chem Soc 116:5011
33. Kern W, Seibel M, Wirth H-O (1959) Makromol Chem 29:164
34. Heitz W, Ullrich R (1966) Makromol Chem 98:29
35. Galda P, Rehahn M (1996) Synthesis 1996:614
36. Gerhardt H (1984) PhD Thesis, Georg-August-Universität Göttingen
37. Kreyenschmidt M (1995) PhD Thesis, Johannes-Gutenberg-Universität Mainz
38. Grimme J, Kreyenschmidt M, Uckert F, Müllen K, Scherf U (1995) Adv Mater 7:292
39. Schumm JS, Pearson DL, Tour JM (1994) Angew Chem 106:1445; Int Ed Engl 33:1360
40. Liess P, Hensel V, Schlüter A-D (1996) Liebigs Ann 1996:614
41. Scherf U, Müllen K (1991) Makromol Chem, Rapid Commun 12:489
42. Stampfl J, Graupner W, Leising G, Scherf U (1995) J Lumin 63:117
43. Tasch S, Niko A, Leising G, Scherf U (1996) Appl Phys Lett 68:1090
44. Huber J, Müllen K, Salbeck J, Schenk H, Scherf U, Stehlin T, Stern R, Acta Polymer 45:244
45. Mahrt RF, Siegner U, Lemmer U, Hopmeier M, Scherf U, Heun S, Göbel EO, Müllen K, Bässler H (1995) Chem Phys Lett 240:373
46. Graupner W, Leising G, Lanzani G, Nisoli M, de Silvestri S, Scherf U (1996) Chem Phys Lett 246:95
47. Köhler A, Grüner J, Friend RH, Müllen K, Scherf U (1995) Chem Phys Lett 243:456
48. Grüner J, Wittmann HF, Hamer PJ, Friend RH, Huber J, Scherf U, Müllen K, Moratti SC, Holmes AB (1994) Synth Met 67:181
49. Mahrt RF, Haring Bolivar P, Pauck T, Wegmann G, Lemmer U, Siegner U, Hopmeier M, Hennig R, Scherf U, Müllen K, Kurz H, Bässler H, Göbel EO (1996) Phys Rev B 54:1759
50. Cimrová V, Neher D, Scherf U (1996) Appl Phys Lett 69:608
51. Graupner W, Eder S, Tasch S, Leising G, Lanzani G, Nisoli M, de Silvestri S, Scherf U (1995) J Fluorescence 7:195s
52. Zenz C, Graupner W, Tasch S, Leising G, Müllen K, Scherf U (1997) Appl Phys Lett 71:2566; Kallinger C, Hilmer M, Haugeneder A, Perner M, Spirkl W, Lemmer U, Feldmann J, Scherf U, Müllen K, Gombert A, Wittwer V (1998) Adv Mater 10:920
53. Grimme J, Scherf U (1996) Macromol Chem Phys 197:2297
54. Keegstra MA, De Feyter S, De Schryver FC, Müllen K (1996) Angew Chem 108:830; Int Ed Eng 35:774
55. Ogliaruso MA, Becker EI (1965) J Org Chem 30:3354
56. Wu R, Schumm JS, Pearson DL, Tour JM (1996) J Org Chem 61:6906
57. Saalbeck J (1996) Ber Bunsenges Phys Chem 100:1667
58. Clarkson RG, Gomberg M (1930) J Am Chem Soc 52:2881
59. Müller M, Mauermann-Düll M, Wagner M, Enkelmann V, Müllen K (1995) Angew Chem 107:1751, Int Ed Engl 34:1583
60. Morgenroth F, Reuther E, Müllen K (1997) Angew Chem 109:647, Int Ed Engl 36:631

61. Kim YH, Webster OW (1990) J Am Chem Soc 112:4592
62. Staab HA, Binnig F (1967) Chem Ber 100:293,899
63. Hara S, Toshima N (1990) Chem Lett 1990:269
64. Percec V, Wang J, Okita S (1992) Polym Prepr (Am Chem Soc Div Polym Sci) 33(1): 225
65. Fahnenstich U, Koch K-H, Müllen K (1989) Makromol Chem, Rapid Commun 10:563
66. Schopov I, Jossifov C (1978) Polymer 19:1449
67. Müller U, Adam M, Müllen K (1994) Chem Ber 127:437
68. Müller U (1993) PhD Thesis, Johannes-Gutenberg-Universität Mainz; Müller U, Baumgarten M (1995) J Am Chem Soc 117:5840
69. Hörhold H-H, Bleyer A, Birckner E, Heinze S, Leonhardt F (1995) Synth Met 69:525
70. Taylor SK, Benett SG, Heinz KJ, Lashley KL (1981) J Org Chem 46:2194
71. Anton U, Müllen K (1993) Makromol Chem, Rapid Commun 14:223
72. Kreyenschmidt M, Baumgarten M, Tyutyulkov N, Müllen K (1994) Angew Chem 106:2062; Int Ed Engl 33:1957
73. Wessling RA, Zimmerman RG (1968) US Pat 3401152; Wessling RA (1985) J Polym Sci, Polym Symp 72:55
74. Hörhold H-H, Palme H-J, Bergmann R (1978) Faserforsch Textiltech, Z Polymerforschung 29:299
75. Denton FR III, Lahti PM, Karasz FE (1992) J Polym Sci A, Polym Chem 30:2223
76. McCoy RK, Karasz FE, Sarker A, Lahti PM (1991) Chem Mat 3:941; Sonoda Y, Kaeriyama K (1992) Bull Chem Soc Jpn 65:853; Garay RO, Karasz FE, Lenz RW, JMS, Pure Appl Chem A 32:905, Zyung T, Kim J-J, Hwang W-Y, Hwang DH, Shim HK (1995) Synth Met 71:2167
77. Denton FR III, Sarker A, Lahti PM, Garay RO, Karasz FE (1992) J Polym Sci A, Polym Chem 30:2233
78. Lahti PM, Sarker A, Garay RO, Lenz RW, Karasz FE (1994) Polymer 35:1312
79. Burroughes JH, Bradley DDC, Brown AR, Marks RN, Mackey K, Friend RH, Burns PL, Holmes AB (1990) Nature 347:539
80. Xia Y, McDiarmid AG, Epstein AJ (1994) Adv Mater 6:293; Garay RO, Baier U, Bubeck C, Müllen K (1993) Adv Mater 5:561
81. Gilch HG, Wheelwright WL (1966) J Polymer Sci A-1 4:1337
82. Swatos WJ, Gordon B III (1990) Polym Prepr (Am Chem Soc Div Polym Chem) 31(1):192
83. Staring EGJ, Demandt CJE, Braun D, Rikken GLJ, Kessener AR, Venhuizen THJ, Wynberg H, ten Hoeve W, Spoelstra KJ (1991) Adv Mater 6:934
84. Doi S, Kuwabara M, Noguchi T, Ohnishi T (1993) Synth Met 57:4174
85. Braun D, Heeger AJ (1991) Appl Phys Lett 58:1982
86. Hsieh BR, Antoniadis H, Bland DC, Feld WA (1995) Adv Mater 7:36
87. Gustaffson G, Cao Y, Treacy GM, Klavetter F, Colaneri N, Heeger AJ (1992) Nature 357:477; Yang Y, Heeger AJ (1994) Appl Phys Lett 64(10):1245
88. Iwatsuki S, Kubo M, Kumeuchi T (1991) Chem Lett 1991:1071; Staring EGJ, Braun D, Rikken GLJA, Demandt RJCE, Kessener YARR, Bauwmans M, Broer D (1994) Synth Met 67:71; Schäfer O, Greiner A, Pommerehne J, Guss W, Vestweber H, Tak HY, Bässler H, Schmidt C, Lussem G, Schärtel B, Stumpflen V, Wendorff JH, Spiegel S, Möller C, Spiess HW (1996) Synth Met 82:1, Vaeth KM, Jensen KF (1997) Adv Mater 9:490
89. Louwet F, Vanderzande D, Gelan J, Mullens J (1995) Macromolecules 28:130
90. Raabe D, Hörhold H-H, Scherf U (1986) Makromol Chem, Rapid Commun 7:613
91. Opfermann J, Scherf U, Raabe D, Nowotny J, Hörhold H-H (1986) Angew Makromol Chem 142:91
92. Raabe D, Hörhold H-H (1992) Acta Polymer 43:275
93. Hsieh BR (1991) Polymer Bull 26:391
94. Hörhold H-H, Gräf D, Opfermann J (1970) Plaste Kautsch 17:84; Hörhold H-H (1972) Z Chem 12:41
95. Lenz R, Handlovitis CE (1960) J Org Chem 25:813
96. Greenham NC, Moratti SC, Bradley DDC, Friend RH, Holmes AB (1993) Nature 365:628

97. Staring EGJ, Demandt RCJE, Braun D, Rikken GLJ, Kessener YARR, Venhuizen AHJ, van Knippenberg MMF, Bouwmans M (1995) Synth Met 71:2179
98. Moratti SC, Cervini R, Holmes AB, Baigent DR, Friend RH, Greenham NC, Grüner J, Hamer PJ (1995) Synth Met 71:2117
99. Hörhold H-H, Opfermann J (1970) Makromol Chem 131:105
100. Koßmehl G, Härtel M, Manecke G (1970) Makromol Chem 131:37
101. Hörhold H-H, Helbig M (1987) Makromol Chem, Macromol Symp 12:229
102. Drefahl G, Plötner G (1961) Chem Ber 94:907; Drefahl G, Kühmstedt R, Oswald H, Hörhold H-H (1970) Makromol Chem 131:89
103. Schenk R, Gregorius H, Meerholz K, Heinze J, Müllen K (1991) J Am Chem Soc 113:2635
104. Stalmach U, Kolshorn H, Brehm I, Meier H (1996) Liebigs Ann 1996:1449; Meier H, Stalmach U, Kolshorn H, Acta Polymer (1997) 48:379
105. Rehahn M, Schlüter A-D (1990) Makromol Chem, Rapid Commun 11:375
106. DeKoninck L, Smets G (1969) J Polym Sci A-1 7:3313
107. Hörhold H-H, Gottschaldt J, Opfermann J (1977) J prakt Chem 319:611
108. Hörhold H-H, Helbig M, Raabe D, Opfermann J, Scherf U, Stockmann R, Weiss D (1987) Z Chem 27:126
109. Damerau T, Hennecke M, Hörhold H-H (1995) Macromol Chem Phys 196:1277
110. Tak Y-H, Vestweber H, Bässler H, Bleyer A, Stockmann R, Hörhold H-H (1996) Chem Phys 212:471
111. Klärner G, Former C, Yan X, Richert R, Müllen K (1996) Adv Mater 8:932
112. Conticello VP, Gin DL, Grubbs RH (1992) J Am Chem Soc 114:9708
113. Miao Y-J, Wong-Foy AG, Bazan GC (1994) Polym Prepr (Am Chem Soc Div Polym Chem) 35(2):538, Miao Y-J, Bazan GC (1994) Macromolecules 27:1063
114. Sun BJ, Bazan GC (1995) Polym Prepr (Am Chem Soc Div Polym Chem) 36(1):253
115. Thorn-Csanyi E, Kraxner P (1995) Macromol Rapid Commun 16:147
116. Greiner A, Heitz W (1988) Makromol Chem, Rapid Commun 9:581; Brenda M, Greiner A, Heitz W (1990) Makromol Chem 191:1083; Klingelhöfer S, Schellenberg C, Pommerehne J, Bässler H, Greiner A, Heitz W (1997) Macromol Chem Phys 198:1511
117. Vestweber H, Greiner A, Lemmer U, Mahrt RF, Richert R, Heitz W, Bässler H (1992) Adv Mater 4:661
118. Pan M, Bao Z, Yu L (1995) Polym Prepr (Am Chem Soc Div Polym Chem) 36(1):618
119. Martelock H, Greiner A, Heitz W (1991) Makromol Chem 192:967
120. Koch F, Heitz W (1997) Macromol Chem Phys 198:1531
121. Remmers M, Schulze M, Wegner G (1996) Macromol Rapid Commun 17:239
122. Remmers M, Neher D, Grüner J, Friend RH, Gelnick GH, Warman JM, Quattrocchi C, dos Santos DA, Bredas J-L (1996) Macromolecules 23:7432
123. Maddux T, Li WJ, Yu LP (1997) J Am Chem Soc 119:844.
124. Gregorius H, Baumgarten M, Reuter R, Tyutyulkov N, Müllen K (1992) Angew Chem 104:1621; Int Ed Engl 31:1635.
125. Baumgarten M, Bunz U, Scherf U, Müllen K (1995) In: Molecular Engineering for Advanced Materials, Becher J, Schaumburg K (eds); Kluwer, Dordrecht, Netherlands, p159.
126. Mauermann-Düll H, Adam M, Böhm A, Reuter R, Müllen K; to be published
127. Stenger-Smith JD, Sauer T, Wegner G, Lenz RW (1990) Polymer 31:1633
128. Antoun S, Gagnon DR, Karasz FE, Lenz RW (1986) J Polym Sci, Polym Lett Ed 24:503
129. Pu L, Wagaman MW, Grubbs RH (1996) Macromol 29:1138
130. Tasch S, Graupner W, Leising G, Pu L, Wagner MW, Grubbs RH (1995) Adv Mater 7:903
131. Ohlemacher A, Schenk R, Weitzel H-P, Tyutyulkov N, Tasseva M, Müllen K (1992) Makromol Chem 193:81
132. Weitzel H-P, Müllen K (1990) Makromol Chem 191:2837
133. Thulin B, Wennerström O (1977) Acta Chem Scand B31:135; Raston W, Wennerström O (1982) Acta Chem Scand B36:655
134. Meier H (1992) Angew Chem 104:1425; Int Ed Engl 31:1399
135. Meier H, Müller K (1995) Angew Chem 107:1598; Int Ed Engl 34:1437
136. Reisch H, Wiesler U, Scherf U, Tuytuylkov N (1996) Macromolecules 29:8204

137. Samoc M, Samoc A, Luther-Davies B, Reisch H, Scherf U (1998) Opt Lett 23:1
138. Chmil K, Scherf U (1993) Makromol Chem, Rapid Commun 14:217
139. Chmil K, Scherf U (1997) Acta Polymer 48:208
140. Steliou K, Salama P, Yu X (1992) J Am Chem Soc 114:1456
141. Toussaint JM, Bredas J-L (1992) Synth Met 46:325
142. Kirstein S, Cohen G, Davidov D, Chmil K, Scherf U, Klapper M, Müllen K (1995) Synth Met 69:415

Author Index Volume 201

The volume numbers are printed in italics

Springer
and the
environment

At Springer we firmly believe that an international science publisher has a special obligation to the environment, and our corporate policies consistently reflect this conviction.
We also expect our business partners – paper mills, printers, packaging manufacturers, etc. – to commit themselves to using materials and production processes that do not harm the environment. The paper in this book is made from low- or no-chlorine pulp and is acid free, in conformance with international standards for paper permanency.